Crystal Growth of Organic Materials

CONFERENCE PROCEEDINGS SERIES

Crystal Growth of Organic Materials

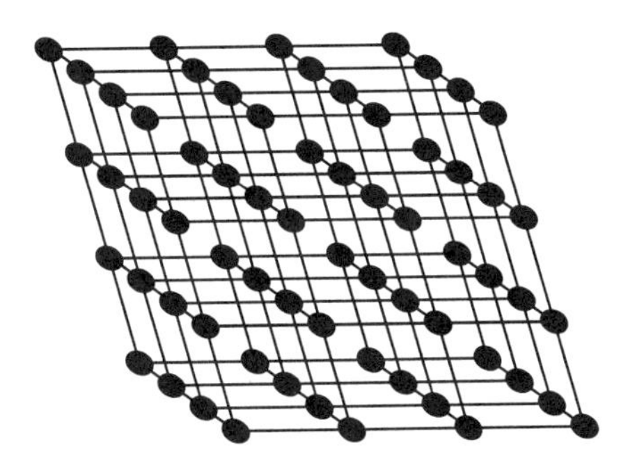

EDITED BY
Allan S. Myerson
Polytechnic University

Daniel A. Green
DuPont

Paul Meenan
DuPont

Proceedings of a conference sponsored
by E.I. du Pont de Nemours and Company,
Biosym/Molecular Simulations,
and American Institute of Chemical Engineers
Washington, D.C.,
August 27–31, 1995

American Chemical Society, Washington, DC

Library of Congress Cataloging-in-Publication Data

Crystal growth of organic materials / edited by Allan S. Myerson, Daniel A. Green, Paul Meenan.

p. cm.—(Conference proceedings series, ISSN 1054–7487)

"Proceedings of a conference sponsored by E.I. du Pont de Nemours and Company, Biosym/Molecular Simulations, and American Institute of Chemical Engineers, Washington, D.C., August 27–31, 1995."

Papers from the Third International Workshop on Crystal Growth of Organic Materials.

Includes bibliographical references and index.

ISBN 0–8412–3382–9 (alk. paper)

1. Crystallization—Industrial applications—Congresses. 2. Chemistry, Organic—Industrial applications—Congresses.

I. Myerson, Allan S., 1952– . II. Green, Daniel A., 1958– . III. Meenan, Paul, 1965– . IV. E.I. du Pont de Nemours and Company. V. Biosym/Molecular Simulations Inc. VI. American Institute of Chemical Engineers. VII. International Workshop on Crystal Growth of Organic Materials (3rd: 1995: Washington, D.C.) VIII. Series: Conference proceedings series (American Chemical Society)

TP156.C57C79 1996
661′.8—dc20

96–1348
CIP

PRINTED IN THE UNITED STATES OF AMERICA

Conference Sponsors

E.I. du Pont de Nemours and Company
Biosym/Molecular Simulations
Co-Sponsored by the American Institute of Chemical Engineers

Organizing Committee and Scientific Committee

Allan S. Myerson
Polytechnic University

Daniel A. Green
DuPont

Paul Meenan
DuPont

Kevin J. Roberts
University of Strathclyde

G. M. Van Rosmalen
Delft University of Technology

P. Bennema
University of Nijmegen

R. W. Rousseau
Georgia Institute of Technology

D. J. Kirwan
University of Virginia

J. M. McBride
Yale University

R. Hartel
University of Wisconsin

D. J. W. Grant
University of Minnesota

K. Berglund
Michigan State University

M. Lahav
Weizmann Institute

L. Leiserowitz
Weizmann Institute

J. Bernstein
Ben Gurion University

C. Sink
Eastman Chemicals

A. Rasmuson
Royal Institute of Technology

J. Ulrich
University of Bremen

R. G. Blanks
Amoco Chemical Company

K. Toyokura
Waseda University

Contents

Preface .... **xi**

MOLECULAR MODELING APPLICATION TO CRYSTALLIZATION

The Application of Computational Chemistry to the Study of Molecular Materials **2**
R. Docherty

On the Crystallographic and Statistical Mechanical Foundations of the Hartman–Perdok Theory **15**
P. Bennema

An Ab Initio Approach to Crystal Structure Determination Using High Resolution Powder Diffraction and Computational Chemistry Techniques. **22**
Paul G. Fagan, Robert B. Hammond, Kevin J. Roberts, Robert Docherty, and Mike Edmondson

An Automated Method for PBC Analyses Applied to Benzoic Acid: A Comparison Between the Monomer and Dimer Analysis **28**
R. F. P. Grimbergen and P. Bennema

Growth Morphology of Crystals and the Influence of Fluid Phase **36**
X. Y. Liu and P. Bennema

HABIT95: A Program for Predicting the Morphology of Molecular Crystals When Mediated by the Growth Environment **43**
G. Clydesdale, K. J. Roberts, and R. Docherty

A Comparison of Binding Energy, Metastable Zone Width, and Nucleation Induction Time of Succinic Acid with Various Additives **53**
Shyh-ming Jang and Allan S. Myerson

Crystal Habit Modification of Caprolactam in the Presence of Carboxylic Acids: Modelling and Verification **59**
Silke Niehörster, Sabine Henning, and Joachim Ulrich

CRYSTALLIZATION IN THE PHARMACEUTICAL INDUSTRY

Regulatory Considerations in Crystallization Processes for Bulk Pharmaceutical Chemicals: A Reviewer's Perspective **66**
Wilson H. DeCamp

Crystal Growth of L-SCMC Seeds in a DL-SCMC Solution of pH 0.5 .... **72**
K. Toyokura, K. Mizukawa, and M. Kurotani

Acetaminophen Crystal Habit: Solvent Effects **78**
Daniel A. Green and Paul Meenan

Use of Single Crystal X-ray Structural Data in Prediction of Solid-State Properties of Alprazolam and Diltiazem Hydrochloride 85
Esa Muttonen, Veli Pekka Tanninen, Len Shields, and Peter York
Incorporation of Structurally Related Substances into Paracetamol (Acetaminophen) Crystals 95
Barry A. Hendriksen, David J. W. Grant, Paul Meenan, and Daniel A. Green
Optical Resolution of DL-SCMC in a Cooled Type Batch Operation 105
K. Toyokura, M. Kurotani, and I. Hirasawa

INDUSTRIAL CRYSTALLIZATION OF ORGANIC MATERIALS

Aspects of Melt Crystallization as a Separation Process 112
Joachim Ulrich
Crystal Growth Kinetics of Complex Organic Compounds 116
Donald J. Kirwan, Ilene B. Feins, and Amarjit J. Mahajan
Precipitation of EDTA from Solution: Effect of Process Conditions on Product Properties 122
R. Guardani and E. P. Fariello
Emulsion Solidification: Purification and Crystallization 127
R. J. Davey, J. Garside, A. M. Hilton, and J. W. Morrison
Solid Layer Melt Crystallization: A Fractionation Process for Milk Fat 137
Michaela Tiedtke, Joachim Ulrich, and Richard W. Hartel
On the Behavior of Adipic Acid Aqueous Solution Batch Cooling Crystallization 145
S. Derenzo, P. A. Shimizu and M. Giulietti
A Fullerene Route to Fuel Additives 151
George W. Schriver, Abhimanyu O. Patil, Kenneth Lewtas, and David J. Martella
Thermodynamic Properties of Supersaturated Solution and Their Use in Determination of Crystal Growth Kinetic Parameters 157
Soojin Kim and Allan S. Myerson
The Migration of Liquid Inclusions in Solid Layers 163
Sabine Henning, Joachim Ulrich, and Silke Niehörster

CRYSTALLIZATION IN THE FOOD INDUSTRY

Controlling Crystallization in Foods 172
Richard W. Hartel
Carminic Acid as a Fluorescent Probe of Sugar Solution Composition 178
Robert J. Richards, Beatrice A. Torgerson, and Kris A. Berglund
Melt Crystallization Behavior of Long Chain Milk Fat Triglycerides 185
Frank Z. Yang, Rebeca Baca-Diaz, Danping Shen, Armand Boudreau, and Joseph Arul
The Effect of Physical State and Glass Transition on Crystallization in Starch 196
Kirsi Jouppila and Yrjö H. Roos

Effect of Impurities on Solubility, Growth Behavior, and Shape of Sucrose Crystals: Experimental and Modelling 200
Masashi Momonaga, Silke Niehörster, Sabine Henning, and Joachim Ulrich

Crystallization and Polymorphism in Cocoa Butter Fat: In-Situ Studies Using Synchrotron Radiation X-ray Diffraction 209
R. N. M. R. van Gelder, N. Hodgson, K. J. Roberts, A. Rossi, M. Wells, M. Polgreen, and I. Smith

Growth and Dissolution Rate Dispersion of Sucrose Crystals 216
Jürgen Fabian, Richard W. Hartel, and Joachim Ulrich

General Papers

Study of Organic Supersaturated Solutions: Theory and Experiment 222
Allan S. Myerson and Alexander F. Izmailov

Growth, Perfection, and Defects in C_{60} and C_{70} Single Crystals 231
K. Kojima, M. Tachibana, Y. Maekawa, H. Sakuma, M. Michiyama, K. Kikuchi, and Y. Achiba

Generation of Chirality in Molecular Crystals Composed of Two Different Achiral Molecules 239
Hideko Koshima, Kuiling Ding, Yosuke Chisaka, Naoya Naka, and Teruo Matsuura

Role of MeOH in Chiral Combination of Host–Guest Molecules in the Inclusion Crystal 246
Koichi Tanaka, Fumio Toda, and Ken Hirotsu

The Effect of Chromium Ion (III) on the Nucleation of Supersaturated Ammonium Sulfate Solutions 249
Wei-Ming Sun and Allan S. Myerson

Crystallization of the Molecular Crystals Between Nitroanilines and Nitrophenols 256
Hideko Koshima, Yang Wang, Teruo Matsuura, Hisashi Mizutani, Hiroyuki Isako, Ikuko Miyahara, and Ken Hirotsu

Gel Growth of the Organomineral Crystal 2-Amino-5-nitropyridinium Dihydrogen Phosphate and the Quadratic Nonlinear Optical Effect 259
N. Horiuchi, Y. Uesu, F. Lefaucheux, and M. C. Robert

A Simple Inexpensive Bridgman–Stockbarger Crystal Growth System for Organic Materials 263
J. Choi, M. D. Aggarwal, W. S. Wang, R. Metzl, K. Bhat, Benjamin G. Penn, and Donald O. Frazier

New and Effective Technique To Study Supersaturated Solutions by the Pulsed (Spin–Echo) NMR Method 266
Alexander R. Kessel, Aleksey N. Temnikov, Alexander F. Izmailov, and Allan S. Myerson

Diffusion Coefficient and Viscosity in Crystal Growth in Microgravity 273
Michael Bohenek and Allan S. Myerson

Index 280

Preface

CRYSTALLIZATION IS AN IMPORTANT SEPARATION AND PURIFICATION TECHNIQUE used in the food, pharmaceutical, and chemical industries. Although much literature exists on the crystallization of inorganic commodity chemicals, until recently the same level of interest has not been devoted to the crystal growth of organic materials. In addition the study of organic crystallization is a multidisciplinary area with chemical engineers, organic solid-state chemists, pharmaceutical scientists, and physicists all involved. To foster interdisciplinary research and provide for information exchange, international workshops on the crystal growth of organic materials have been organized. The first two workshops were held in Japan (1989) and Scotland (1992). The Third International Workshop on Crystal Growth of Organic Materials was held August 27–31, 1995, in Washington, D.C.

The goal of the workshop was to focus on research in emerging areas as well as to provide a forum for the presentations in areas of interest to the food and pharmaceutical industries and to commodity and specialty chemical manufacturers. The main conference topics were molecular modeling: application to crystallization, crystallization in the pharmaceutical industry, industrial crystallization of organic materials, crystallization in the food industry, and general subjects. This volume is organized in the same fashion and contains selected oral and poster presentations from the meeting.

Acknowledgments

We thank all the conference participants for making the conference a scientific and social success. In addition, we thank the scientific committee members for their help in reviewing the papers that appear in this volume and in aiding in the development of the program. We also thank the staff in the chemical engineering department at Polytechnic University for their help in all aspects of the conference organization. We gratefully acknowledge the financial support of DuPont and Biosym/Molecular Simulations Inc.

ALLAN S. MYERSON
Department of Chemical Engineering
Polytechnic University
Brooklyn, NY 11201

DANIEL A. GREEN
DuPont Experimental Station
Building 304
Wilmington, DE 19898

PAUL MEENAN
DuPont Experimental Station
Building 304
Wilmington, DE 19898

January 25, 1996

Molecular Modeling Application to Crystallization

The Application Of Computational Chemistry to the Study Of Molecular Materials.

R. Docherty, Zeneca Specialties Research Centre, Hexagon House, Blackley, Manchester England M9 8ZS.

An understanding of the specific arrangements of molecules in the solid state allows the chemist to manipulate the solid state to optimise the performance characteristic of interest. In this paper the application of molecular modelling and computational chemistry techniques to the study of molecular materials will be described including lattice energy calculations, crystal shape prediction property estimation methods and crystal structure prediction/determination. The potential and limitations of these methods will be discussed.

Introduction

An understanding of the specific arrangements adopted by molecules within the crystal lattice allows the solid state chemist to manipulate the crystal chemistry to optimise the performance characteristic of interest.

Given the importance of crystal engineering and polymorph control to the development and production of a vast range of speciality chemicals (i.e. pharmaceuticals, agrochemicals, pigments, dyes, opto-electronic materials and explosives) it is not surprising that many techniques are applied to improve our understanding of the solid state structure of such materials. Over recent years molecular modelling and computational chemistry have played and increasingly important role in this field.

The solid state arrangement(s) adopted by a molecules depends on the subtle balance of intermolecular interactions that it can achieve for a given conformation in a particular packing arrangement. Crystallisation and the properties of the solid state are dependent on a process which is essentially molecular recognition on a grand scale. Polymorphism and changes in properties are due to the recognition of different balances of these subtle interactions. In this paper the use of molecular modelling to establish the link between molecular structure, intermolecular interactions packing motifs and solid state property will be illustrated.

Molecular Modelling

Molecular modelling and computational chemistry methods operate on three levels. Firstly molecular modelling is almost unique in it's ability in allowing the examination of the detailed, complex and often elegant arrangements of molecules, proteins, fibres, polymers, surfaces and solid state structures. Secondly in conjunction with computational chemistry molecular modelling permits the determination of structure activity relationships (SAR's) linking calculated properties and hence molecular structure to performance characteristics. Ultimately it enables (based in these SAR's) the design of novel molecular/solid state structures with improved properties and performance characteristics.

The Crystal Chemistry Of Molecular Materials

The structures and crystal chemistry of molecular materials are often classified into different categories according to the type of intermolecular forces present. These include;

* simple Van der Waals attractive interactions,
* classical hydrogen bonding (Taylor, 1982),
* electrostatic interactions,
* C-H:::::O non classical hydrogen bonding,
* short directional contacts (Desiraju, 1989).

Molecules can essentially be regarded as impenetrable systems whose shape and volume characteristics are governed by the molecular conformation and the radii of their constituent atoms. The general uneven, awkward shape of molecular structures tends to result in unequal unit cell parameters being adopted during crystallisation. The vast majority of the structures

1054–7487/96/0002$12.00/0

reported prefer the triclinic, monoclinic and orthorhombic crystal systems.

A useful parameter for judging the efficiency of a molecule for using space in a given solid state arrangement is the packing co-efficient (PC). This model assumes that the molecules within the crystal will attempt to pack in a manner such as to minimise the amount of unoccupied space (Kitaigorodskii, 1973).

In general there is a rough correlation between higher PC values and increasing size of large flat aromatic molecules i.e perylene (0.8). Once even slight deviations from planarity are introduced the effective packing ability falls. For benzophenone, an aromatic ketone ($Ph_2C{=}O$) the phenyl groups are twisted to 54° with respect to each other and the PC falls to 0.64. One of the most interesting features of such a model is the low PC for hydrogen bonded systems such as urea (0.65) and benzoic acid (0.62). This rather surprising feature is due to the rather open architecture of hydrogen bonded structures which is the result of a need to adopt particular arrangements to maximise the hydrogen bonded network (Etter, 1991).

In general it is probably safe to suggest that in the majority of cases with molecular materials it is the desire to pack efficiently is the single biggest driving force towards selected structural arrangements. The notable exceptions will be in cases where the need to form complex hydrogen bonding networks will override this need. Weaker interactions such as special hydrogen bonds and polar interactions are probably not primary movers in the arrangements adopted but will tend to be optimised within a given efficient arrangement.

Lattice Energy Calculations

In order to understand the principles which govern the wide variety of solid state properties and structures of organic materials it is important to describe both the energy and nature of interactions in specific orientations and directions. As a result of the pioneering work of Williams (1966) and Kitaigordskii (1968) in the development of atom-atom potentials it is now possible to interpret packing effects in organic crystals in terms of interaction energies. The basic assumption of the atom-atom method is that the interaction between two molecules can be considered to simply consist of the sum of the interactions between the constituent atom pairs.

The lattice energy E_{latt} (often referred to as the crystal binding or cohesive energy), can for molecular materials be calculated by summing all the interactions between a central molecule and all the surrounding molecules. If there are n atoms in the central molecule and n' atoms in each of the N surrounding molecules then lattice energy can be calculated by Equation (1).

$$E_{latt}=1/2\sum_{k=1}^{N}\sum_{i=1}^{n}\sum_{j=1}^{n'}V_{kij} \qquad (1)$$

V_{kij} is the interaction between atom i in the central molecule and atom j in the k'th surrounding molecule. Each atom-atom interaction pair consists of a Van der Waals attractive and repulsive interaction, an electrostatic interaction and in some special cases a hydrogen bonding potential. Figure 1 shows the profiles of the calculated lattice energy as a function of summation limit for α-glycine, anthracene, ß-succinic acid and urea. These plots show the same general trend, on increasing the summation limit there is an initial increase in the lattice energy is recorded. This is followed by the reaching of a plateau region beyond 20Å. Further increase in the summation limit has no effect on the calculated lattice energy.

The validity of the potentials can to some extent be tested by comparing the theoretical values against the experimental sublimation enthalpy. Table 1 contains a selection of calculated lattice energies and experimental sublimation enthalpies. Figure 2 shows a plot of calculated against experimental lattice energies for a range of around eighty compounds. The molecular classes reported include a wide range of molecular materials. The excellent agreement between theory and experiment is clear, the mean error is 1.5 kcal/mol and the maximum error 3.5 kcal/mol. The average difference between calculated and experimental less than 6%.

Intermolecular Interactions

A particular advantage of the calculated lattice energy is that it can be broken down into the specific interactions along particular directions and further partitioned into the constituent atom-atom contributions.

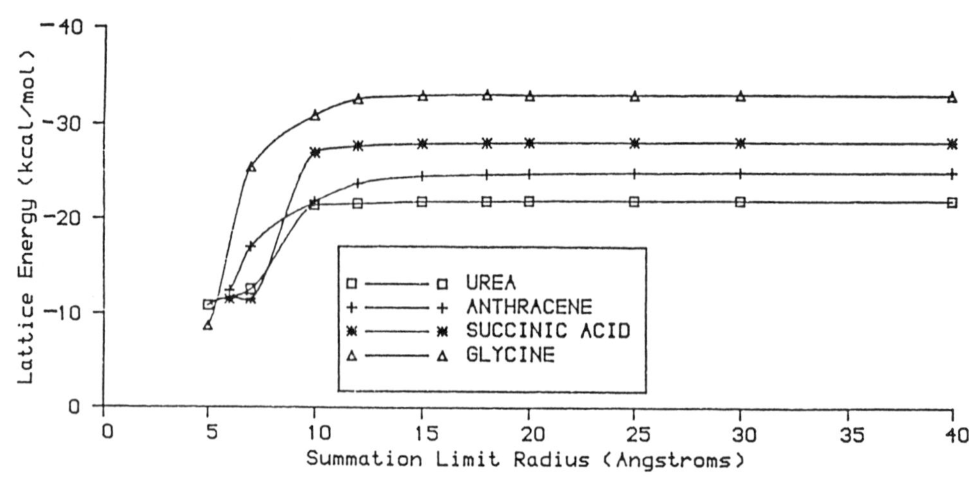

Figure 1 The calculated lattice energy as a function of summation limit for anthracene, urea, succinic acid and glycine

Table 1 Calculated and 'experimental' lattice energiess for a range of molecular materials. This is a subset of the data presented in Figure 2.

Material	Lattice Energies (kcal/mol) Calculated	Experimental
n-Octadecane	-35.2	-37.8
Biphenyl	-21.6	-20.7
Napthalene	-19.4	-18.6
Anthracene	-24.9	-26.2
Perylene	-32.5	-31.0
Benzophenone	-24.5	-23.9
Trinitrotoluene	-25.1	-24.4
Glycine	-33.0	-33.8
L-alanine	-33.3	-34.2
Benzoic acid	-20.4	-23.0
Urea	-22.7	-22.2
ß-succinic acid	-30.8	-30.1

As a result it is possible to build up an understanding of the interactions which contribute to particular packing motifs. The study of the strength and geometry of intermolecular interactions remains an area of active research as it is a key element in molecular solid state chemistry (as described in the elegant work by Etter, 1991), in the design of molecular aggregates and in the understanding and construction of molecular recognition complexes for biologically interesting substrates (Chang and Hamilton, 1988).

Urea (see Fig 3) has a three dimensional arrangement of hydrogen bonds where each urea molecule is surrounded by six other urea molecules. This cluster is responsible for 85% of the total lattice energy. The important intermolecular interactions are given in Table 2.

The calculated intermolecular interactions, in particular the weaker ones, can be further examined to determine their relative importance, geometry and strength by using the vast amount of experimental data available in the Cambridge Crystallographic Database.

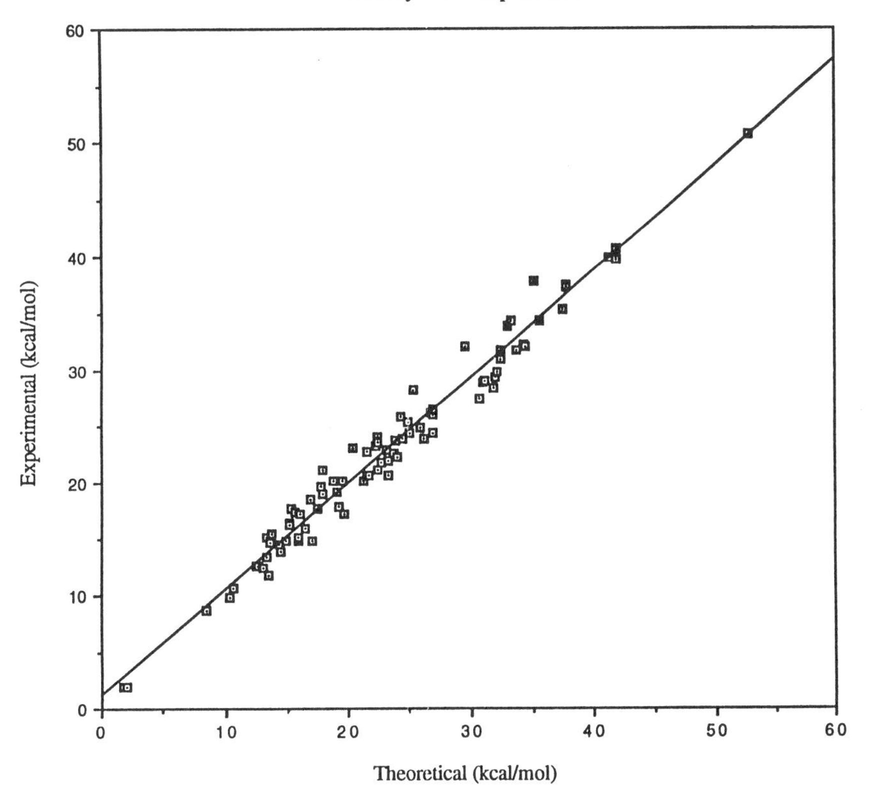

Figure 2 The calculated lattice energy against the experimental lattice energy derived from the experimental sublimation enthalpy

The Cambridge Crystallographic Database

Over the past ten years total number of structures stored in the Cambridge Crystallographic Database has increased from 40,000 to over 130,000. There has also been a noticeable improvement in the quality of the structures reported and in the complexity of the structures solved. The crystallographic discrepancy factor R has fallen from an average of 12% in 1965 to a present value of 5%. At the same time the number of atoms in the structures being investigated has increased from about 20 to 50 (Allen et al , 1983).

The database stores and can be searched on, three main categories of information for each entry including the bibliographic summary, the two dimensional chemical structure, and the full three dimensional structure details. The full three dimensional crystal structure information includes the unit cell dimensions, the space group symmetry and the individual atomic fractional co-ordinates. A specific full three dimensional query can now be introduced. This allows particular intermolecular bonding patterns to be examined both within molecules and between molecules. Clearly this is a powerful search facility which increases value of the information already within the database.

The occurrence, relative importance, strength and geometry of different intermolecular interactions can be assessed. An example of such an analysis concerns the role of weaker interactions such as special hydrogen bonds (e.g. C-H:::O=C interactions). They have been considered in detail (Taylor and Kennard, 1984 Desiraju, 1989). These 'special' hydrogen bonds are generally weaker than traditional hydrogen bonds and spectroscopic, crystallographic and theoretical investigations continue to probe their magnitude and directionality. An analysis of the database suggests that the strength of the

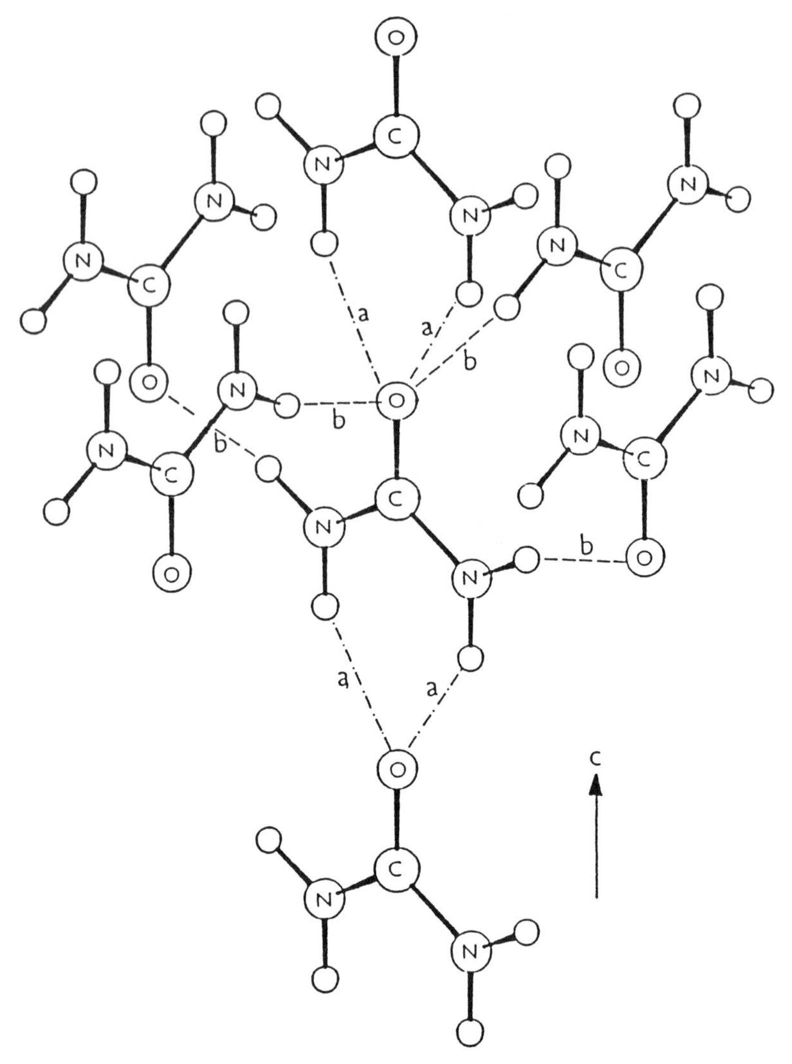

Figure 3 The three dimensional structure of a urea cluster involving six surrounding molecules

Table 2 The important intermolecular interactions in urea.

Position UVW	Interaction Energy Z	(kcal/mol)	Bond Type	Fraction (%)
001	1	-3.64	a	16.1
00-1	1	-3.64	a	16.1
100	2	-3.00	b	13.2
-110	2	-3.00	b	13.2
000	2	-3.00	b	13.2
010	2	-3.00	b	13.2

interaction depends on the acidity of the hydrogen involved and so the strength order C(sp)-H > C(sp^2)-H > C(ar)-H > C(sp^3)-H (Desiraju, 1989).

Crystal Shape Calculation

Early crystallographers were fascinated by the flat, plane and symmetry related external faces in both natural and synthetic crystallised solids. This led them to postulate that the ordered external arrangement was a result of an ordered internal arrangement.

Morphological simulations based on crystal lattice geometry were proposed by Bravais, Friedel, Donnay and Harker (1937). This work spans over seventy years but is often quoted collectively by modern workers as the BFDH law. Hartman and Perdok (1955) were the first to extensively quantify the crystal morphology in terms of the interaction energies between crystallising units. They used the assumption by that the surface energy is directly related to the chemical bond energies and identified chains of 'strong' intermolecular interactions called periodic bond chains (PBC). The strength of a PBC is determined by the weakest link in that chain. They also characterised flat, stepped and kinked faces according to the number of PBC's that are present in these faces.

Attachment and slice energies can be determined from a PBC analysis or calculated directly from the crystal structure by partitioning the lattice energy calculated from each symmetrically independent molecule in the unit cell into slice and attachment energies (Berkovitch-Yellin, 1985). The slice energy is defined as the energy released on the formation of a growth slice of a particular thickness (Hartman and Bennema, 1980). The attachment energy is defined as the energy released on the attachment of a particular growth slice onto the surface in a specified orientation. Faces with the lowest attachment energies will be the slowest growing and therefore will be the morphologically most important.

Figure 4 shows a precursor in the manufacture of the agrochemical paclobutrazol (Black et al, 1990). The molecule crystallises in the orthorhombic space group $P2_12_12_1$ with four molecules in a unit cell of dimensions a=5.799, b=13.552 and c=19.654Å. The packing of the molecule is dominated by chains of C-H::::O

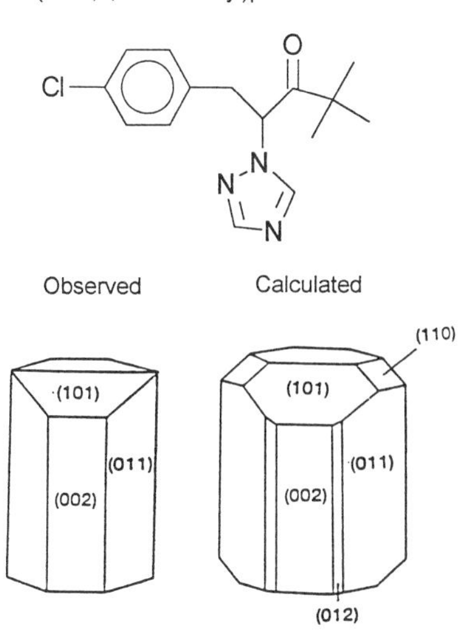

Figure 4 The molecular structure and experimental and calculated morphologies for the intermediate in the manufacture of paclobutrazol.

interactions running in the b-direction and a cascade of edge to face interactions running down the a-direction. The calculated crystal shape based on attachment energies shows excellent agreement with the observed morphology as shown in Figure 4. The only major differences is the prediction of the (012) faces which are not usually observed.

It has been established over a number of years that impurities can have drastic effects on crystal shape. The elegant work of the Weizmann Institute (Lahav et al, 1985) has shown that for molecular crystals this can be accounted for through structural explanations.

Tailor made additives can be broken down into two different categories, the 'disruptive and 'blocker' types (Clydesdale et al, 1994). Disruptive additives are generally smaller than the host system. They are structurally similar to the host and use this similarity to enter a surface site where there difference is not recognized. Additional on-coming molecules encounter a normal surface except in an area on a particular face where the disruptive molecule is situated. This is shown in Figure 5. For disruptive additive this results in a failure to complete the proper intermolecular bonding sequence normally

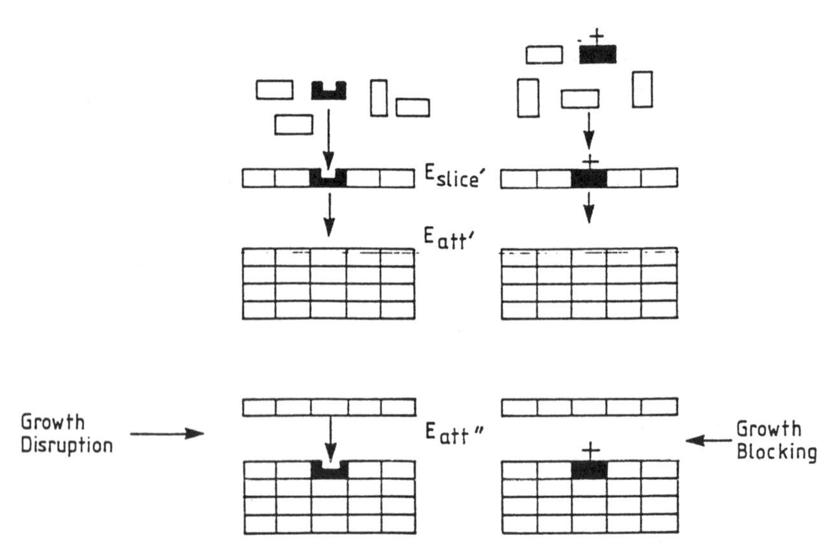

Figure 5 Schematic representation of the growth of crystals in the presence of tailor made disruptive and blocker additives.

adopted by the pure host system. Altering a particular bonding sequence in a certain direction affects the growth rate along that direction relative to the unaffected faces and consequently alters the relative growth rates.

Blocker type additives are usually bigger than the host system. Again the additive is structurally similar to the host but normally with an end group that differs significantly from the host structure. The part of the additive that is structurally similar to the host is then accepted into certain faces. These faces again recognise the similarity regions of the impurity and not the differences. The differing units in the blocker then sit on the respective crystal faces and prevent on-coming molecules getting into their rightful positions at the surface. This is shown in Figure 5.

In order to model these changes in morphology induced by these additives a number of new parameters must be calculated. The difference in binding energy for the additive compared to the host molecule (in a particular site) can be used as a measure of the ease of incorporation of the impurity. The effect of the impurity on subsequent growth can then be considered.

This can be illustrated by considering the effect of toluene on the morphology of benzophenone (Roberts et al, 1994). Figure 6 shows the morphology of benzophenone grown from the melt and from a toluene solution. The most dramatic effect is the considerable increase in the importance of the {021} face. Figure 6 also shows the benzophenone structure down the a-axis with a toluene molecule fitted on the {021} plane.

Sublimation Enthalpy Estimation

Measurement of thermochemical data such as the sublimation enthalpy can be a costly and time-consuming process. Prediction of solid state properties before a compound is synthesised would obviously be of considerable use.

Recent applications of statistical methods including neural networks include the determination of structure-activity relationships in drug design and the classification of spectra. Thermochemical data such as solubilities and boiling points of organic heterocycles have also been the subject of investigation. There are also a number of reviews of the use NNs in chemistry (Lacy, 1990 and Zupan and Gasteiger, 1993).

The work of a number of authors including (Gavezzotti, 1991) has shown that the lattice energy (and consequently the sublimation enthalpy) can be predicted based upon linear regression studies on molecular crystals. The heat of sublimation can be estimated from the packing potential energy (PPE) (Gavezzotti, 1991) which correlates with molecular descriptors such as molecular weight, Van der Waals' surface, volume and molecular outer surface.

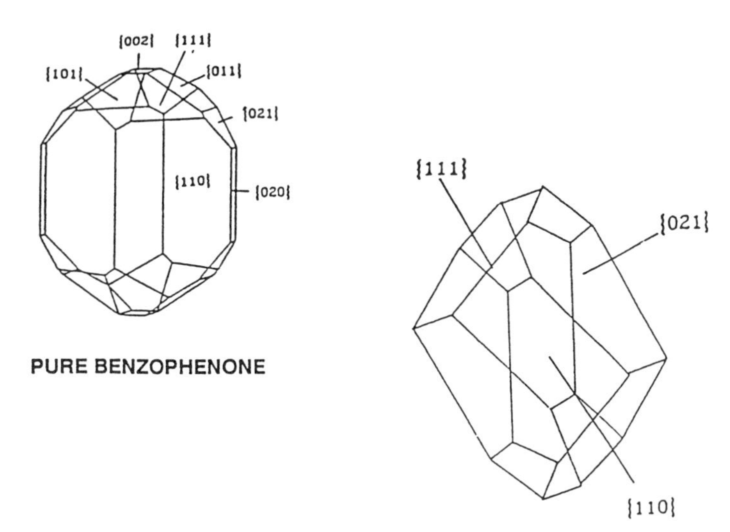

BENZOPHENONE GROWN WITH TOLUENE

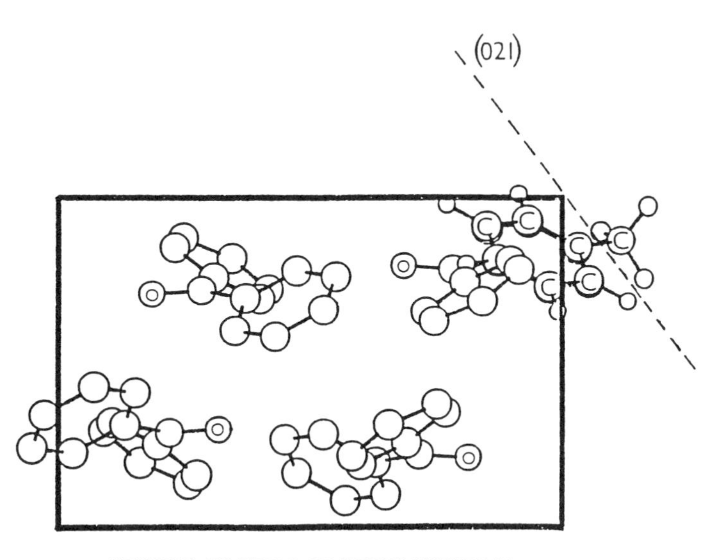

CRYSTAL PACKING OF BENZOPHENONE

Figure 6 The morphology of pure benzophenone, benzophenone grown in toluene and the packing of benzophenone

Recently a feed forward Neural Network (NN), has been trained to reproduce sublimation enthalpies for 62 molecules including aliphatic hydrocarbons through to amino acids (Charlton et al 1995). The results are compared both with calculated values from traditional crystal packing techniques and with a multilinear regression analysis (MLRA) model. The molecular descriptors input to the NN was kept as simple as possible to facilitate the prediction of values for novel molecules. The parameters are the number of carbon atoms (C), the number of hydrogen atoms (H), the number of nitrogen atoms (N), the number of oxygen atoms (O), the number of π-atoms (PI), the number of hydrogen bond donors (HBD) and the number of hydrogen bond acceptors (HBA). The MLRA uses the same input data as the NN, although some covariance was found between the input parameters, indicating that the number of inputs could be reduced. It is therefore concluded that a simple 3-parameter MLRA model fitted the data used. A plot of calculated values of sublimation enthalpy based on Equation 2 is shown in Figure 7. The mean and maximum errors are given in Table 3 for the various methods.

$$SE = 3.47 + 1.41C + 4.55HBD + 2.7HBA \quad (2)$$

Crystal Structure Prediction

The generation by computational methods of reliable solid state structural details and solid state properties on novel materials from only molecular descriptors remains both a major scientific goal and the subject of some controversy (Maddox, 1988 and Gavezzotti, 1994). It is often difficult/impossible to obtain crystals of sufficient size and quality, in the

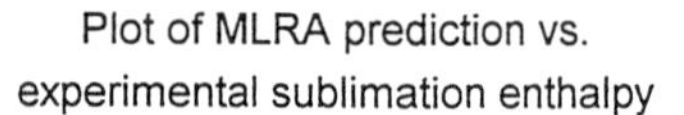

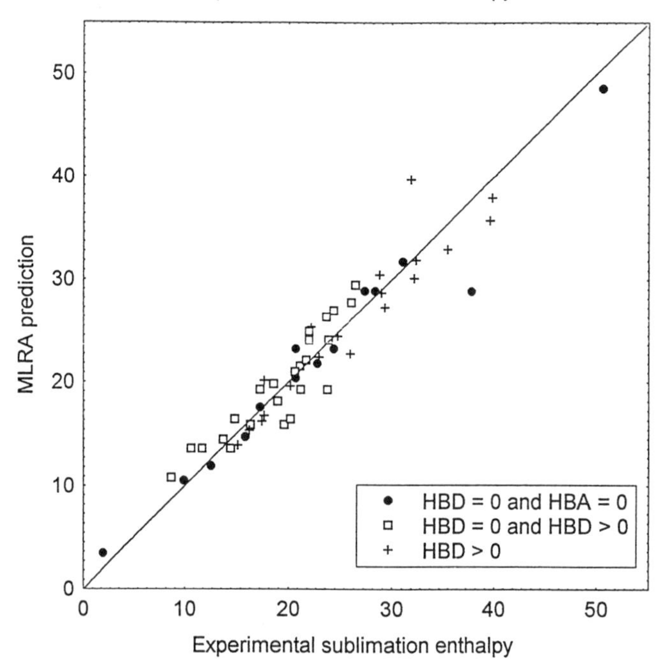

Figure 7. The plot of the sublimation enthalpy determined from equation (2) against experiment.

Table 3 A summary of the predicted sublimation enthalpy results

	Theory[a]	NN[b]	MLRA[c]
Max. Error	3.5	10.1	8.9
Mean Error	1.4	2.5	1.8
r^2	0.97	0.87	0.92

[a] Theoretical prediction using crystal packing.
[b] Neural network prediction (Leave One Out Experiment).
[c] Multilinear Regression Analysis prediction

desired polymorph in order to study a structure by conventional single crystal methods.

The lack of a general approach enabling the prediction of the solid state structure solely from molecular structure has been described as ***'one of the continuing scandals in the physical sciences'*** (Maddox, 1988). Progress has been hindered due to problems with global minimisation and force-field accuracy. The electrostatic component of the force-field has received much consideration (Price and Stone, 1984). These problems have limited the routine application of such methodologies to industrially important molecules. The methods being developed for structure prediction usually involve three stages;

- trial structure generation,
- clustering of similar trial structures,
- refinement of trial structures.

A systematic efficient search/structure generation algorithm is needed to generate all the potential structures. It is important that a sufficient sample of structural space is covered so that not only local minima are located but the global minimum as well. A reduction in these trials is achieved by either a similarity clustering procedure or filtering approach to remove the similar or unlikely candidates. Finally refinement of the potential structures is carried out using lattice energy minimisations. During refinement the lattice energy is refined with respect to the structural variables. The unit cell dimensions (**a,b,c,** α**, ß and** γ) are allowed to alter along with the molecular orientation and translational parameters.

Despite the inherent difficulties, predictions from first principles remains an admirable scientific goal and the subject of much investigation reflected in the elegant work of Gavezzotti (1992), Holden (1993), Perlstein (1994) and Karfunkel (1992).

Gavezzotti (1992) and his co-workers have pioneered the concept of the 'molecular nuclei'. Given a molecular structure clusters of molecules (nuclei) are built using various symmetry operators. The relative stability of each of these nuclei is appraised through a calculation of it's intermolecular energy. When these clusters (nuclei) have been selected a full crystal structure can be generated using a systematic translational search. The translational search is carried out in a systematic manner with lines, layers and full three dimensional structures being built. A further selection process is used based on packing co-efficient and packing density limits to filter out the most likely structures. These are then refined using lattice energy minimisation techniques. The results for this approach are very impressive.

A similar process was described by Holden (1993) who considered an approach in terms of molecular co-ordination spheres. From a detailed examination of the Cambridge Crystallographic Database they identified co-ordination numbers for various space groups. They restricted the search to molecules containing just C,H,N,O and F with only one molecule in the asymmetric unit (i.e. no solvates or complexes).

The construction of possible full crystal packing patterns follows in three stages. Initially a line of molecules is established by moving a second molecule towards the central molecule until a specified interaction criteria is meet. A two dimensional grid is then organised by moving a line of molecules towards the central line. The final step involves moving a two dimensional grid parallel to the central grid. The orientation of the molecules within the two dimensional grids and the three dimensional packing arrangement depends on the symmetry operations within the space group being investigated. In a number of cases the lowest energy theoretical structure is in good agreement with the experimental arrangement.

In recent publications Perlstein (1994) has used a Monte Carlo cooling approach to describe the packing of one dimensional stacked aggregates including perylenedicarboximide pigments. Starting aggregates are generated using five identical molecules randomly orientated at set separation distances along a defined axis. The Monte Carlo cooling is carried out from 4000K to 300K. Each Monte Carlo step involves randomly rotating the central molecule and regenerating the other molecules along the axis at the set distances. The energy of this arrangement is then computed and compared against the previous energy and either accepted or rejected based on the acceptance criteria of the Monte Carlo algorithm. This process is repeated twenty times for both rotational angles and then the separation distance r. The temperature is then cooled by 10% and the process repeated. At a temperature of 300K the lowest energy structure found is accepted as a local minima. This whole procedure is repeated until a number of local minima (sufficient to include the global minimum) have been found.

The results show good agreement between predicted and experimental values for the interplanar separation distance, and the longitudinal and the transverse displacement. Although this is not three dimensional structure prediction it is a technique which has potential to contribute to the problem of predicting three dimensional structures. It is of particular interest that these results have been obtained for molecules with flexible end groups with non-bonded interaction and torsional terms taken into account during the simulation.

Recently Karfunkel (1993) has described an approach to the crystal packing problem based around a modified Monte Carlo simulated annealing process. The modification proposed by the authors involves the separation of the variables that define the spatial extensions of the crystal from those concerned with the angular degrees of freedom. This removal of the variables responsible for spatial extension of the crystal from the Monte Carlo search space reduces the tendency for the crystal to remain as a gas. Essentially each Monte Carlo step chooses a set of independent angular variables. For each set of angular variables the parameters affecting the spatial extension of the crystal are optimised. This is initially done by relieving any bad contacts followed by optimisation with respect to the crystal energy.

The results of this methodology are extremely promising. The postulated structures show excellent agreement with experiment, the more sophisticated the charge description the better the description of the structure (Leusen, 1995).

Although these methods are promising they do have a number of inherent assumptions. They all tend to use rigid probes as their input structures and so conformational flexibility remains problematic. They are all at present restricted to one molecule in the asymmetric unit. It is interesting to note that the difference between the most stable and other potential structures is usually only a fraction of a kcal This is a common feature amongst all the prediction methods. The magnitude of the differences might be a consequence of difficulties in the method, in particular the accuracy of force-fields, or the shallowness of the potential surface being examined. It may also be due to the existence of as yet unknown and/or unstable polymorphs.

Structure Solution From Powders

Clearly the limitations described above has hindered the routine application of any one of the structure prediction methods to commercially important materials. Although it is sometimes difficult to get a single crystal of sufficient size and quality for traditional single crystal studies it is usually possible to get a powder of the material of interest. Traditionally structure solution using the Rietveld method (Rietveld, 1969) has been restricted to inorganic structures but recently there has been an increase in the number of studies on molecular materials. Starting structures for refinement have conventionally been generated using single crystal techniques which involve the separation and integration of individual peaks. These approaches tend to struggle with low symmetry systems where peak overlap is significant. Computational chemistry methods are currently being developed allow better trial structure generation using a combination of molecular orbital calculation and crystal packing calculations (Fagan et al, 1994) as well as Monte Carlo algorithms (Harris et al 1994) and Newsam (1993).

The solution of a structure by the Rietveld method can be considered as a three stage problem;

- data collection, index, space group
- trial structure generation
- Rietveld Refinement

The first stage involves the collection of the data, the indexing of the resulting pattern and the determination of the unit cell dimensions and the space group. The third stage involves the actual Rietveld refinement stage where slight changes to the packing arrangement/molecular structure are carried out so that the simulated pattern from the proposed structure gets closer to the observed pattern. The second stage involves trial structure generation and both crystal packing calculations and Monte Carlo algorithms have been employed. The validity of the crystal packing/lattice energy method has been demonstrated by a powder determination for an unknown molecule with an independent single crystal study. Amongst the

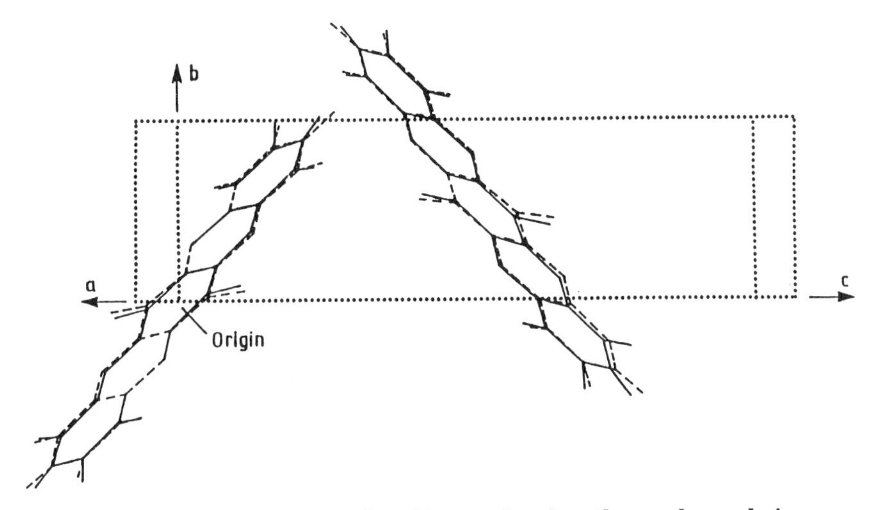

Figure 8. The single crystal packing overlayed on the powder analysis.

molecules studied was the commercially important triphendioxazine systems. 6,13-dichlorotriphendioxazine ($C_{18}H_8N_2O_2Cl_2$), is the basic chromophore unit of a number of commercially important dyestuffs. Figure 8 shows the overlay of the unit cell for the structure solved by powder diffraction with the packing of the arrangement determined by single crystal methods. The structures are very similar with only slight deviations. The structure of 6,13-dichlorotriphendioxazine is held together through π-π interactions down the b-axis. Non bonded Cl--H contacts at 2.8Å form dimers in the ac-plane.

The difference in with the ab-initio prediction methods is that in this approach the simulated/experimental powder trace is used as a final check of any trial/optimised structure and so concerns over global minimisation, force field accuracy and the description of the electrostatics is reduced.

Conclusions

Molecular modelling and computational chemistry techniques are playing an increasingly important role in the study of molecular materials. Through the use of lattice energy calculations it is possible to interpret solid state structures in terms of intermolecular interactions and hence molecular structure.

Ab-initio structure prediction remains an admirable long term goal and much progress has been made. At present the predictive methods use fixed molecular conformations and the unification of conformational search routines within crystal packing algorithms remains a major difficulty.

The use of a combination of high resolution x-ray diffraction and theoretical methods can provide structural information on molecular materials comparable with traditional single crystal methods. The refinement to the diffraction trace mitigates concerns over force field accuracy and global minimisation in crystal packing calculations.

References

Allen, F.H.; Kennard, O.; Taylor, R. *Acc. Chem. Res.* **1983** 16, 146.

Berkovitch-Yellin, Z.; J. Amer. Chem. Soc., **1985**, 107, 8239.

Black, S.N.; Williams, L.J.; Davey, R.J.; Moffat, F.; McEwan, D.M.; Sadler, D.E.; Docherty, R.; Williams, D.J. *J. Phys. Chem.* **1990**, 94, 3223

Chang, S.; Hamilton, A.D. *J. Amer. Chem. Soc.* **1988** 110, 1318.

Charlton, M.H.; Docherty, R.; Hutchings, M.G. *Faraday Trans.* **1995** (in press).

Clydesdale, G.; Roberts, K.J.; Docherty, R. *J. Crystal Growth*, **1994**, 135, 331

Desiraju, G.R., *Crystal Engineering - The Design Of Organic Solids*, Elsevier, Amsterdam, **1989.**

Desiraju, G.R.; Gavezzotti, A., *Acta Crystallogr. Sect B*, **1988** 44, 427

Desiraju, A.R.; Jagarlapudi, A.R.P., *Chem. Soc. Perkin Trans* **1987**, 2, 1195.

Docherty, R.; Roberts, K.J.; Saunders, V.; Black, S.N.; Davey, R.J., *Faraday Discuss*, **1993** 95, 11.

Docherty, R.; Clydesdale, G.; Roberts, K.J.; Bennema, P., *J. Phys. D. Applied Physics* **1991** 24 89.

Donnay, J.D.; Harker D., *Amer. Miner.* **1937, 22**, 446.

Etter, M.C., *J. Phys. Chem.* **1991** 95, 4601.

Fagan, P.G.; Roberts, K.J.; Docherty R.; Chorlton, A.P.; Jones, W.; Potts, G.P. *Chem. Comm* (in press)

Gavezzotti, A., *J. Phys. Chem.*, **1991**, 95, 8948.

Gavezzotti, A., *Acc Chem. Res*, **1994** ,27, 309.

Gavezzotti, A., *J. Amer. Chem. Soc.* **1981** 113, 4622.

Harris, K.D.M.; Tremayne, M.; Lightfoot, P.; Bruce, P.G.C. *J. Amer. Chem. Soc.* **1994, 116** 3543.

Hartman, P.; Bennema, P., *J. Cryst. Growth*, **1980**, 49, 145.

Hartman, P.; Perdok, W.G., Acta Crysttogr., **1955** 8 49.

Holden, J.R.; Du, Z.; Ammon, L. *J. Computational Chemistry* **,1993**, 14, 422.

Hunter, C.A.; Sanders, J.K.M., J. Amer. Chem Soc. **1990** 112, 5525.

Karfunkel,R.; Gdanitz, R.J., *J. Computational Chemistry*, **1992** 13, 1771 .

Kitaigordskii, A.I.; Mirskaya, K.V.; Tovbis, AB, *Sov. Phys. Crystallogr.* **1968** 13, 176.

Kitaigorodskii, A.I. *Molecular Crystals and Molecules* , Academic Press, New York, **1973**

Lacy,M.E, *Tetrahedron Computer Methodology*, **1990**, 3, 119.

Lahav, M.; Berkovitch-Yellin, Z.; Van Mil, J.; Addadi, L.; Idelson, M.; Leiserowitz, L., *Israel J. Chem*, **1985**, 25, 353.

Leusen, F.J.J.; Pinches, M.S.; Lovell, R.;

Karfunkel, H.R.; Paulus, E.F., *Chem Comm.* (in press)

Maddox, J., *Nature*, **1988**, 335, 201

Maginn, S.J.; Compton, R.G.; Harding, M.; Brennan, C.M.; Docherty, R. *Tetrahedron Letters*, **1993,** 34, 4349.

Newsam, J.M.; Deem, M.W.; Freeman, C.M. *Accuracy In Powder Diffraction II, NIST*, May **1992**.

Perlstein, J. *Chem. Mater.* **1994,** 6, 319.

Price, S.L.; Stone A.J. *Molecular Physics* **1984** 51, 569.

Rietveld, H.M. *J. Applied Crystallography* **1969** 2, 65 .

Roberts, K.J.; Sherwood, J.N.; Yoon, C.S.; Docherty, R., *Chem. Of Materials*, **1994**, 6, 1099.

Taylor, R.; Kennard, O. *Acc Chem. Res*, **1984** 17, 320

Taylor, R.; Kennard, O. *J. Chem. Soc.* **1982**, 104, 5063.

Williams, D.E., *J. Chem. Phys.* **1966** 45, 3770, 1966.

Zupan, J.; Gasteiger, J. *Neural Networks for Chemists.* VCH, Weiheim, Germany. **1993**.

On the Crystallographic and Statistical Mechanical Foundations of the Hartman-Perdok Theory

P. Bennema, RIM Laboratory of Solid State Chemistry, Faculty of Science, University of Nijmegen, Toernooiveld, 6525 ED Nijmegen, The Netherlands

Abstract

A survey is given of the integrated Hartman-Perdok roughening transition theory, developed during the last 15 years. Special attention is paid to a new extension of this theory taking the fluid part of the interface into account.

1.1 BFDH theory

According to what is now called the BFDH theory (Bravais, Friedel, Donnay, Harker theory) the morphology of crystals can be described on basis of the following parallel relation:

$$d_{h_1k_1l_1} > d_{h_2k_2l_2} \Rightarrow MI_{h_1k_1l_1} > MI_{h_2k_2l_2} \qquad (1)$$

(1) implies that the larger the interplanar distance between the netplanes d_{hkl} (corrected for the extinction conditions of the space group of the crystal structure of a given crystal structure) the larger the MI (Morphological Importance) of growth forms of crystals. MI is a statistical measure for the relative size and frequency of occurrence of a face (hkl). The larger the relative size and/or the frequency of occurrence the larger MI (Burke, 1966, Friedel, 1911, Donnay et al. 1937). In order to get an impression of the shape of crystals according to the BFDH law (1), the distances of faces (hkl) referenced to an origin can be taken inversely proportional to d_{hkl}. Using a computer program like HABIT a crystal form can be constructed.

In passing it is interesting to note that the last 15 years faces to be indexed with four integers (hklm) on so called modulated crystals and faces to be indexed with six integers (hklmno) on quasi crystals respectively were discovered (Janner et al., 1980, Dam et al., 1985, Balzuweit 1993, Janssen et al. 1987, Bennema et al. 1991). Recently a theory was developed to give a first explanation of these findings (Bennema et al., 1993, Kremers et al. to be published, Kremers, 1995, Bennema et al. 1994, Heijmen et al. 1995, van Smaalen, 1993).

1.2 Hartman-Perdok theory and recent developments

The BFDH law gives to some extend a satisfying description of the morphology of crystals. However, the BFDH theory lacks a physical chemical foundation.

Forty years ago Hartman and Perdok published three papers, which can be considered as a breakthrough in the subscience of crystal growth: morphology (Hartman et al., 1955). In this now called Hartman-Perdok (HP) theory, the concept of Period Bond Chain (PBC) was introduced. A PBC is an uninterrupted chain of first nearest neighbour bonds with a periodicity of the lattice [uvw]. According to the HP theory, three types of faces can be distinguished:

(1) F faces parallel to connected nets consisting of at least two different sets of PBC's,

(2) S faces parallel to only one PBC,

(3) K faces not parallel to any PBC,

Connected nets are defined as nets, with an overall thickness d_{hkl}, according to BFDH, of which all units are connected to each other. They consist of at least two different sets of connected PBC's. In order to disconnect a connected net parallel to PBC's [uvw] and other directions [uvw] which are coplanar with the face (hkl) such

1054–7487/96/0015$12.00/0

that hu+kv+lw=0, a positive cut energy is necessary. Nets of S and K faces are not connected. Within the framework of the Hartman-Perdok theory it is assumed (and this can be justified) that the relative rate of growth of a face (hkl), R_{hkl}, runs parallel to the attachment energy. In order to construct growth forms R_{hkl} is taken proportional to E^{att}_{hkl} (This can be justified to some extend). E^{att}_{hkl} is defined as the energy to remove a growth unit from the surface (hkl). This energy corresponds to the broken bonds, making an interface. (It is complementary to E^{slice}_{hkl}. E^{att}_{hkl} and E^{slice}_{hkl} give together the crystallization energy E^{cr}.) It follows from the HP theory that crystals will be bounded by F faces having the lowest R_{hkl} or the lowest E^{att}_{hkl} and the highest E^{slice}_{hkl}. Connected nets (hkl) with the highest E^{slice}_{hkl} will also have in most cases the highest d_{hkl} (Hartman 1955, Bennema et al. 1987, Bennema, 1993, Bennema, 1992, Docherty et al. 1991).

2. Application of modern morphological theories to predict morphology

2.1 Crystallographic rules to determine connected nets from a crystal graph based on physical principles

In modern theories of crystal morphology the concept of connected net plays a key role. Preceeding a brief treatment of a statistical mechanical interface model, we will first focus the attention on the determination of connected nets from a crystal structure or crystal graph.

From a modern point of view the determination of all possible connected nets and the relevant surviving ones can be carried out also using recently developed software:

(1) determine the growth units of the crystal structure under consideration,

(2) calculate overall bond energies Φ_i^{ss} between growth units referenced to vacuum. These bond energies Φ_i^{ss} may be calculated from atom-atom potentials of neighbouring growth units. Overall bond energies are then calculated using for example the best available Buckingham potentials applied to precisely determined distances between atoms of neighbouring molecules, resulting from an X-ray or neutron ray analysis. Recently also molecular mechanical force fields are used from for example Cerius 2 software, made available by Molecular Simulations in Cambridge, U.K. to calculate overall bond energies. Those force fields are selected, which yield a stable crystal structure corresponding to a minimum in crystal energy of the real crystal structure,

(3) make a selection of strongest bonds from the calculated bond energy Φ_i^{ss}, which will determine the crystal graph,

(4) reduce growth units to centres of gravity and define the crystal graph. A crystal graph is defined as an infinite set of points, related with bonds to each other, fulfilling for a monomer crystal graph the space group symmetry of the original crystal structure,

(5) determine all connected nets or unambiguous connected nets according to one of the following methods (or a combination of these methods):

(a) visual inspection of a crystal structure using (computer made) projections of the structure in certain crystallographic directions and projections of connected nets seen from above,

(b) a computer based method, the so called flat box method. According to this method the crystal graph is partioned in flat boxes with a thickness d_{hkl} (according to BFDH) and it is checked whether in these boxes a connected can be identified,

(c) the direct method developed by Strom using a computer (Strom, 1980, 1981, 1985). In this method all possible PBC's of a crystal graph are identified and it is checked which PBC's combine to a connected net. Method (c) is the most complete method,

(6) construct growth or equilibrium forms. For growth forms different logical recipes can be used. In most cases the relative rate of growth R_{hkl} is taken proportional to $E^{a\ tt}_{h\ kl}$ [17-20,9a,b]. It is well known that for an equilibrium form the distance of a face (hkl) to the centre of a coordinate system is proportional to the surface (free) energy.

Note that concerning (1) different solvents may give rise to different growth units (monomers or dimers). This will give rise to different "monomer or dimer graphs", having a different bond structure and symmetry. The larger the growth units, the less bonds occur in the crystal graph and the less connected nets occur. An example will be treated in this book by Grimbergen and Bennema.

2.2 Actual bond energies at the interface

As discussed above the original Hartman-Perdok theory is based on bond energies Φ_i^{ss} referenced so to say to vacuum. This is because bond energies Φ_i^{ss} are defined as energies resulting from a thought experiment where growth units in the crystal structure are brought infinitely far apart from each other. As shown above in previous papers (Bennema et al., 1987, Bennema, 1993, Bennema, 1992, Docherty et al. 1991) actual bond energies occurring at the interface between crystal and mother phase have the shape

$$\Phi_i = \Phi_i^{sf} - \frac{1}{2}(\Phi_i^{ss} + \Phi_i^{ff}) \qquad (2)$$

where sf, ss, ff refer to solid-fluid, solid-solid and fluid-fluid bonds respectively. Φ^{sf} etc. are the corresponding bond energies respectively. As discussed in section 2.1 (see (2)) Φ_i^{ss} bond energies can be calculated. However, Φ_i^{sf} and $_i\Phi$ bond energies are not known. In order to make an estimate of Φ_i^{ff} very often the so called proportionality relation is introduced:

$$\Phi_1 : \Phi_2 : \ldots : \Phi_i \ldots = \Phi_1^{ss} : \Phi_2^{ss} \ldots : \Phi_i^{ss} \ldots \qquad (3)$$

Using (3) in principle vacuum morphology and mother phase morphology become equal.

2.3 Roughening temperature

Assuming that the dimensionless roughening temperature θ^R and the order-disorder phase transition temperature θ^C are close (this can be justified to some extend (Bennema et al. 1987, Bennema, 1993) it follows from the theory of roughening transition in a so called SOS (solid-on-solid) interface cell model and the generalized theory of Onsager for an order disorder phase transition in a two dimensional spin up spin down or solid fluid model that

$$
\begin{aligned}
{}^{C}_{hkl}(\approx\theta^{R}_{hkl}) &\rightarrow \gamma_{step}(hkl) \\
{}^{C}_{hkl}(\simeq\theta^{R}_{hkl}) &\rightarrow \gamma_{step}(hkl)
\end{aligned}
\qquad (4)
$$

γ_{step}(hkl) is the step free energy of a step within the connected net (hkl). γ^R_{hkl} and γ^C_{hkl} are dimensionless critical temperatures defined as:

$$\theta^{R}_{hkl} \simeq \theta^{C}_{hkl} = \left(\frac{2kT}{\Phi_{str}}\right)^{C} \qquad (5)$$

and θ the dimensionless temperature of the growing crystal

$$\theta = \frac{2kT}{\Phi_{str}} \qquad (6)$$

Note that the choice of the factor 2 and the choice of the strongest bond energy of the crystal graph (Φ_{str}) are just conventions.

Using the theory of Rijpkema, Knops, Bennema and van der Eerden (Rijpkema et al., 1982) for each connected net, which is a real connected net (without crossing bonds) θ^C can be calculated exactly (within the precision of the numerical solution of a matrix equation). In case connected nets have a few crossing bonds an upper and a lower bound of θ^C_{hkl} may be calculated. If connected nets become very complex (consisting for example of two or more connected subnets ("floors")), calculation of θ^C_{hkl} may become impossible (Bennema et al., 1987, Bennema, 1993, Bennema, 1992, Docherty et al. 1991, Rijpkema et al. 1982).

It follows from this subsection that in the concept of connected net the crystallographic morphological theory of Hartman-Perdok and the statistical mechanical theory of roughening transition or order-disorder phase transition Rijpkema et al. meet. The implication of (4) is that if $\theta < \theta^R_{hkl}$ the face (hkl) will grow as a flat face by a layer growth mechanism (2D nucleation or spiral growth) and if $\theta \geq \theta^R_{hkl}$ as a disoriented rough (macroscopically) rounded off face (Bennema et al., 1987, Bennema, 1993, Bennema, 1992).

2.4 Methods to determine Φ_i's from observed roughening phenomena

Assume that from the faces 1,2, ... i, i+1, etc. corresponding to connected nets of the crystal graph under consideration with decreasing order disorder dimensionless order phase transitions temperatures, $\theta_i^C > \theta_2^C ... > \theta_i^C > \theta^C_{i+1}$, etc. the faces 1 to i are observed as flat growing faces, but the faces i+1 etc. not. We may then conclude that for the actual θ

$$\theta^C_{i+1} < \theta < \theta^C_i \tag{7}$$

A good approximation may be

$$\theta \simeq \frac{\theta^C_i + \theta^C_{i+1}}{2} \tag{8}$$

If θ is known, it follows from (6) that since the temperature T at which the crystal is growing is known, Φ_{str} can be calculated and hence also all Φ_i's, using (3).

If the actual roughening temperature T^R can be determined experimentally like in the case of {110} faces of paraffin (Bennema, 1993, Bennema, 1992) it follows again from (6) that from the calculated θ^C_{hkl}, Φ_{str} and hence using (3) all Φ's can be calculated.

2.5 Models to calculate Φ_i's

In order to calculate from a theory Φ_i's, X.Y. Liu developed the following inhomogeneous cell model (Liu et al. to be published, Liu et al., 1994, Liu et al., to be published, Liu et al., submitted). The idea that the fluid phase is homogeneous up to the crystal surface is abandoned.

So it is assumed that for growth from solution, in case of thermodynamic equilibrium with a saturated solution, that a profile $X_{hkl}(Z)$ for the face (hkl) of solute molecules occurs. That means that going from the crystal surface at Z=0 to the bulk along the Z axis perpendicular to a face (hkl), the fraction of solute molecules X(Z) varies with Z. Sufficiently far from the interface (say ten or more interplanar distances d_{hkl}) X(Z)=X(bulk) of the saturated solution. Now according to the inhomogeneous cell model the fluid phase adjacent to a connected net (hkl) is partioned in parallel slices with a thickness d_{hkl} parallel to the orientation (hkl). It is assumed that each of these layers is a homogeneous subphase. Applying thermodynamics and/or regular solution theory to each of these layers separately the following relations exist between $X_{i(hkl)}$ the fraction of solute molecules and the corresponding enthalpy of dissolution Δh_i^{diss} of the ith layer in the interface

$$X_{i(hkl)} = \exp - \frac{\Delta h_i(hkl)^{diss}}{k T_M^2} (T_M - T_S) \tag{9}$$

$\Delta h_{i(hkl)}^{diss}$ is the enthalpy of dissolution per molecule in the ith layer of the fluid interface for the face (hkl). T_M is the melting temperature and T_S the saturation temperature. If we now take the extremes $X_{i(hkl)}=X(0)_{hkl}$ at the interface and in the bulk $X_{i(hkl)}$, we can from (9) derive expression (10). Since T_M and T_S are the same for all layers of the fluid interface the relation (10) holds:

$$\frac{\ln X(0)_{hkl}}{\ln X(b)} = \frac{\Delta h(0)_{hkl}^{dis}}{\Delta h_b^{diss}} \tag{10}$$

This means that if $X(0)_{hkl}$ is known, $\Delta h(0)_{hkl}^{diss}$ can be calculated and can be compared with Δh_b^{diss}, the bulk enthalpy of dissolution.

Now the sum of all bond energies Φ_i of all the bonds originating from a growth unit corresponds for crystal structures with one type of growth unit to the bulk enthalpy of dissolution or

$$\sum \Phi_i = \Delta h_b^{diss} \tag{11}$$

We have to take for the face (hkl)

$$(\sum \Phi_i)_{hkl} = \Delta h(0)_{hkl}^{diss} \tag{12}$$

$X_{hkl}(Z)$ and $X_{hkl}(0)$ can be obtained from statistical mechanical models for the interface. Recently the inhomogeneous cell model was applied to explain the needle shape of urea crystals growing from an aqueous solution and the very thin platy shape of paraffin crystals growing from the hexane solution. In the first case profiles were obtained from extensive MD simulations (Liu et al., 1995, Liu et al. to be published) and in the second case from a kind of mean field regular solution like theory developed for polymers and applied to paraffins by Scheutjens and Fleer (Liu et al., 1995, Liu et al. to be published).

We note that an example to derive the morphology of naphthalene and biphenyl in a nutshell on basis of the theories presented above inclusive the fluid part will be given in this volume (Liu and Bennema, this volume).

3. Problems to be solved in the near future and conclusions

3.1 Development of software to determine all relevant connected nets

In the near future user-friendly software to determine all connected nets of a given crystal graph on basis of the theory or method developed by Strom (Strom 1980, 1981, 1985) will be developed. Such software can be attached to molecular mechanical software of for example Cerius 2 developed by MSI in Cambridge U.K. Since there are no artificial constraints, such as flat boundaries for the flat box method indeed all possible connected nets are developed by the graph theoretical method by Strom. The method Strom is based on checking all possible combinations of all possible PBC's to connected nets.

But now a new problems arises. Even for not very complex crystal graphs a sometimes bewildering number of connected nets are produced. To mention just one example. A paper was published on the morphology of paraffins (see also case study (Bennema, 1993)). Connected nets were determined by visual inspection (See 5.A section 2.1). Strom found many more connected nets (work in progress). So the claim in (Bennema, 1993) that all connected nets were found was wrong. (For the interpretation of experimental data the above reference is fine, because only the first two or three strongest nets were below the roughening temperature). For example for the faces {110} two different connected nets were identified. However, Strom found these 2 plus 4 plus 8 alternative connected nets. So during visual inspection 12 alternative connected nets for the orientations {110} were overlooked. We are now in the process to develop crystallographic criteria to select from the many possible alternative connected nets for one orientation (hkl) the relevant connected nets. It can be shown that the four plus eight connected nets assuming certain hypotheses of interaction of fluid phase with the particular connected nets, have to be rejected (work in progress).

A general criterion is to select the strongest slice with the highest E^{slice} or the highest θ^{C}_{hkl}. These connected nets are in most cases also the most symmetrical ones. We then introduce the hypothesis that the connected net with the highest E^{slice} will be the dominant one during crystal growth. However, before applying this criterion we will wherever possible apply to be developed crystallographic criteria to discriminate genuine from non genuine (apparant) connected nets (work in progress).

3.2 Reasons for the occurrence of different connected nets

From studying different crystal graphs we found the following reasons, that for one orientation (hkl) sometimes quite a lot of connected nets occur.

(1) Very often one connected net occurs at a certain height and an alternative one with a height $\frac{1}{2}d_{hkl}$ shifted in reference to the connected net just mentioned.

(2) In case of strong nets with a high d_{hkl} very often a connected net consists of 1,2,3 or more connected subnets. So one connected net consists as it were of 1,2,3 etc. "floors" of subnets. In case of 2 connected subnets two alternative cuts can be made. This leads to two different real connected nets. In case of n alternative connected subnets n different real multifloor connected nets occur.

(3) It is also possible that connected nets can be made consisting of a connected core with so called attached growth units. These growth units are connected to a core connected net by loose bonds. For the calculation of the relative roughening temperature (see (5)) only the growth units contributing to the connected core net are relevant.

The attached growth units can be omitted for a calculation of θ^{C} (5).

Referring to the case (2) just mentioned it can be imagined that a single core net can be selected from an overall net consisting of n floors of connected subnets. To this single net loose growth units often can be attached to the chosen core net.

So it can be imagined why for thick multifloor connected nets many alternative genuine connected nets are possible. Of course the thinner the core net the lower its E^{slice}_{hkl} and θ_{hkl} the lower the chance that such a net will occur on the outside of the orientation (hkl). Work is now underway to formulate general rules to select relevant connected nets when alternative connected nets occur.

3.3 M.D studies of the fluid part and prediction of morphology

It is desirable to carry out MD studies of the fluid part of the interface. An example is found in recent papers on the morphology of urea (Liu et al. 1995, Liu et al. to be published).

The prediction of a more realistic morphology can then be based on

(i) most reliable $(\Phi_i)_{hkl}$ bond energies. (Note that the $(\sum\Phi_i)_{hkl}$ may vary from one face $(h_1k_1l_1)$ to another $(h_2k_2l_2)$),

(ii) rate of growth versus supersaturation functions $R_{hkl}(\sigma)$ based on kinetic theories.

(i) and (ii) will give the most reliable prediction of the morphology, see for example (Liu et al. 1995, Liu et al. to be published, Liu et al., to be published, Liu et al., 1994, Liu et al., to be published, Liu et al., submitted).

References

Balzuweit, K.; Thesis, University of Nijmegen (1993).

Bennema, P.; Balzuweit, K.; Dam, B.; Meekes, H.; Verheijen, M.A.; Vogels, L.J.P.; J.Phys. D: Appl. Phys. 24 (1991) 186.

Bennema P.; van der Eerden, J.P.; Crystal graphs, connected nets, roughening transition and the morphology of crystals, Chapter 1 in: Morphology of Crystals, ed. I. Sunagawa, TERRAPUB, Tokyo (1987) 1-75.

Bennema, P.; Handbook of Crystal Growth, ed. D.T.J. Hurle, Elsevier, Amsterdam (1993) Vol. 1, Chapter. 7.

Bennema, P.; KONA 10 (1992) 25.

Bennema, P.; Morphology of crystals: past and future, Science and Technology of Crystal growth (1995) 159, Summerschool ISSCGIX, Papendal, 1995, Kluwer Academic Publishers

Bennema, P.; On the crystallographic and statistical mechanical foundations of the forty year old Hartman-Perdok theory, to be published in J. Crystal Growth (1996)

Bennema, P.; Kremers, M.; Meekes, H.; Balzuweit K.; Verheijen, M.A.; Faraday Discuss. 95 (1993) 3.

P. Bennema, P.; Kremers, M.; Meekes H.; Verheijen, M.; Phys. Stat. Sol. (a) 146 (1994) 13.

Burke, J.G.; Origins of the Science of Crystals, University of California Press, Berkeley, 1966.

Dam, B.; Janner A.; Donnay, J.D.H.; Phys. Rev. Lett. 21 (1985) 2301.

Docherty, R.; Clydesdale, G.; Roberts, K.J.; Bennema, P.; J. Phys. D: Appl. Phys. 24 (1991) 89.

Donnay, G.D.H.; Harker, D.; Am. Mineral. 22 (1937) 446.

Friedel G.; Leçon de Cristallographie, Paris, Hermann (1911).

Grimbergen, R.F.P.; Bennema, P.; this volume.

Hartman P.; Perdok, W.G.; Acta Cryst. 8 (1955) 49.

Hartman P.; Perdok, W.G.; Acta Cryst. 8 (1955) 521.

Hartman P.; Perdok, W.G.; Acta Cryst. 8 (1955) 525.

Hartman P.; Crystal Growth, an Introduction, ed. P. Hartman, North Holland, Amsterdam, 1973, 367-402.

Hartman, P.; Morphology of crystals, Part A, ed. I. Sunagawa, Terra Scientific Publishing and D. Reidel, Dordrecht, 1987, pp. 269-319.

Heijmen, T.G.A.; Kremers M.; Meekes, H.; Phil. Mag. B 71 (1995) 1083.

Janner, A.; Rasing, Th.; Bennema P.; van der Linden, W.H.; Phys. Rev. Lett. 45 (1980) 1700.

Janssen, T.; Janner, A.; Adv. in Physics 36 (1987) 519.

Kremers, M.; Meekes, H.; Bennema, P.; Verheijen M.A.; van der Eerden, J.P.; to be published in Acta Cryst. A.

Kremers, M.A.; Thesis, Nijmegen University (1995).

Liu X.Y.; Bennema, P.; Nature 374 (195) 342.

Liu X.Y.; Bennema, P.; survey paper on inhomogeneous cell model, to be published.

Liu X.Y.; Bennema, P.; Phys. Rev. B49, (1994) 765.

Liu X.Y.; Bennema, P.; Phys. Rev. B, to be published.

Liu X.Y.; Bennema, P.; Phys. Rev. E, submitted.

Liu X.Y.; Bennema, P.; this volume.

van Smaalen, S.; Phys. Rev. Lett. 70 (1993) 2419.

Rijpkema, J.J.M.; Knops, H.J.F.; Bennema, P.; van der Eerden, J.P.; J. Crystal Growth 61 (1982) 295.

Strom, C.S.; Z. Kristallogr. 153 (1980) 99.

Strom, C.S.; Z. Kristallogr. 154 (1981) 31.

Strom, C.S.; Z. Kristallogr. 172 (1985) 11.

An *Ab Initio* Approach to Crystal Structure Determination Using High Resolution Powder Diffraction and Computational Chemistry Techniques

Paul G. Fagan, Robert B. Hammond, Kevin J. Roberts Department of Pure and Applied Chemistry, University of Strathclyde, Glasgow, UK
Robert Docherty, Mike Edmondson, Zeneca Specialities, Blackley, Manchester, UK

Despite the importance of molecular engineering concepts in the development of novel speciality effect chemicals our knowledge of the solid-state structure of such materials is surprisingly limited. This has been due, in part, to difficulties inherent in the preparation of crystals of sufficient size and quality for single crystal structure determination. As the result of this, the generation of structural data *ab initio* has become, in recent years, both a significant scientific goal and the subject of some controversy. In this paper we report the application of a combination of high resolution powder diffraction and computational chemistry techniques as a new *ab initio* approach to crystal structure determination. We demonstrate the method through an application to the crystal structure determination of the commercially important organic pigment 6,13-dichlorotriphendioxazine.

1. Introduction

In recent years the developed world has seen much of the quality end of the chemical product market moving away from capital-intensive bulk chemicals towards the manufacture of high value-added speciality materials such as pharmaceuticals, dyes & pigments, additive & effect chemicals. Factors such as material stability, environmental concerns and energy efficiency requirements often mean that the processing of such systems is likely to involve dealing with products in solid form. The fact that chemical production processes involving solid phases have tended to result in more plant scale difficulties than those involving gaseous or liquid phases, has produced a growing awareness of the need to understand and define solid-state factors associated with the preparation and optimisation of particulate solids. Given this scenario, crystal structure data are often vital if molecular, surface and crystal engineering approaches are to be used to manipulate particle properties. However, whilst crystal structure data-bases contain ca. 10^5 solved structures, the number of industrially significant entries tends to be rather limited as only ca. < 1% of speciality chemical products are known to have a solved crystallographic structure. This lack of crystal structural data is due to difficulty in growing crystals of sufficient size and quality for conventional X-ray analysis as such methods demand single crystals ca. > 50µm in size which is not always possible.

In this paper we report an alternative method for the determination of the crystal structure of particulate solids which uses a combination of high resolution powder diffraction and computational chemistry techniques to produce a new *ab initio* approach to crystal structure determination. We demonstrate the utility of the method through its application to the commercially important organic pigment 6,13-dichlorotriphendioxazine. The specific aim of this study was to assess the reliability of the technique in cases where good single crystals cannot easily be obtained and to examine alternatives to other, equally viable, techniques.

2. Basic Approach and Methodology

Our overall approach to crystal structure determination using the new route involves essentially four main stages, each well

1054–7487/96/0022$12.00/0 © 1996 American Chemical Society

established but never before utilised together in such a way to provide a reliable route to crystal structure solution.

Stage 1. Using standard modelling approaches, with the aid of routinely available software packages, molecular arrangements are obtained for the molecule under consideration. In turn the structure is optimised and refined using molecular mechanics and semi-empirical quantum chemistry techniques to obtain precision in bond lengths and angles. These calculations allow determination of the molecular geometry with respect to these two parameters with a high degree of accuracy. By way of structure validation a comparison with a table of standard bond lengths and angles[1] is performed. However, in cases where the molecule possesses internal conformational flexibility the modelling of torsion bonds confers a non-trivial component. The use of an existing database of structures may provide an alternative route to the initial structure but cannot be considered as a routine approach to more complex systems and is thus not pursued in this paper.

Stage 2. High resolution (synchrotron) powder diffraction data are collected on the sample material. Following normalisation, to account for beam decay, using the PODSUM[2] program, individual peaks are identified visually before being modelled mathematically using PKFIT[2] according to a prescribed peak shape function. The peak positions so obtained are passed to a series of indexing programs which should in turn proffer the unit cell dimensions for the system under test. From systematic absences the symmetry is obtained and from density the number of molecules in the unit cell calculated. Synchrotron radiation is crucial in providing the peak resolution required (generally taken as $< 0.03°$) for indexing in systems of low symmetry characteristic of organic materials.

Stage 3. Using the information obtained from the above stages it is possible to construct a series of starting crystal models for the system under test. Initially, with appropriate graphical display this can be achieved visually to obtain a sensible structure for energy minimisation. After manual, but stepwise and systematic, variation of molecular orientation within the unit cell, by applying rotations and then translations to the rigid molecular framework, and adjusting torsion angles in flexible molecules, atom-atom potential minimisation methods, using such programs as PCK83[3] are used to calculate a series of possible low energy crystal structures for further consideration. In order to ascertain the feasibility of one structure with respect to another the powder diffraction profile for each is simulated and compared directly with the experimental data with respect to peak intensities - positions by virtue of indexing techniques must concur. This is done quantitatively by refining the scale factor and observing the difference spectrum. On comparing the theoretical structures in such a way we obtain a rapid and reliable way of filtering out improbable structures. Given that all the trial packing configurations have a low potential energy, a structure is only considered for final refinement if the powder diffraction simulation presents a good match with experimental data.

Stage 4. In this, the final stage, Rietveld method[4] is used to refine the theoretically produced structure. In this well documented technique, a calculated diffraction profile according to the proposed structure is fitted to the experimental data in a series of least squares minimisations. Both structural (atomic positions, lattice and thermal) parameters and parameterised profile functions are treated in the refinement.

3. Application to 6,13-Dichlorotriphendioxazine

6,13-Dichlorotriphendioxazine, ($C_{18}H_8N_2O_2Cl_2$, MW 355.2, see inset to figure 1, hereafter referred to as DCTPD), is the basic chromophore unit of a number of commercially important dyestuffs. It was prepared as a highly crystalline powder following recrystallisation from nitrobenzene and ground to a particle size of c.a. 45μm.

High resolution powder diffraction data using a Debye-Scherrer scattering geometry were taken on beamline 2.3[5] at the Synchrotron

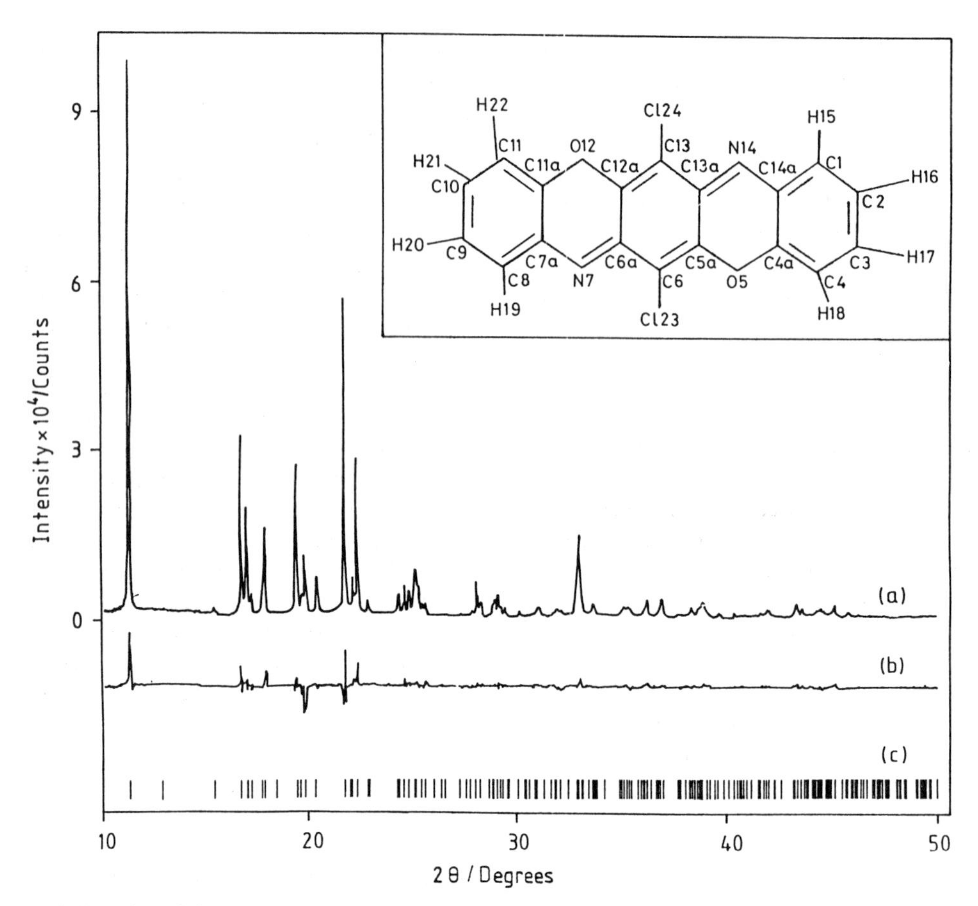

Figure 1. Results of the Rietveld Refinement of DCTPD: (a) experimental high resolution powder diffraction pattern; (b) residual plot showing the difference between experimental and theoretically modelled diffraction patterns; (c) marker points showing the location of diffraction points. The inset shows the molecular structure of DCTPD.

Radiation Source (SRS) at CCRL Daresbury Laboratory in the UK. The storage ring operated at an energy of ca. 2 GeV with a stored beam current of ca. 175 mA. The incident beam wavelength was selected using a Ge (111) monochromator to provide photons at a wavelength of 1.20229 Å. Data were collected using an angular scanning range of 2θ angle from 2° to 50° at a step size of 1 mdeg using a counting time of 1 s/point.

Unit cell dimensions were obtained from the first thirty reflections using the indexing programs of Werner, Visser and Louer[6] and refined using REFCEL[2]. DCTPD was found to crystallise in a monoclinic structure with a=8.717(1)Å, b=4.887(1)Å, c=17.147(2)Å, β =97.865(3)°, unit cell volume=723.59 $Å^3$, density= 1.646 mg cm^{-3}). The space group ($P2_1/c$) was determined by consideration of systematic extinction conditions. This unit cell was consistent with a bimolecular unit cell forcing the molecule to lie on a centre of symmetry and thus halving the number of atoms involved in structure refinement.

An approximate molecular model of DCTPD was generated via building modules within standard software, optimised using the molecular mechanics package MM2[7] and refined using semi-empirical molecular orbital methods[8]. The optimised molecular conformation was placed in a proposed lattice from the unit cell dimensions and by using molecular packing minimisation methods together with the atom-

atom technique a series of optimised lattice systems postulated. For these calculations we used the universal force field[9] together with charges calculated by the equilibration method[10]. Because of symmetry considerations there were no translational elements involved in the minimisation.

The hypothetical crystal structures derived from the molecular packing calculations were then used to simulate the powder diffraction patterns[11] which were, in turn, compared with the experimental data. In this way, through the on-line manipulation of the trial structures, a model structure having the best fit to the experimental data was obtained. For example, a structure of feasible lattice energy of -46.9 kcal/mol) was obtained in the first set of minimisations. Comparing the diffraction simulation data with the real profile (figure 2(a)) revealed reasonable agreement across the two theta range with only a few significant peaks missing. The most significant of these $(2\ 0\ \bar{1})$ should have occurred at approximately 13°, 2θ. The exaggerated presence of the (2 0 1) peak led the authors to rotate the minimised structure at 90° relative to the b-axis. Subsequent minimisation of the new arrangement led to a new structure of -47.2 kcal/mol which on pattern simulation presented a very close match (figure 2(b)) and was thus selected, on both criteria, for Rietveld refinement. The program DBWS[12] was used and yielded a structure solution with a final R-factor of 0.1369 where R is defined by:

$$R = \Sigma \mid Y_{obs} - Y_{cal} \mid / \Sigma \mid Y_{obs} \mid$$

where Y is diffracted intensity. The structural parameters, at 298K, are given in Tables 1 and 2. The mean isotropic temperature factor, based on all the atoms in the molecule, was refined to be 4.14(2) Å^2. The resulting Rietveld fit to the data is shown in Figure 1.

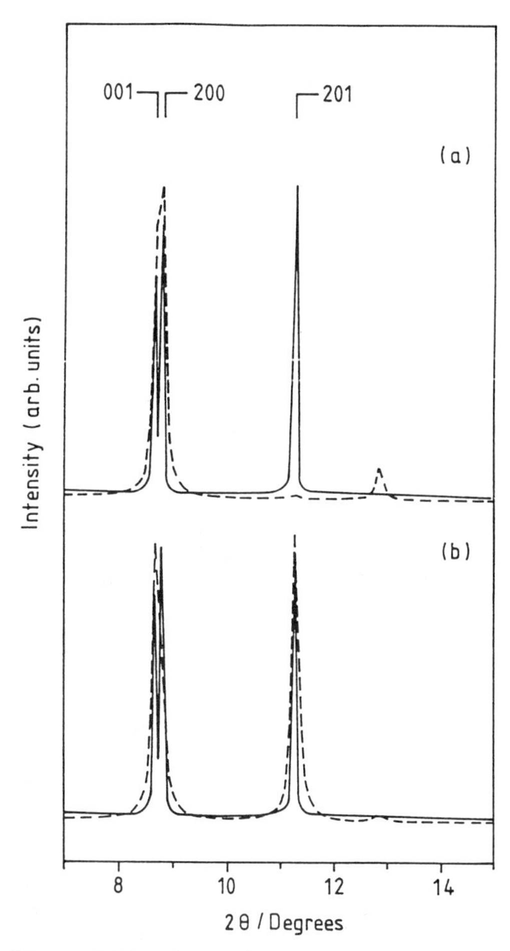

Figure 2. Experimental (_____) and theoretical (- - - - -) powder diffraction patterns illustrating how comparison between two packing motifs (a and b) reveals a suitable starting structure (b) for subsequent Rietveld refinement.

4. Conclusions

Although the molecular and crystallographic structures of DCTPD are constrained via the ring systems and space-group symmetry there is much we can conclude from this work with a view to the application of the technique for more complex systems. Despite the fact that *ab initio* prediction of crystal structures from calculations based purely on molecular modelling calculations remains a long term goal, the use of high resolution X-ray diffraction provides, perhaps, one of the key experimental benchmarks needed in the molecular modelling calculations to mitigate concerns over global minimisation and force field accuracy. This work demonstrates the utility of combining molecular modelling and diffraction techniques to solve molecular structure *ab initio*, particularly in the case of low symmetry organic structures. There

Table 1. List of non-hydrogen atomic co-ordinates for DCTPD derived from this study

Atom	x/a	y/b	z/c
C1	0.1422(15)	0.7244(21)	0.2040(7)
C2	0.2782(13)	0.8774(21)	0.2226(9)
C3	0.4053(13)	0.8428(22)	0.1818(7)
C4	0.3994(11)	0.6534(22)	0.1220(7)
O5	0.2665(8)	0.3128(17)	0.0414(3)
C13	0.1351(11)	-0.0244(20)	-0.0391(5)
N14	0.0000(9)	0.3727(18)	0.1219(4)
C14a	0.1338(11)	0.5300(22)	0.1426(6)
C4a	0.2643(12)	0.4986(21)	0.1024(6)
C13a	0.0006(10)	0.1967(20)	0.0644(7)
C5a	0.1336(11)	0.1567(22)	0.0208(5)
Cl24	0.2897(4)	-0.0562(9)	-0.0881(2)

Table 2. Bond lengths (/Å) of DCTPD obtained from this study.

Atom-Atom Interaction	Bond Length / Å
Cl24-C13	1.69
C13-C12a	1.36
C12a-C6a	1.48
C6a-C6	1.47
C6a-N7	1.31
N7-C7a	1.40
C7a-C8	1.41
C8-C9	1.40
C9-C10	1.40
C10-C11	1.38
C11-C11a	1.40
C11a-O12	1.39
O12-C12a	1.39

are many examples within speciality fine chemicals, such as pharmaceutical materials, where single crystals of sufficient quality and size cannot be obtained. It is in the examination of such materials that our approach offers the most promise. Work is currently in hand to extend the methodology developed towards an approach for the routine consideration of conformationally flexible molecules.

Results from this will be reported in future papers.

Acknowledgements.

We are grateful to; the SERC and Zeneca Specialities for the financial support of a CASE studentship to one of us (PGF); to CCRL Daresbury Laboratory for beamtime on the SRS; and to the SERC/EPSRC for a research grant (GR/H/40891) and current support of a senior fellowship (KJR).

References

1. Allen, F.H.; Kennard, O.; Watson, D.G.; Brammer, L.; Orpen, A.G.; Taylor, R., *J. Chem. Soc. Perkin Trans. II,* **1987,** 12, S1 - S19.
2. Part of the Powder Diffraction Program Library, SERC Daresbury Laboratory, Warrington, UK.
3. Williams, D.J., PCK83, Quantum Chemistry Programme Exchange programme number 548.
4. Cernik, R.J.; Murray, P.K.; Pattison, P.P.; Fitch, A. N., *J. Appl. Cryst.,* **1990,** 23, 292.
5. Rietveld, H. M., *J. Appl. Cryst.,* **1969,** 2, 65.
6. Werner, P.E.; Eriksson, L.; Westdahl, M., *J. Appl. Cryst.,* **1985,** 18, 367: Visser, J.W., *J. Appl. Cryst.,* **1969,** 2, 89: Boulif, A.; Louër, D., *J. Appl. Cryst.,* **1991,** 24, 987.
7. Sprague, J.T.; Tai, J.C.; Yuh, Y.; Allinger N.L., *J. Compt. Chem.,* **1987,** 8, 581:

Liljefors, T.; Tai, J.C.; Shusen L.; Allinger, N.L., *J. Compt. Chem.,* **1987,** 8, 1051.

8. Stewart, J.P.P., MOPAC 6.0, Quantum Chemistry Programme Exchange programme number 455.
9. Mayo, S.L.; Olafson, B.D.; Goddard, W.A., *J. Phys. Chem.,* **1990,** 94, 8897.
10. Rappe, A.K.; Goddard, W.A., *J. Phys. Chem.,* **1991,** 95, 58.
11. Abrahams, S.C.; Keve, E.T., *Acta Cryst.,* **1972,** A27, 157.
12. Sakthivel, A.; Young, R.A., DBWS, School of Physics, Georgia Institute of Technology, Atlanta, GA 30332, USA.

An Automated Method for PBC Analyses applied to Benzoic Acid. A Comparison between the Monomer and Dimer Analysis.

R.F.P. Grimbergen and P. Bennema. RIM, Laboratory of Solid State Chemistry, University of Nijmegen, Toernooiveld, 6525 ED Nijmegen, The Netherlands.

In this paper a recently developed computer program for automatic Periodic Bond Chain (PBC) analyses is presented. The program is a user-friendly implementation of the so-called Flat Box Method and searches automatically for connected nets within a slice thickness d_{hkl}. For benzoic acid the connected nets were derived using either monomers or dimers as growth units. Subsequently attachment energy calculations enabled us to construct a theoretical morphology for these different cases. The dimer morphology is in good agreement with the experimentally observed crystals grown from the vapour.

Crystallization is an important method for purification of all kinds of compounds. During the last few years many industries have been trying to optimize their crystallization processes because of energy consumption and environmental reasons. In this optimization process the morphology plays an important role. Therefore, many people are interested in methods able to predict crystal shapes.

A well-established method to predict the morphology of a crystal is the Periodic Bond Chain (PBC) analysis, developed by Hartman and Perdok (Hartman, 1955). In this approach the assumption is made that there is a high growth rate along directions of strong uninterrupted chains of bonds between the growth units in the lattice.

If a crystal is cut along a crystallographic direction (hkl), it is possible to distinguish three types of faces:

(1) There is no PBC in the slice (hkl). The corresponding face is defined as a K(kinked)-face. The growth rate in a direction perpendicular to these faces will be large, because there is no PBC to stabilize the face.

(2) There is one PBC in the plane. This is defined as a S(stepped)-face.

(3) There are two or more non-parallel intersecting PBCs in the plane (hkl). Hartman en Perdok defined this situation as an F(flat)-face.

The F-faces determine the crystal morphology and also to a large extent the growth kinetics if the crystal faces grow below their roughening temperature.

The derivation of connected nets from a crystal structure is a difficult and time consuming job. Therefore we automated this procedure. In the literature an automated method to derive all F-slices has already been described (Strom, 1985). This method is based on graph theoretical principles. In this work we chose to implement the so-called Flat Box Method. The procedure for derivation of F-slices from a crystal structure using this method will be outlined below.

The Flat Box Method

In a PBC analysis, first the growth units of the crystal have to be defined and all bonds between the growth units in the unit cell and the neighbouring ones. Next the growth units are reduced to their corresponding centers of gravity and the bonds between the growth units are represented by bonds between the centers of gravity. This set of centers of mass and bonds we define as the *Crystal Graph*. The Crystal Graph contains all information to derive the F-faces.

In the Flat Box Method a slice of thickness d_{hkl} is cut along a crystallographic orientation (hkl). Then all centers of gravity within the slice have to be determined. Finally it is checked whether all centers are mutually connected via bonds defined in the crystal graph or not. If this is the case, we define the slice as a connected net. A connected net corresponds to an F-slice if it is stoichiometric and not affected by the extinction conditions of the space group of the structure.

The C-program F_GRAPH is a user-friendly implementation of the Flat Box Method. It searches for connected nets in a given range of orientations (hkl) and checks for stoichiometry. Because of the use of the Cambridge Structure Search and Retrieval (CSSR) file format, the results of this analysis can be directly imported for visual inspection in any molecular modeling package like Cerius^2 (MSI, 1994). In Fig. 1 is an example of one of the connected nets (210) of benzoic acid. After application of the extinction conditions, the resulting connected nets are all F-slices.

If the F-slices are known, the morphological importance (MI_{hkl}) for each F-slice (hkl) has to be determined. A very simple method for this purpose is the Bravais Friedel Donnay Harker (BFDH) method which uses the interplanar distance d_{hkl}, corrected for the extinction conditions of the space group, as a direct measure for the MI_{hkl}. The larger d_{hkl}, the larger MI_{hkl}. Later Hartman and Bennema (Hartman, 1980) showed that the proportionality between the growth rate and the attachment energy E_{att} may be used for the derivation of growth forms. The attachment energy is defined as the energy released per molecule when a new layer is attached to the surface of the crystal. The larger E_{att}^{hkl}, the smaller MI_{hkl}. A third method is the use of the critical roughening temperature of a face T_C^{hkl} as a measure for the MI_{hkl}. In this case a crystal face (hkl) is more important when T_C^{hkl} is higher.

In the next section, we will describe the procedure for determining all F-faces applied to the case of benzoic acid. The analysis was carried out assuming either monomers or dimers to be the growth units. Subsequently we have performed attachment energy calculations for all connected nets using different force fields. We used the attachment energies to construct theoretical growth forms and compared them to experimentally observed morphologies.

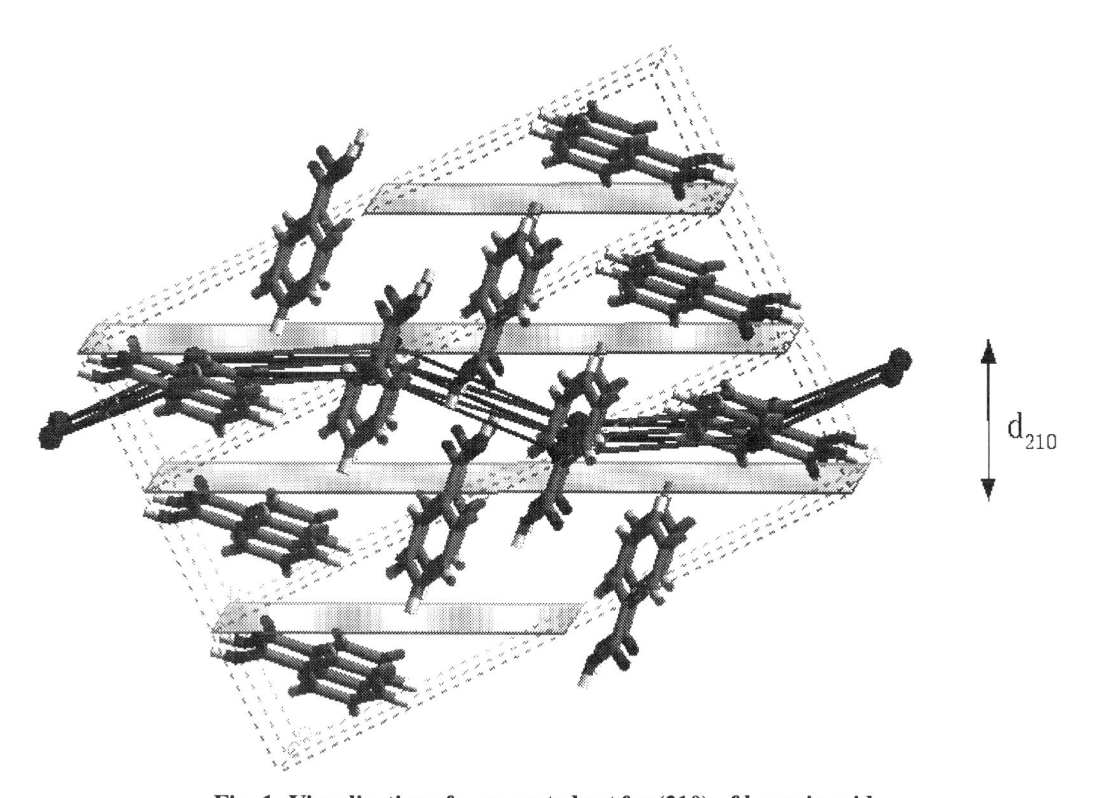

Fig. 1. Visualization of a connected net for (210) of benzoic acid.

Computation of the F-faces of benzoic acid

The structural information of benzoic acid was retrieved from the Cambridge Structural Database (CSD). Instead of the conventional cell parameters a_c=5.4996 Å, b_c= 5.1283 Å, c_c=21.9500 Å and β_c=97.37° in space group $P2_1/c$, the following cell was chosen: $\mathbf{a}=\mathbf{c}_c$, $\mathbf{b}=\mathbf{b}_c$ and $\mathbf{c}=-\mathbf{a}_c$ and β=82.63°. The new cell is in the setting $P2_1/a$. The benzoic acid molecule is always hydrogen bonded to another molecule in the crystal structure. Such a pair we will consider a growth unit in the dimer analysis. Thus there are 4 growth units in the unit cell for the monomer case and 2 growth units in case of dimers (Fig. 2). The positions of the centers of mass of the molecules are summarized in Table 1.

In order to construct the crystal graph, all bonds between the molecules have to be determined. In the case of monomers, we determined all important bonds and found a coordination number of 16. In the dimer case, the crystal graph corresponds to the general case for organic compounds having two centrosymmetric molecules in the unit cell. Hartman did already treat this class of structures and he found that because of the special positions of the centers of mass, an "extra extinction" condition $h+k=2n$ has to be applied (Hartman, 1991). This is an example of a crystal graph with a higher symmetry than the crystal structure. We applied this a extinction condition to the connected nets of benzoic acid dimers. However, we added to the crystal graph an extra bond between CM1 in the unit cell and CM1 with translation (011) (bond 1-1(011)).

Table 1 Positions of the centers of mass in the unit cell in fractional coordinates.

Name	X	Y	Z
Monomer			
CM1	0.88621	0.51973	0.19938
CM2	0.61379	0.01973	0.80062
CM3	0.11379	0.48027	0.80062
CM4	0.38621	0.98027	0.19938
Dimer			
CM1	0.00000	0.00000	0.00000
CM2	0.50000	0.50000	0.00000

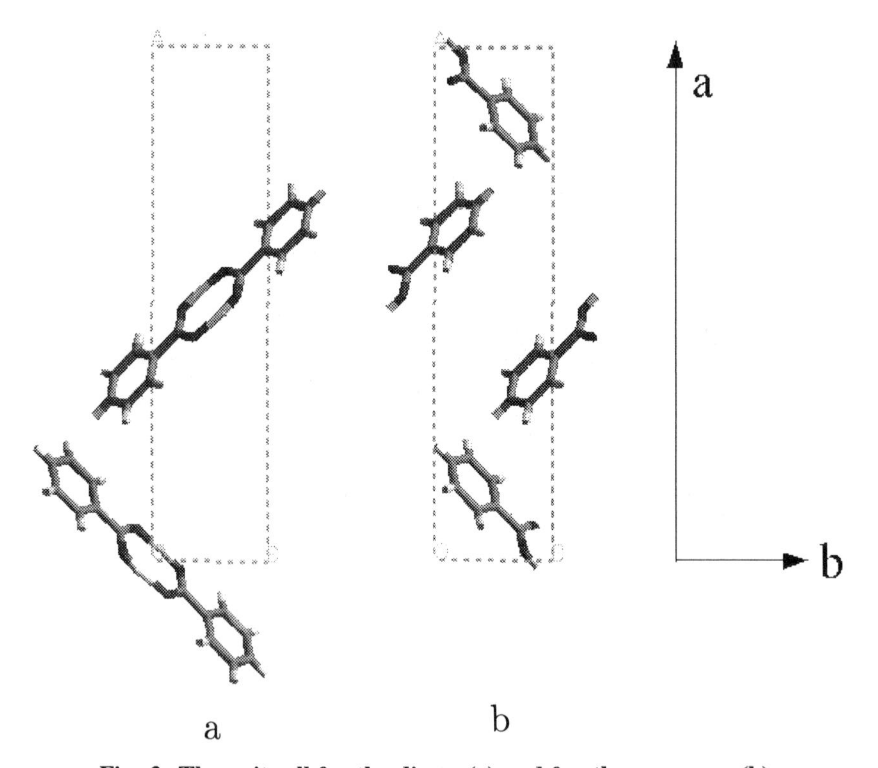

Fig. 2. The unit cell for the dimer (a) and for the monomer (b).

The bonds used to construct the crystal graph for the monomer and dimer case are given in Table 2. It is important to stress here that the choice for another growth unit usually results in a completely different crystal graph as shown here for the case of benzoic acid (Fig. 3). The bonding pattern of the monomers is, because of the larger number of growth units in the unit cell, much more complicated than the pattern of the dimers. Therefore, we expect that there will be more alternative connected nets for the monomer case for some orientations.

The F_GRAPH program was used to calculate all the connected nets for the monomer and the dimer case using the appropriate crystal graphs. The results of these computations after correction for extinction conditions are summarized in Table 3.

In the fourth and fifth column the number of alternative F-faces has been printed. It is immediately clear that the more complicated monomer graph results in much more F-faces than the dimer graph.

Table 2 Summary of the bonds in the crystal graph for monomers and dimers.

Monomer				Dimer	
Nr.	Bond	Nr.	Bond	Nr.	Bond
1	1-1(001)	7	1-3(110)	1	1-1(001)
2	1-1(010)	8	1-3(1$\bar{1}$0)	2	1-1(010)
3	1-1(011)	9	1-3(10$\bar{1}$)	3	1-1(011)
4	1-2	10	1-3(11$\bar{1}$)	4	1-1(01$\bar{1}$)
5	1-2(00$\bar{1}$)	11	1-3(1$\bar{1}\bar{1}$)	5	1-2
6	1-3(100)			6	1-2(00$\bar{1}$)

The results of the dimer graph are in agreement with the results of Hartman (Hartman, 1991). For monomers the number of F-slices per orientation is in many cases larger than one, while the program for dimers exclusively generates singlets. It is well-known that in many structures multiple connected nets appear for some orientations. Some are completely different and others are mutually related by some symmetry element. The latter ones are so-called

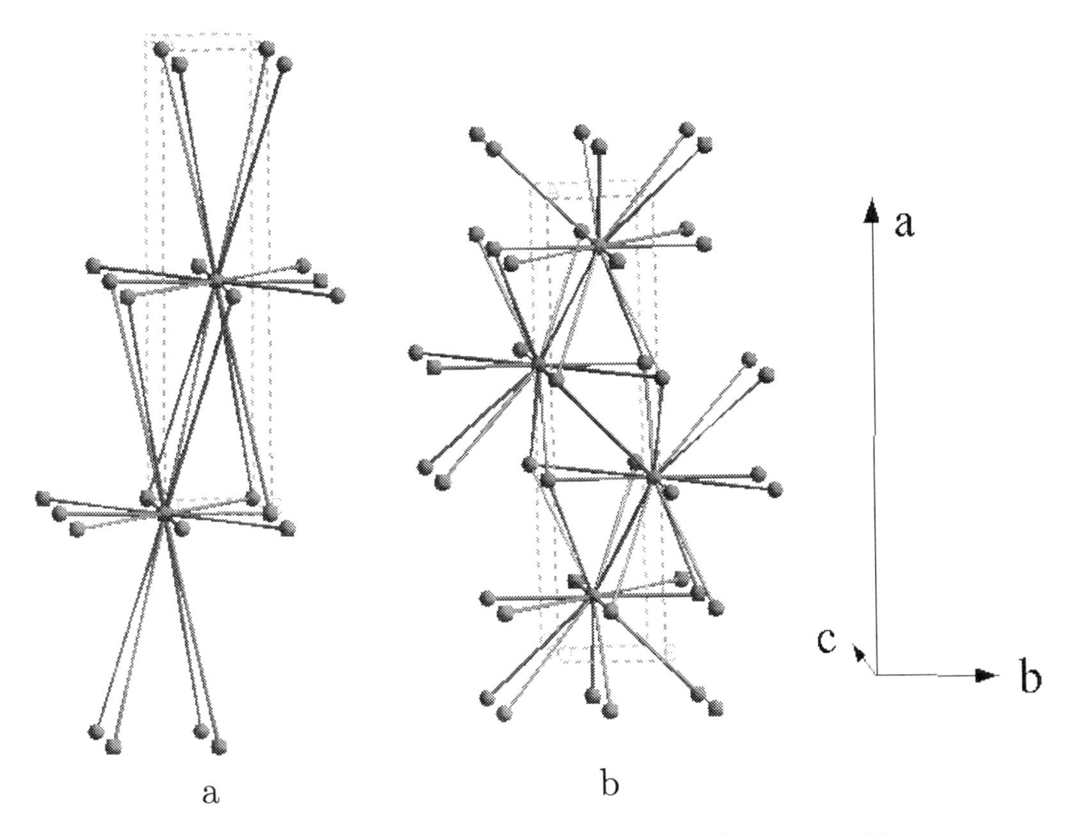

Fig. 3. The crystal graph for the dimer (a) and for the monomer (b).

Table 3 Summary of F-faces and attachment energies calculated for benzoic acid with either monomers or dimers as growth units.

Form	d_{hkl} (Å)	N_m	N_d	E_m^{drei}	E_d^{drei}	E_m^{lif}	E_d^{lif}	E_m^{mom}	E_d^{mom}	E_m^{sch}	E_d^{sch}
{200}	10.8843	2	1	12.09	12.08	13.01	13.01	12.68	12.67	15.59	15.59
{001}	5.4542	2	1	27.92	27.90	29.32	29.32	27.38	27.37	32.44	32.43
{201}	5.1479	1	1	24.44	24.41	29.07	29.05	27.27	27.25	32.24	32.22
{110}	4.9917	2	1	45.07	38.37	38.66	37.73	55.28	39.25	62.03	46.02
$\{20\bar{1}\}$	4.6435	1	-	39.54		34.26		32.52		38.78	
{210}	4.6392	2	-	42.82		38.29		50.56		56.78	
{310}	4.1880	1	-	45.43		42.96		55.50		62.29	
{111}	3.7375	1	1	50.01	44.43	46.65	42.81	61.82	45.00	70.21	53.08
{011}	3.7361	4	-	46.61		43.92		56.98		64.69	
{211}	3.6332	4	-	51.09		47.58		71.60		79.39	
$\{11\bar{1}\}$	3.6295	2	1	51.17	46.38	46.53	44.93	59.86	47.24	67.75	55.90
{311}	3.4473	2	1	52.53	45.00	48.52	44.72	62.31	47.12	70.62	55.72
$\{21\bar{1}\}$	3.4421	4	-	51.42		47.64		61.01		69.20	
$\{31\bar{1}\}$	3.2089	2	-	52.53		48.74		62.16		70.65	
{112}	2.4234	1	-	52.26		53.10		66.88		76.07	
{212}	2.4092	2	-	51.84		50.96		63.60		72.47	
{012}	2.4078	4	-	55.13		53.93		66.98		76.20	
{312}	2.3668	3	-	51.42		52.77		67.05		76.27	
$\{11\bar{2}\}$	2.3641	1	-	55.69		53.48		66.38		75.46	
$\{21\bar{2}\}$	2.2968	4	-	56.12		53.64		66.49		75.60	
$\{31\bar{2}\}$	2.2116	3	-	56.22		53.74		66.56		75.69	

N_m and N_d are the number of different connected nets in each orientation found for monomers and dimers, respectively. E_m^{drei}, E_m^{lif}, E_m^{mom} and E_m^{sch} are the calculated attachment energies for monomers using the Dreiding, Lifson, Momany and Scheraga force fields. E_d^{drei}, E_d^{lif}, E_d^{mom} and E_d^{sch} is the same as the previous for dimers. All attachment energies are in -kcal/mol.

multiplets. Singlets and multiplets of F-slices have been studied extensively for sodiumoxalate in a recent paper (Strom, 1995). Work on this subject is still in progress. For the moment we calculate the energetically most favourable F-slice and consider this as the actual growth layer.

Calculation of the attachment energies

For the calculation of attachment energies we made use of the morphology module of the commercially available molecular modeling package Cerius[2] (MSI, 1994). This module is based on the method described by Berkovitch-Yellin (Berkovitch-Yellin, 1985). The attachment energy of an F-face E_{att}^{hkl} is computed as

$$E_{att}^{hkl} = E_{latt} - E_{slice}^{hkl} \qquad (1)$$

E_{latt} is the lattice energy of the crystal and E_{slice}^{hkl} is the energy released on the formation of a growth slice of thickness d_{hkl}.

The slice energy for an F-slice (hkl) is calculated as follows. A reference molecule is chosen and all interactions of molecules lying within a radius R from this molecule are calculated. The radius R has to be chosen in such a way that the total interaction energy has converged to a constant value. Then the total interaction energy is equal to the lattice energy. Subsequently a slice with thickness d_{hkl} is stepped through the crystal in a direction normal to the slice. E_{slice}^{hkl} is calculated for each step by summation of all interactions of molecules within the slice with the reference molecule. The attachment energy is now calculated from the highest E_{slice}^{hkl} using relation 1. This procedure is repeated for all growth units in the unit cell and for all orientations (hkl).

For the calculation of the attachment energies we needed point charges for the benzoic acid molecule. The charges were calculated using the Cerius[2] interface to AMPAC (MSI, 1994). With the MNDO semi-emperical Hamiltonian the Electrostatic Potential-Derived (ESPD) point charges were calculated. All subsequent E_{att}^{hkl} calculations made use of these charges.

We calculated the attachment energies for the monomer and dimer case using four different force fields. The Dreiding, Lifson, Momany and Scheraga force field parameters were available in Cerius[2]. The results of the calculations are listed in the last eight colums of Table 3.

Construction of the theoretical morphologies

In order to construct a theoretical crystal habit, we need center-to-face distances for all F-faces. The calculated attachment energies E_{att}^{hkl} were used as a direct measure for the center-to-face distance. Thus, assuming that the proportionality between the growth rate of face (hkl) and E_{att}^{hkl} holds (Hartman, 1980), the constructed morphologies are growth forms. For illustration, two representative examples of habits were constructed using the data from Table 3. Fig. 4 shows the result for the Dreiding force field and Fig. 5 for the Momany force field. The theoretical morphologies are so-called vacuum

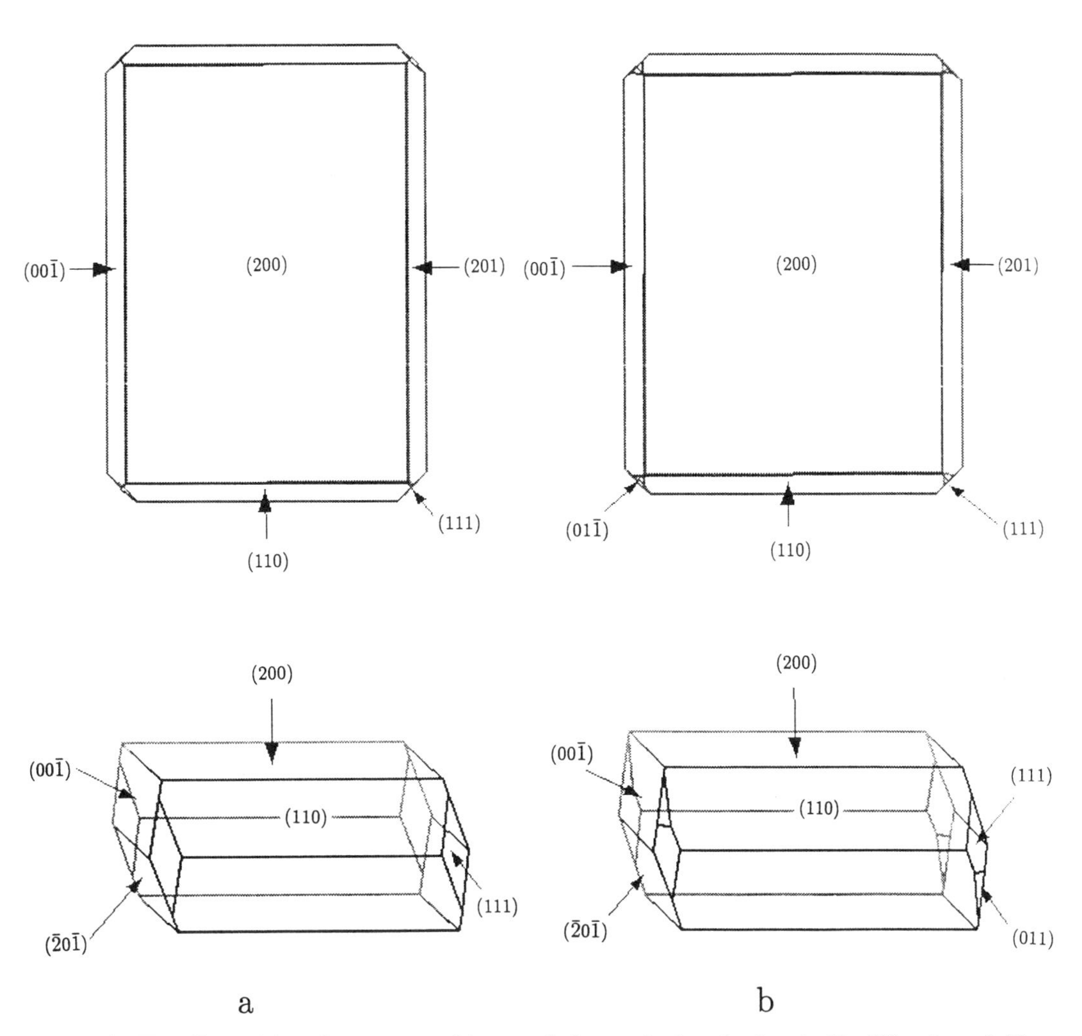

Fig. 4. The dimer (a) and monomer (b) morphology calculated using the Dreiding force field.

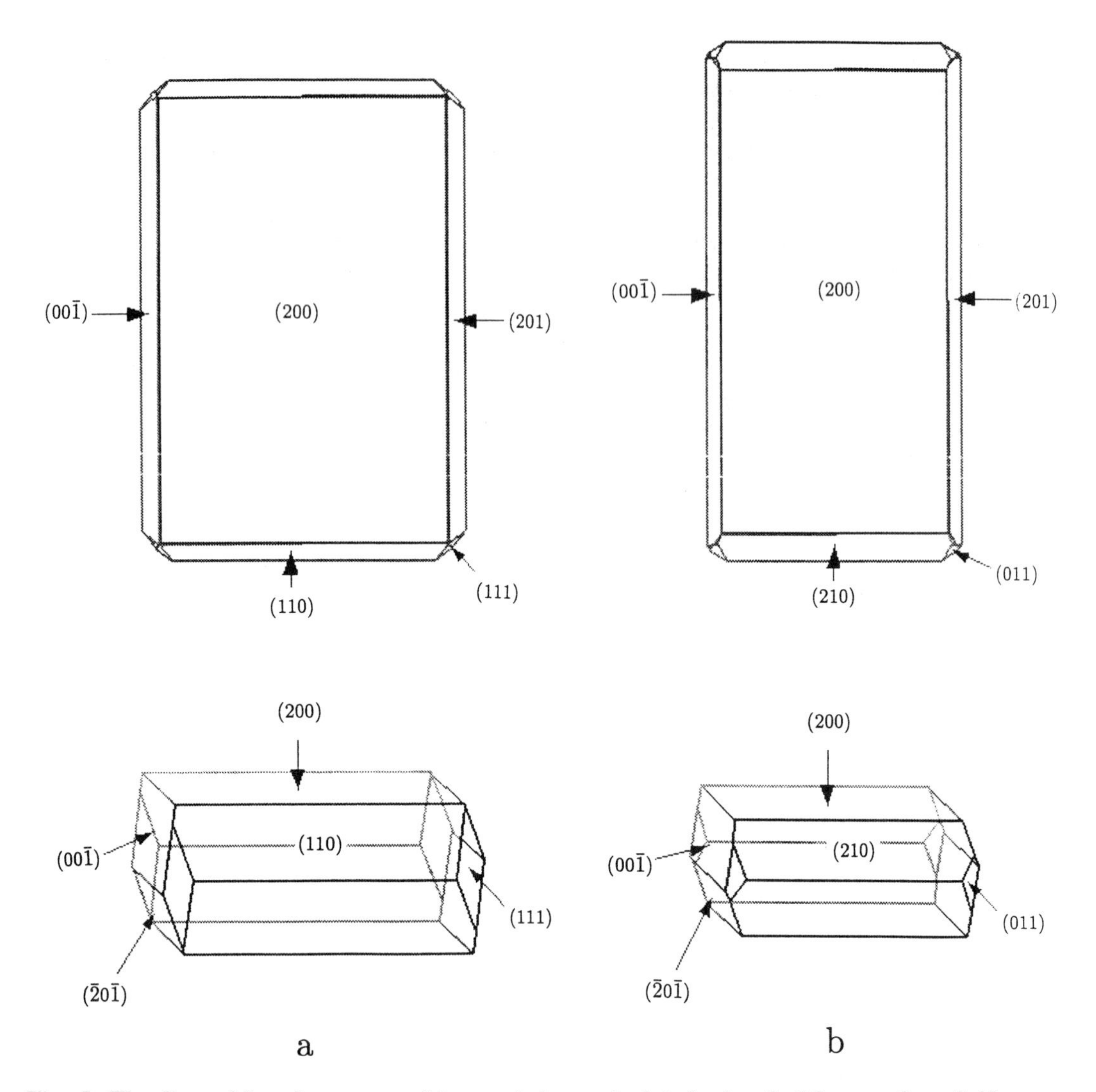

Fig. 5. The dimer (a) and monomer (b) morphology calculated using the Momany force field.

morphologies, because all energies are calculated in reference to single molecules in vacuo. For this reason the predicted growth forms have to be compared with crystals grown from the vapour. As an experimental reference we used the experimental results of Berkovitch-Yellin (Berkovitch-Yellin, 1985), which is bounded by $\{200\}$, $\{201\}$, $\{001\}$ and $\{110\}$.

Discussion and conclusions

If we compare the theoretical crystal habit based on the dimer PBC analysis, we observe that the $\{111\}$ form appears, while it is absent in the experimental habit. Furthermore, $\{001\}$ is somewhat overestimated compared to $\{201\}$. These effects are independent of the force field parameters we used and might be caused by the choice of the atomic point charges. Because of the extra extinction condition $h + k = 2n$, the forms $\{210\}$ and $\{011\}$ are not permitted and do not show up. However, the attachment energy of $\{011\}$ would cause this face to appear in the growth morphology.

For the monomer habit, we found for the Momany, Lifson and Scheraga force field, the $\{210\}$

form. This form is not observed in experimentally grown crystals. For all four force fields, the attachment energies for both the $\{210\}$ and $\{110\}$ are very close to the threshold value for appearance in the morphology. Therefore, they are very sensitive for changes in the force field parameters. In case of the Momany and Scheraga force field, the $\{110\}$ does not show up, because of the large MI_{210}. For the Lifson force field, we found that $MI_{210} \simeq MI_{110}$. The Dreiding force field yielded $\{110\}$ and $\{210\}$ was absent. The crystal shape was in case of the Momany, Lifson and Scheraga force field, elongated along the b-axis compared to the dimer growth form. This is in agreement with experimental crystal habits.

We might conclude that in general the theoretical crystal habits based on the dimer PBC analysis, are in good agreement with the experimental morphology. For the monomers, in many cases the $\{011\}$ and $\{210\}$ appear, which are not observed on the experimental habit.

Acknowledgements

This work was supported by the Dutch Organisation for Technical Sciences (STW). The author acknowledges the valuable discussions with C.S. Strom and H. Meekes.

References

Berkovitch-Yellin, Z. *J. Am. Chem. Soc.* **1985**, *107*, 8239-8253.

Hartman, P.; Perdok, W.G. *Acta Cryst.* **1955**, *8*, 49-52.

Hartman, P.; Bennema, P. *J. Cryst. Growth* **1980**, *49*, 145-156.

Hartman, P. *J. Cryst. Growth* **1991**, *110*, 559-570.

Molecular Simulations Incorporated, Cerius^2 release 1.6, **1994**.

Strom, C.S. *Z. Krist.* **1985**, *172*, 11-24.

Strom, C.S.; Grimbergen, R.F.P.; Hiralal, I.D.K.; Koenders, B.G.; Bennema, P. *J. Cryst. Growth* **1995**, *149*, 96-106.

Growth Morphology of Crystals and the Influence of Fluid Phase

X.Y. Liu and P. Bennema, RIM, Laboratory of Solid State Chemistry, Faculty of Science, University of Nijmegen, Toernooiveld, 6525 ED Nijmegen, The Netherlands

Based on the kinetics of crystal growth, the growth morphology of crystals can be predicted by a conventional **PBC** (**P**eriodic **B**ond **C**hain) analysis, a so-called **IS** (**I**nterfacial **S**tructure) analysis and/or scaling experiments. The PBC analysis is carried out to obtain internal habit controlling factors while the IS analysis and/or scaling experiments are carried out to obtain external habit factors. In this paper, our concentration is focused on the analysis of fluid phase effect. The growth morphology of some organic crystals is predicted using our approach and the results are compared with experiments.

The shape of a crystal is determined by the distances from the center of the crystal to the respective crystal faces {*hkl*}. Obviously, these distances are proportional to the relative growth rates R_{hkl}^{rel} of the crystal faces for a growing crystal. This implies that the prediction of the growth morphology is equivalent to predict the relative growth rate in different crystallographic orientations.

In the world of crystal growth, the Hartman-Perdok (HP) theory (Hartman, 1987) is often applied to predict the growth morphology of crystals. It is assumed by this theory that a growing crystal is bounded by F (or flat) faces (Hartman, 1987), and the relative growth rate R_{hkl}^{rel} of an F face is directly proportional to its attachment energy E_{hkl}^{att}, which corresponds to the energy released per structural unit as the crystal slice (*hkl*) attaches to the crystal surface from infinite distances. This energy is related to the 3D crystallization energy E^{cr} and the slice energy E_{hkl}^{slice} by

$$E^{cr} = E_{hkl}^{slice} + E_{hkl}^{att}. \qquad (1)$$

(E_{hkl}^{slice} denotes the 2D lattice energy of the crystal slice (*hkl*) having a thickness of d_{hkl}.) However, in the case of growth of crystals from solutions or from the melt the discrepancies occur frequently, comparing the predicted morphology with observed one.

According to Bennema (Bennema, 1993), the concept of F faces has its physical origin from modern statistic physics. However, the relation between R_{hkl}^{rel} and E_{hkl}^{att} has an *ad hoc* character. In addition, the influence of the fluid phase is not considered. These are the major drawbacks of the HP theory.

The growth of crystals is governed by kinetics. This implies that the growth kinetics should be first taken into account to predict R_{hkl}^{rel} in different orientations. Based on this idea, we arrived at the relation between R_{hkl}^{rel} and habit controlling parameters for the case of crystals grown from solutions and the growth governed by the screw dislocation mechanism (Liu *al et.*, 1995; Liu & Bennema, 1995), as

$$R_{hkl}^{rel} \sim d_{hkl} n_{hkl} X_A^{C^*_{\ell(hkl)}} [\xi_{hkl} C^*_{\ell(hkl)}]^{-1} \times \exp(-C^*_{\ell(hkl)} \xi_{hkl} \Delta H^{diss} / n_{hkl} kT). \qquad (2)$$

Here d_{hkl} is the interplanar distance of the crystallographic orientations {*hkl*}, X_A is the concentration of the solute in the bulk, ΔH^{diss} is the molar enthalpy of dissolution, k is Boltzmann's constant, T is the temperature, n_{hkl} is a constant which is approximately equal to the coordination number of a structural unit within the crystal slice (*hkl*), ξ_{hkl} is the crystallographic orientation factor, defined as

$$\xi_{hkl} = E_{hkl}^{slice} / E^{cr} \qquad (3)$$

and $C^*_{\ell(hkl)}$ denotes the surface scaling factor, defined as the ratio between the local enthalpy of dissolution at the crystal surface (*hkl*) ΔH_{hkl}^{diss} and ΔH^{diss} (Liu & Bennema, 1993).

We notice that the factors d_{hkl}, n_{hkl} and ξ_{hkl} are determined by the structure of crystals. Therefore these factors can be considered as the internal habit controlling factors. These factors can be calculated from the crystal structure or a conventional PBC analysis (Bennema, 1993). On

1054–7487/96/0036$12.00/0 © 1996 American Chemical Society

the other hand, X_A, ΔH^{diss} and $C^*_{\ell(hkl)}$ are strongly determined by the ambient phase and growth conditions. Therefore these factors are the external habit-controlling factors. It can be seen from Eq.(2) that due to the exponential relation, $C^*_{\ell(hkl)}$ strongly affects R^{rel}_{hkl} for a given crystal-solution system. According to our recent investigation (Liu, 1995) its value depends strongly on the concentration and the composition of solutions, crystallographic orientations and other crystal growth conditions. Henceforth, this factor can be regarded as one of the key factors to characterize the influence of the fluid on the morphology of crystals. To investigate the effect of the fluid, one should know $C^*_{\ell(hkl)}$ under the crystal growth condition.

The goal of this paper is to discuss the way of the calculation or the estimation of $C^*_{\ell(hkl)}$ for a given crystal solution system. To demonstrate these approaches, the growth morphology of some organic crystals grown from solutions will be predicted.

1. Surface Scaling Factor

As mentioned in the beginning, the surface scaling factor is expressed as

$$C^*_{\ell(hkl)} = \Delta H^{diss}_{hkl} / \Delta H^{diss} \approx \phi_i/\Phi_i, \quad (4)$$

with $$\Delta H^{diss}_{hkl} = \sum_i^m \phi_i \quad (5)$$

and $$\Delta H^{diss} = \sum_i^m \Phi_i \quad (6)$$

where ϕ_i denotes the exchange bond energy between a solute unit and a solid unit in direction i at the crystal surface (hkl) and Φ_i is the corresponding bond energy in the bulk phase. Suppose that there are m bonds connecting a structural unit with neighboring units. It follows from the proportionality condition (Liu & Bennema, 1994) that ratio between the bond energies in different directions remains the same in different situations. Once the ratio is calculated from solid-solid interaction energies, Φ_i can then be derived from ΔH^{diss} according to Eq.(6). Obviously, the remaining problem for the estimation of $C^*_{\ell(hkl)}$ is to measure ϕ_i. This can be accomplished from roughening transition or scaling experiments.

Alternatively, $C^*_{\ell(hkl)}$ can also be calculated theoretically. In the case of regular solutions, $C^*_{\ell(hkl)}$ can be related to the concentration of the solute in different regions as follows (Liu *al et.*, 1995; Liu & Bennema, 1993)

$$C^*_{\ell(hkl)} \approx \ln X^{eff}_{A(hkl)} / \ln X_A, \quad (7)$$

where $X^{eff}_{A(hkl)}$ is the concentration of effective growth units at the crystal surface. $X^{eff}_{A(hkl)}$ can be obtained from a so-called interfacial structure (IS) analysis. This implies that $C^*_{\ell(hkl)}$ can be calculated according to Eq.(7).

In the following, we will discuss the aforementioned two approaches.

1.1 Interfacial structure analysis

Due to the influence of the crystal surface, fluid units at the solid-fluid interface reveal a certain degree of ordering. This implies that growth units at the crystal surface will preferentially adapt some orientations and molecular conformations in order to lower the free energy of the system. The ordering leads to a consequence that some adsorbed growth units can easily be incorporated into the crystal structure at kink sites while others show insurmountable free energy barriers to enter kinks of steps. The so-called effective growth unit conceptually corresponds to those adsorbed growth units which can effectively participate in the process of crystal growth at the crystal surface. To explain unambiguously this term, we need to have a further analysis.

Due to the ordering, some adsorbed growth units have roughly the same orientations and the conformations as demanded for the growth of the crystal surface (hkl) (see Fig.1). These adsorbed growth units are defined as F_1-units. Other adsorbed growth units whose orientations and conformations are different from those as demanded for the growth are defined as F_2-units (Fig.1). In order to be transformed into a solid unit (S_I) at a kink, an F_1-unit needs to surpass a barrier ΔG^*_{kink} (> 0, corresponding to desolvatation free energy), as

$$F_1 \underset{I}{\overset{\Delta G^*_{kink}}{\rightleftharpoons}} S_I \quad .$$

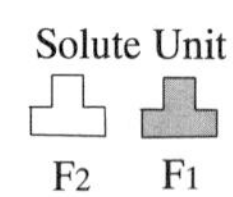

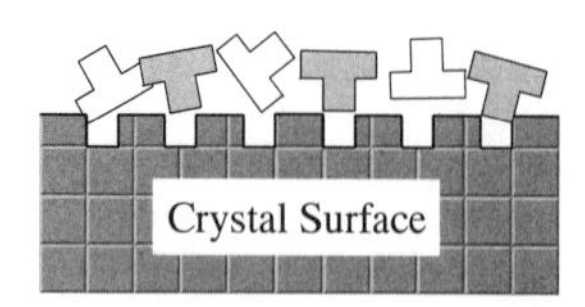

Fig.1 Schematic illustration of F_1 and F_2 units at the crystals surface. F_1 units are those adsorbed solute units having the roughly same orientations and the conformations as demanded for the growth of the crystal surface; F_2 units are those whose the orientations and conformations are different from those as demanded for the growth.

It follows that the relaxation time for the transformation from an F_1 to an S_I unit is

$$\tau_1 \sim 1/\nu_1 \exp(\Delta G^*_{\text{kink}}/kT) \tag{8}$$

(ν_1 denotes the frequency of thermal vibration of F_1 units.) On the other hand, the transformation from an F_2 to an S_I unit is somewhat different. Before an F_2 unit can be incorporated into the crystal at a kink they need to be transformed into an F_1-unit, for which they have to surpass a barrier ΔG^* (≥ 0), as

$$F_2 \underset{\text{II}}{\overset{\Delta G^*}{\rightleftharpoons}} F_1 \underset{\text{I}}{\overset{\Delta G^*_{\text{kink}}}{\rightleftharpoons}} S_I.$$

The relaxation time for the transformation is then given by

$$\tau_2 \sim 1/\nu_2 \exp(\Delta G^*_{\text{kink}}/kT) + 1/\nu_1 \exp(\Delta G^*/kT). \tag{9}$$

(ν_2 denotes the frequency of thermal vibration of F_2 units.) Obviously, $\tau_2 \geq \tau_1$. This implies that in dynamic equilibrium with solid units at the surface, some F_2 units might behave in a similar way as F_1 units whereas some will react very slowly with solid units. In order to measure the effecivity of F_2 units, we take the transition rate of F_1 units as a reference to define a effective coefficient of F_2 units:

$$\varsigma_{F2} \approx (1/\tau_2)/(1/\tau_1) \approx \exp(\Delta G^*_{\text{kink}}/kT) / [\exp(\Delta G^*_{\text{kink}}/kT) + \exp(\Delta G^*/kT)]. \tag{10}$$

(To obtain Eq.(10), we assume that ν_2 and ν_1 have the same magnitude.) Based on this definition, an adsorbed growth unit can be "cut" into two parts: one part ς_{F2} can be regarded the same as a fraction of an F_1 unit, and the other part ($1 - \varsigma_{F2}$) can be regarded as a fraction of a "solvent" or "impurity" unit which may not participate in the crystal growth process (c.f. Fig.2(a)). Adding these small pieces of quasi "F_1" units together with genuine F_1 units, we then define the concentration of effective growth units, as

$$X^{\text{eff}}_{A(hkl)} = \sum_{q=F_1,F_2} \left[\varsigma(q) X^q_{A(hkl)}\right] = \delta X_{A(hkl)}, \tag{11}$$

with
$$\delta = \sum_{q=F_1,F_2} \left[\varsigma(q) P(q)\right], \tag{12}$$

$$P(q) = X^q_{A(hkl)} / X_{A(hkl)}. \tag{13}$$

From a point of view of statistic physics, this implies that we can "detect" a molar fraction $X^{\text{eff}}_{A(hkl)}$ of "genuine growth units" at the crystal surface which will "react" with solid units at the crystal surface (c.f. Fig.2(b)).

Based on the above-mentioned principles, the interfacial structure analysis can be generalized as follows:

a) Apply computation or simulation techniques, such as Molecular Dynamic or Monte Carlo simulations, self-consistent field lattice model calculations, etc., to calculate the detailed structure of the crystal-solution interface. The essential information, like density distributions, orientation and conformation distributions, should be obtained from the calculations.
b) Define the orientations and the conformations required for growth units to be incorporated into the crystal in a given crystallographic orientation. In other words, define F_1 units according to the structure of crystal surface.
c) Calculate, according to Eqs.(11)-(13), $X^{\text{eff}}_{A(hkl)}$ from the density distributions, orientation and conformation distributions obtained from step a). $C^*_{\ell(hkl)}$ is then calculated from $X^{\text{eff}}_{A(hkl)}$ and X_A (c.f. Eq.(7)).

1.2 Scaling experiments

It can be visualized from Eq.(4) that $C^*_{\ell(hkl)}$ can be obtained once ϕ_i is available. In the case of

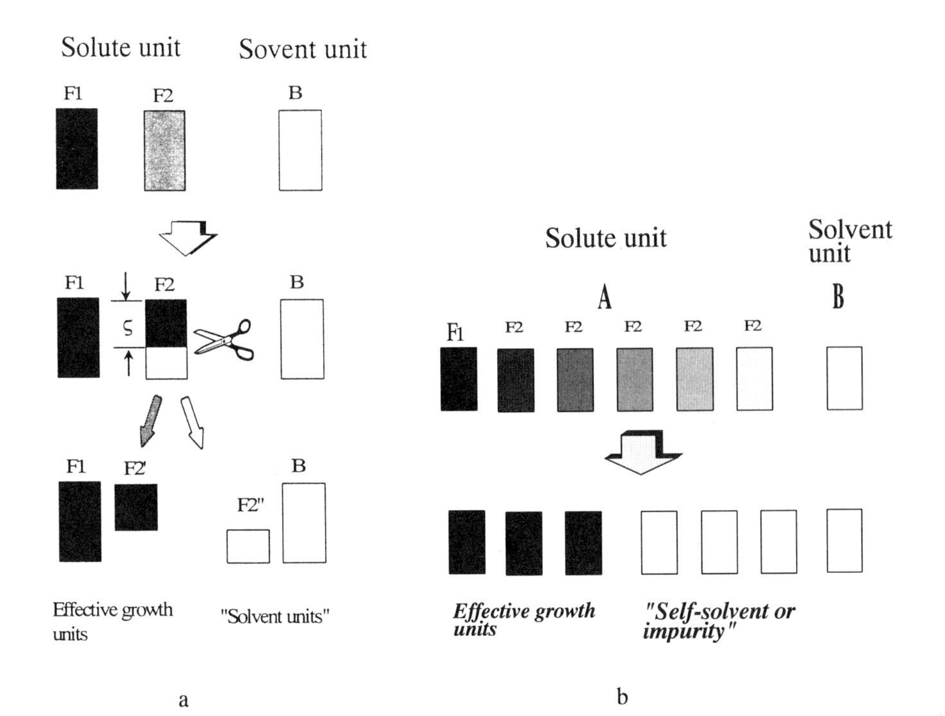

Fig.2 Schematic illustration of effective growth units. (a) In reference to F_1-units, each F_2-unit can be separated into an effective part ς and an ineffective part $(1 - \varsigma)$ according to Eq.(10). (b) Effective growth units consist of F_1-units and the effective parts of all F_2-units.

surface roughening experiments, it is possible to measure ϕ_i.

The roughening transition is a phase transition occurring at a flat crystal face (Bennema, 1993). If the temperature T is lower than the roughening temperature T^R of a crystal face, the face will maintain flat on molecule scale. If $T \geq T^R$ the crystal surface will become rough. In terms of the dimensionless roughening temperature θ^R_{hkl}, T^R and ϕ_i can be related in the following way:

$$\theta^R_{hkl} = \frac{2kT^R}{\phi_i}. \qquad (14)$$

For a given crystal face θ^R_{hkl} is constant and can in principle be estimated (Bennema, 1993). The implication of (14) is that in case T^R can be measured from experiments, ϕ_i may subsequently be obtained for the face (hkl) under consideration. Odd n-paraffin crystals are one of the examples where ϕ_i for the {110} faces can be measured from the surface roughening experiments (Liu, 1995).

We notice that for the faces occurring on a crystal, there are only some faces whose ϕ_i (and $C^*_{\ell(hkl)}$) can be measured under experimental conditions. For other faces, their $C^*_{\ell(hkl)}$ could be estimated from the following relation (Liu & Bennema, 1994; Liu & Bennema, to be published):

$$(1 - \xi_{hkl})\, C^*_{\ell(hkl)} \sim (1 - \xi_{h'k'l'})\, C^*_{\ell(h'k'l')}. \qquad (15)$$

This relation implies that if $C^*_{\ell(hkl)}$ is available for one orientation, $C^*_{\ell(hkl)}$ for other orientations may also be estimated. This empirical relation was obtained from our calculations (Liu & Bennema, 1993; Liu & Bennema, to be published) and can be approximately applied to crystals with structurally simple or sphere-like structure units.

In many cases, the roughening temperature is not measurable under crystal growth conditions. Then a scaling approximation can be made to estimate $C^*_{\ell(hkl)}$ for different crystallographic orientations. Assume that under the crystal growth condition, the actual temperature is in between the roughening temperature the crystal faces $\{h_1k_1l_1\}$ and that of $\{h_2k_2l_2\}$ as

$$T^{R}_{h_1k_1l_1} > T > T^{R}_{h_2k_2l_2},$$

or $$\theta^{R}_{h_1k_1l_1} > \theta > \theta^{R}_{h_2k_2l_2},$$

with $$\theta = \frac{2kT}{\phi_i}. \qquad (16)$$

In other words, the crystal faces $\{h_1k_1l_1\}$ grow in a flat mode and the crystal faces $\{h_2k_2l_2\}$ grow in a rough mode. We may theoretically create an artificial face $(h^ok^ol^o)$ whose roughening temperature and ξ_{hkl} are also in between those of $\{h_1k_1l_1\}$ and of $\{h_2k_2l_2\}$, and we may approximate $\theta^{R}_{h^ok^ol^o} \cong \theta$. It follows from (16) and (4) that ϕ_i and $C^{*}_{\ell(hkl)}$ for the artificial face can be estimated. Consequently, $C^{*}_{\ell(hkl)}$ for other faces on the crystals can also be obtained from (15).

2. Prediction and Discussion

2.1 Growth morphology of urea crystals grown from aqueous solutions

Urea $[O{=}C(NH_2)_2]$ crystallizes in the tetragonal space group $P_{\bar{4}2_1m}$ (a = 5.661Å, c = 4.712Å) (Worsham *al et.*, 1957). In growing from aqueous solutions, the crystals have a long needle-like shape (Davey *al et.*, 1986). However, applying the HP theory (Boek *al et.*, 1991) it is predicted that the crystals have a block shape, which is completely different from the observations.

In order to predict the growth morphology of urea crystals properly, we have carried out an IS analysis (Liu *al et.,* 1995). The analysis is based on the results of molecular dynamics computer simulations. We obtained from this analysis the long needle shape of urea crystals (Liu *al et.,* 1995), in excellent agreement with experiments (Davey *al et.*, 1986). This is the first example of successfully applying the IS analysis to the prediction of the growth morphology of crystals. For more details, refer to Liu *al et.*, 1995.

2.2 Growth morphologies of biphenyl and naphthalene crystals grown from toluene solutions

Both biphenyl and naphthalene crystals belong to the monoclinic structure of $P_{2_1/a}$ (Trotter, 1961; Hargreaves & Rizvi, 1962; Cruickshank, 1966). For biphenyl crystals, the cell parameters are a = 8.12Å, b = 5.46Å, c = 9.47Å, β = 95.4° and for naphthalene crystals, a = 8.24Å, b = 6.00Å, c = 8.65Å, β = 122.9°. According to the results of the PBC analysis carried out by Human *al et.* and Jetten *al et.* (Human *al et.*, 1981; Jetten *al et.*, 1984), the important F-faces are $\{001\}$, $\{110\}$, $\{20\bar{1}\}$, $\{100\}$ and $\{11\bar{1}\}$. In the following, we will apply the scaling approach to estimate $C^{*}_{\ell(hkl)}$ for these faces.

For biphenyl crystals grown from toluene solutions, the $\{110\}$ faces are faceted and the $\{20\bar{1}\}$ faces are rough at low supersaturations (Human *al et.*, 1981). This implies that the actual temperature is in between T^{R}_{110} and T^{R}_{201}. Applying the principle introduced in section 1.2, we approximate the surface scaling factor for these faces as $C^{*}_{\ell(001)} \cong 3.2$, $C^{*}_{\ell(20\bar{1})} \cong 0.97$, $C^{*}_{\ell(110)} \cong 1.19$, $C^{*}_{\ell(100)} \cong 0.914$ and $C^{*}_{\ell(11\bar{1})} \cong 1.03$. In addition, other habit controlling factors are obtained as follows: $n_{001} \cong 6.03$, $n_{20\bar{1}} \cong 4.38$, $n_{110} \cong 4.33$, $n_{100} \cong 4.44$, $n_{11\bar{1}} \cong 4.72$; $X_A = 0.43$, $\Delta H^{diss}/kT \cong 7.43$ (Jetten *al et.*, 1984). d_{hkl} can be easily calculated from the cell parameters and ξ_{hkl} has been obtained from the PBC analysis (Human *al et.*, 1981). Based on these factors, R^{rel}_{hkl} is then calculated from Eq.(2) for these orientations and the growth morphology of biphenyl crystals is constructed using the Wulff plots (Bennema, 1993). (See Fig.3(a).) In order to make a comparison, the growth morphology constructed based on the HP theory is also presented in Fig.3(b).

One can visualize from Fig.3(a) that according to our approach biphenyl crystals have a very flat, lozenge shape, bounded by the $\{001\}$ and the $\{110\}$ faces according to our formalism. In contrary, the HP theory predicts a rather thick growth habit, bounded by the faces of $\{001\}$, $\{20\bar{1}\}$, $\{110\}$ and $\{11\bar{1}\}$. According to Jetten *al et.* (Jetten *al et.*, 1984) and Human *al et.* (Human *al et.*, 1981), the prediction from our

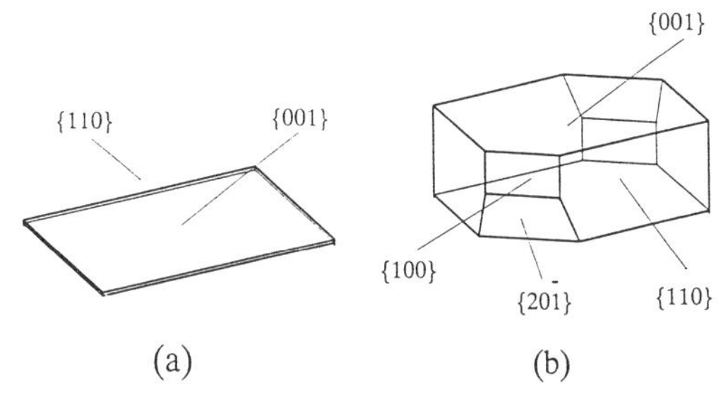

Fig.3 Wulff construction of the growth morphology of biphenyl crystals. (a) The growth morphology predicted according to our formalism [Eq.(2)]. The crystals possess a very thin, lozenge shape. (b) The growth morphology predicted according the HP theory. The crystals show the very thick growth habit with the hexagonal shape.

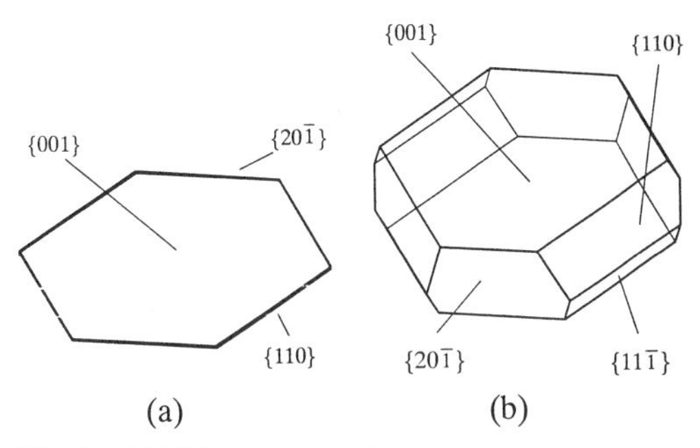

Fig.4 Wulff construction of the growth morphology of naphthalene crystals. (a) The growth morphology predicted according to our formalism [Eq.(2)]. The crystals possess a very thin, hexagonal shape. (b) The growth morphology predicted according the HP theory. The crystals show the much thicker growth habit than that shown in (a).

formalism is in excellent agreement with the observations.

In the case of naphthalene crystals grown from toluene solutions, the actual temperature is somewhat below $T^{R}_{20\bar{1}}$ and in between $T^{R}_{20\bar{1}}$ and T^{R}_{100} under the growth condition. Applying the scaling approximation, we obtain for these faces the surface scaling factor $C^{*}_{\ell(001)} \cong 3.35$, $C^{*}_{\ell(20\bar{1})} \cong 1.07$, $C^{*}_{\ell(110)} \cong 1.12$, $C^{*}_{\ell(100)} \cong 0.995$ and $C^{*}_{\ell(11\bar{1})} \cong 0.994$. Other habit controlling factors are: $n_{001} \cong 6$, $n_{20\bar{1}} \cong 4.36$, $n_{110} \cong 4.31$, $n_{100} \cong 4.44$, $n_{11\bar{1}} \cong 4.7$; $X_A = 0.53$, $\Delta H^{diss}/kT \cong 7.43$ (Jetten *al et.*, 1984). Similarly, d_{hkl} and ξ_{hkl} can obtained the crystal structure data and from the PBC analysis (Jetten *al et.*, 1984), respectively. R^{rel}_{hkl} is then calculated from these factors, according to Eq.(2). The growth morphology of naphthalene crystals is then constructed using the Wulff plots, and is shown in Fig.4(a). The crystals reveal a very thin, hexagonal shape, bounded by the {001}, the {110} and the {20$\bar{1}$} faces (see Fig.4(a)). This is in very good agreement with the observations (Jetten *al et.*, 1984). In contrary, the growth morphology of the crystals predicted according to the HP theory are somewhat too thick and shows too many faces Fig.4(b). This discrepancy is somewhat similar to the case of biphenyl crystals (c.f. Fig.3).

It can be seen from all three examples that our predictions are in excellent agreement with observations. This is because first our prediction is based on the well established relation between the growth rate and habit-controlling factors; secondly, the effect of the fluid phase is properly taken into account. Therefore, the major drawbacks of the HP theory are avoided in our approaches.

3. Summary

We have presented in this paper the principles of prediction of the growth morphology of crystals, in particular by taking into account the effect of the fluid phase. The influence of the fluid phase can be characterised by the surface scaling factor $C^{*}_{\ell(hkl)}$. An interfacial structure analysis and scaling experiments can be carried out to evaluate $C^{*}_{\ell(hkl)}$. The essential of an IS analysis is to identify the effective growth units at the crystal surface for the crystallographic orientations under the study, and to calculate their concentration. $C^{*}_{\ell(hkl)}$ can be then obtained based on Eq.(7). The scaling experiments are to measure or estimate the exchange bond energy at the given crystal surface, and to calculate or approximate $C^{*}_{\ell(hkl)}$ for different orientations based on (4) and (15).

The growth morphology of urea crystals, of biphenyl crystals and of naphthalene crystals are predicted according to the above-mentioned approaches, which are compared with those obtained from the HP theory and the experiments. Our predictions are in all cases in very good agreement with the experiments.

References

Bennema, P. in *Handbook on Cryst. Growth:* Hurle, D.T.J., Ed.; North-Holland, Amsterdam, 1993; Vol. 1, pp. 477-581.

Boek, E.S.; Feil, D.; Briels, W.J.; Bennema, P. *J. Crystal Growth* **1991**, *114*, 389-410.

Cruickshank, D.W.J. *Acta Cryst.* **1966**, *19*, 3770.

Davey, R.; Fila, W.; Garside, J. *J. Crystal Growth* **1986**, *79*, 607-613.

Hargreaves, A.; Rizvi, S.H. *Acta Cryst.* **1962**, *15*, 365.

Hartman, P. in *Morphology of Crystals*: Sunagawa, I. Ed.; Terra, Tokyo, 1987; Vol. 1, pp.269-319.

Human, H.J.; van der Eerden, J.P.; Jetten, L.A.M.J.; Odekerken, J.G.M. *J. Crystal Growth* **1981**, *51*, 598-600.

Jetten, L.A.M.J.; Human, H.J.; Bennema, P.; van der Eerden, J.P. *J. Crystal Growth* **1984**, *68*, 503

Liu, X.Y. *J. Chem. Phys.* **1995,** *102,* 1373-1384.

Liu, X.Y.; Bennema, P. *J. Chem. Phys.* **1993**, *98*, 5863-5875.

Liu, X.Y.; Bennema, P. *Phys. Rev.* **1994**, *B49*, 765.

Liu, X.Y.; Bennema, P. *Phys. Rev.* **1995**, to be published.

Liu, X.Y.; Bennema, P. *J. Crystal Growth*, to be published.

Liu, X.Y.; Boek, E.S.; Briels, W.J.; Bennema, P. *Nature* **1995**, *374*, 342-345.

Trotter, J. *Acta Cryst.* **1961**, *14*, 1135.

Worsham, J.E.; Levy, J.H.A.; Peterson, S.W. *Acta Cryst.* **1957**, *10*, 319.

HABIT95 - A Program for Predicting the Morphology of Molecular Crystals when Mediated by the Growth Environment

G. Clydesdale, K.J. Roberts,* Department Of Pure And Applied Chemistry, University of Strathclyde, Glasgow G1 1XL, UK
* also CCL-Daresbury Laboratory, Warrington WA4 4AD, UK
R. Docherty, ZENECA Specialties, Hexagon House, P.O. Box 42, Blackley, Manchester M9 3DA, UK

From a knowledge of the internal structure the computer program HABIT95 uses the atom-atom approximation to determine the intermolecular interactions in a molecular crystal. Summing these for faces (hkl) gives attachment energies, which are used as a measure of the relative growth rates to simulate the growth morphology. The effect of solvents or impurities is considered by calculating habit-modified morphologies, based on modified attachment energy terms. The program allows detailed analysis of the intermolecular and interatomic bonding and is written in standard FORTRAN 77.

HABIT95 is a computer program designed to aid in morphological investigations on molecular materials, and represents a major development of an earlier program HABIT (Clydesdale et al., 1991). It predicts the external crystal shape (often referred to as habit or morphology) based on a knowledge of the internal structure. On the whole the predicted and observed morphologies have shown good agreement (Clydesdale et al., 1994abc). Given the importance of the crystallisation of speciality molecular materials, such as pharmaceuticals, agrochemicals, dyestuffs, explosives and non-linear optical systems to the chemical industry, a theoretical approach which allows morphological investigations to be carried out on a routine basis is likely to prove extremely valuable; whereas a purely experimental approach is likely to be both time-consuming and costly. Simulations can give considerable insight into the crystallisation of molecular materials and recent developments have allowed the effects of solvent and impurities/additives to be predicted, aiding the optimisation of growth conditions to produce a required crystal habit.

Program Features

The new morphological prediction capabilities within HABIT95 include understanding the effects of either blocking or tailor-made additives (Clydesdale et al., 1994bc); modelling polar morphology (Docherty et al., 1993); modelling conformational changes during growth (Roberts et al., 1994) and predicting the morphology of molecular complexes.

Further capabilities include an interface with the crystal drawing program SHAPE (Dowty, 1980) which allows the simulated morphologies to be viewed and, to allow the effect of the additive to be seen graphically via molecular graphics systems, a data file can be written containing the atomic coordinates of an additive molecule in a host crystal. Data files containing a single intermolecular bond may also be written. In addition there is an option to select the most stable growth slice and a number of routines have been added which improve the examination of the important intermolecular interactions within molecular materials aiding the understanding of crystallisation, the effects of impurties/solvent and polymorphism as a molecular recognition phenomena.

The program is written in standard FORTRAN 77 and designed to run on unix-based workstations. It is supplied as source code along with a detailed manual and a number of test cases (input and output files) which illustrate the various operational modes. The source code of HABIT95 is available from the Quantum Chemistry Program Exchange.

Modelling Crystal Morphology Using the Attachment Energy Model

Morphological modelling techniques are based upon correlating the bulk structure in some way with the crystal morphology via a knowledge of the intermolecular forces involved in the crystallisation process. Solid-state calculations are used to determine the structural preference, in terms of (hkl), of a crystal growth layer (thickness d_{hkl}) to adsorb upon a growing crystal surface. This is achieved by

1054–7487/96/0043$12.00/0

calculation of the surface attachment energy (E_{att}) (Hartman and Perdok, 1995) which they defined as the energy released on the addition of a growth slice to the surface of a growing crystal. It is related to the crystallisation energy or lattice energy (E_{cr}, approximately equal to the negative of the sublimation enthalpy) by:

$$E_{cr} = E_{sl} + E_{att} \qquad - (1)$$

where E_{sl} is the intermolecular bonding energy contained within the surface growth slice (see Fig. 1(a)). The growth rate of a given crystal face (hkl) is proportional to E_{att} and hence inversely so to the morphological importance (Hartman and Bennema, 1980).

The lattice energy (E_{cr}) can be considered to consist of the summation of all the atom-atom interactions between a central molecule and all the surrounding molecules within a summation radius limit, beyond which the lattice energy increases by a negligble amount. From this it follows that E_{sl} can be calculated by summing all the atom-atom interactions within a layer of thickness of d_{hkl}, whilst E_{att} is simply the sum of those interactions which fall outwith this growth layer. By adjusting the position of the growth slice within the crystal structure the most energetically stable slice (i.e. that with the highest value for E_{sl}) can be determined (the program will do this automatically, if required). For structures having a number of symmetrically independent asymmetric units in the unit cell and/or an asymmetric unit containing more than one molecule, the calculations are repeated over all molecular sites and an average value obtained (Berkovitch-Yellin, 1985).

The interaction energy between two non-bonded atoms i and j, can be calculated using an atom-atom approach which assumes that the interaction between two molecules consists of the sum of all the interactions between the constituent atoms. Calculation of these interactions is achieved using, for example, a Lennard-Jones potential to calculate the van der Waals attractive and repulsive contributions and a coulombic potential to calculate the electrostatic contribution as shown in equation (2), where V is the value of the potential, A and B are constants, r is the interatomic distance, q_i and q_j are the partial charges on the atoms which are calculated using the semi-empirical quantum chemistry program MNDO (Dewar and Theil, 1979).

$$V = -A / r^6 + B / r^{12} + q_i \cdot q_j / r \qquad - (2)$$

The Bravais- Friedel- Donnay -Harker (BFDH) approach (Donnay and Harker, 1937) is used to identify likely growth faces. This assumes that, after allowing for the reduction of the growth slice thickness from space group symmetry considerations, the most morphologically important forms (hkl) and hence those with the lowest growth rates are those having the greatest interplanar spacings d_{hkl}. Once the attachment energy has been calculated for these faces the resulting predicted crystal morphology can be simulated using Wulff's (1901) classical polar plot by assuming that E_{att} is directly proportional to the centre-to-face distance. For our work we use the computer program SHAPE, which is interfaced directly, to do this (see below).

Habit Modification via Blocker and Disruptor Tailor-made Additives

Computer modelling techniques have been developed for the simulation of the crystal morphology of molecular materials in the presence of "tailor-made" additive molecules, which are used to selectively habit-modify a pure system in order to obtain a specific habit. They are "designer" impurities which resemble the host system with differences which affect the crystal growth processes in a number of ways. Mechanistically this process is essentially a case of "molecular trickery" in which the hetero-molecule mimics, to a degree, the surface chemistry of the host system thus "fooling" the crystal surface into accepting it as a pure host molecule. Once in the surface of the growing crystal the impurity affects the growth rate of individual crystal faces. It does this either by modifying properties of the crystallising solution, or by affecting the solid-state structure of the precipitated material by preventing further molecules getting to their rightful sites on the surface.

In the latter case additive molecules affect the growth rate of crystal surfaces either by blocking the movement of surface step/kink terraces (referred to as blocking modifiers, Fig. 1(b)) or by incorporating in the solid-state and disrupting the intermolecular bonding networks (disrupting modifiers, Fig. 1(c)). This causes a change in the value of the attachment energy associated with the adsorption of subsequent growth layers and affects the crystal habit.

In order to assess the structural ramifications of the additive adsorption we need to calculate the modified slice and attachment energies with the additive present (Berkovitch-Yellin, 1985); these are illustrated schematically in Fig. 1. $E_{sl'}$ is the slice energy calculated with an additive at the centre of the slice; $E_{att'}$ is the attachment energy of a growth slice

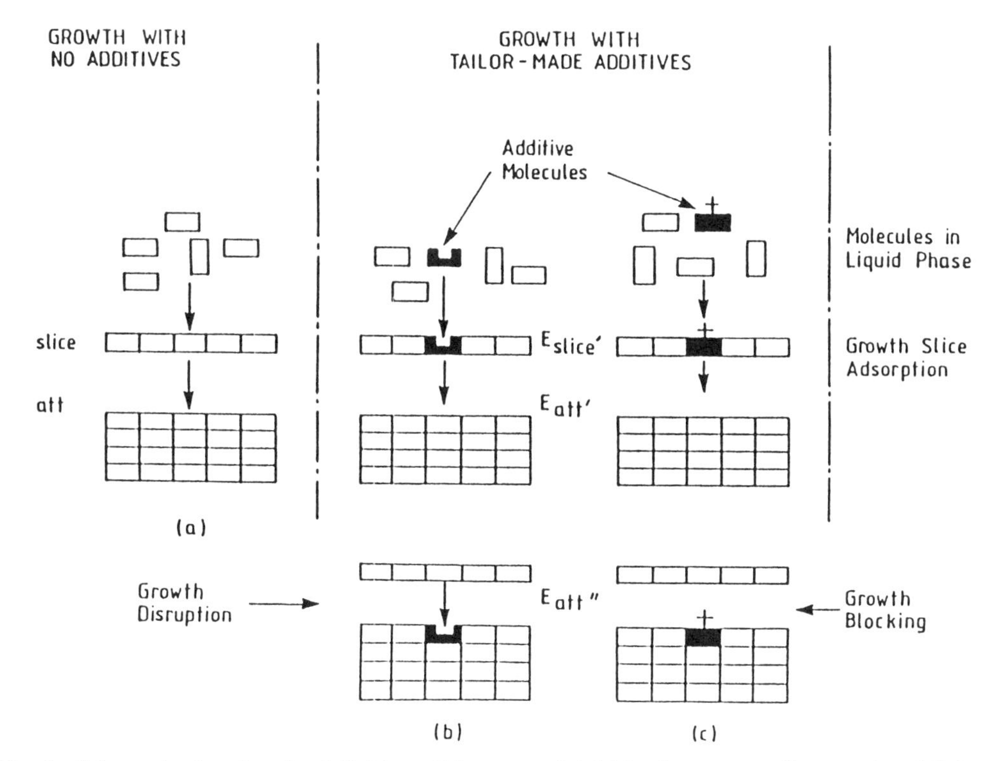

Fig. 1. Schematic showing the definition of the energy terms E_{sl}, E_{att}, $E_{sl'}$, $E_{att'}$ and $E_{att''}$ used in morphological modelling for (a) pure systems and systems having (b) disruptive-type and (c) blocker-type tailor-made additives. Reproduced with permission from Clydesdale et al. (1994c). Copyright 1994 Elsevier/North-Holland.

containing an additive onto a pure surface; $E_{att''}$ is the attachment energy of a pure growth slice onto a surface containing an additive. In order to assess how easily an additive will adsorb on a given crystal surface (hkl) the parameters E_{sl} and E_{att} can be compared to $E_{sl'}$ and $E_{att'}$ and in doing so probe the molecular compatibility of the host and additive system. This is done by defining the relative binding energy (Δb) (Lahav et al., 1985) as the difference between the binding energy of the additive ($E_{b'}$) and the pure material (E_b) where:

$$\Delta b = E_{b'} - E_b = (E_{sl'} + E_{att'}) - (E_{sl} + E_{att}) \quad - (3)$$

From this we can see that crystal faces where there is minimum change in the binding energy are where the additives are likely to incorporate. If Δb is strongly dependant upon crystal orientation then the incorporation will be specific to one crystal face and vice-versa. In our approach we propose the use of the further parameter $E_{att''}$ which reflects the energy released on the addition of a pure growth slice onto a surface on which an additive has adsorbed. This additional parameter can be used as a direct measure of the growth rate of a crystal face "poisoned" with a "tailor-made" additive molecule thus enabling calculation of the crystal morphology with an additive present.

Disruptor Additive Calculation: Benzamide with Benzoic Acid as Additive. Modelling the effect of tailor-made additives using this method has proved successful for a number of systems (Lahav et al., 1985; Clydesdale et al., 1994b). For example, the effect of the disruptor additive benzoic acid on benzamide (described in detail by Clydesdale et al. (1994b)) is given in Table 1 which lists the change in binding energy (Δb) for each likely growth form along with the host and additive attachment energies. Note that for pure systems the attachment energies (E_{att}) are averaged over all symmetry-independent sites in the unit cell whereas for additive systems the additive binding site is important and values for $E_{att''}$ and Δb are given for all sites.

The simulated morphologies (Wulff plots) were viewed using the program SHAPE:

Table 1. Changes in binding energy (Δb) and attachment energies (in kcal mol^{-1}) for benzoic acid additive in benzamide lattice. Z defines the central molecule in the unit cell: Z=1 (x, y, z); 2 (-x, y+1/2, 1/2-z); 3 (-x, -y, -z); 4 (x, 1/2-y, z+1/2). (av) denotes results averaged over all Z. For brevity only observed faces are shown.

Face (hkl)	Z	Δb	$E_{att''}$	E_{att}
(0 0 2)	1	1.73	-3.30	
	2	1.74	-3.30	
	3	1.74	-3.30	
	4	1.73	-3.30	
	(av)	1.74	-3.30	-3.31
(1 0 0)	1	2.56	-9.63	
	2	2.17	-9.63	
	3	2.17	-9.63	
	4	2.56	-9.63	
	(av)	2.37	-9.63	-8.64
(0 1 -1)	1	1.42	-8.45	
	2	1.41	-8.44	
	3	-0.25	-8.45	
	4	-0.26	-8.44	
	(av)	0.58	-8.45	-10.89

HABIT95 writes files to be directly read into this program. For additive calculations the program writes two files which allows direct comparison of the undoped and doped crystal morphologies. Fig. 2(a) shows the simulated morphology of pure benzamide using E_{att} values as a measure of growth rate, Fig. 2(b) simulates the observed effect of benzoic acid (using $E_{att''}$ values, as there is low loss of binding energy for all faces) which is to reduce growth in the b-direction due to replacing a host-host hydrogen-bond with an additive-host oxygen-oxygen repulsion subsequently blocking growth along the [010] direction (Lahav et al., 1985).

The Vacancy Approach for Blocker Tailor-made Additives

Considering blocker additives in the same way as disruptors gives rise to large, positive $E_{att''}$ values for faces where the blocking part of the additive hinders adsorbing molecules. These attachment energies with additive present cannot be utilised to produce a morphology but serve only as a guide of the "blocking ability" of the additive at each face. We improved (Clydesdale et al., 1994c) our calculation of $E_{att''}$ by introducing "vacancies" in cases where

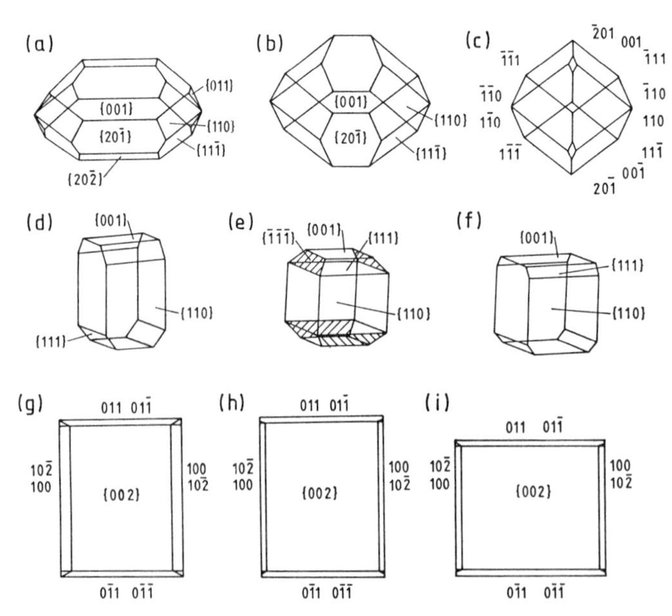

Fig. 2. Theoretical morphologies calculated using HABIT95: (a) benzamide, attachment energy model; (b) benzamide with benzoic acid additive; (c) naphthalene, attachment energy model; (d) naphthalene with biphenyl additive; (e) urea, classical simulation with the host and attaching molecules having bulk atom charges; (f) urea, polar simulation with host atoms having {111} and {-1-1-1} surface charges and the attaching molecules having isolated molecule charges.

the intermolecular interactions are unfavourable for steric reasons. This is achieved by calculating each intermolecular interaction and omitting its contribution to $E_{att''}$ (equivalent to removing the molecule) if it is considered to result from the blocking nature of the additive. This vacancy concept exhibits some of the main effects of blocker-type additives: a lowering in $E_{att''}$ (and hence an increase in morphological importance) and the loss of host molecules from adsorbing due to additive molecules blocking surface sites. The modified $E_{att''}$ parameter may now be used to obtain a morphological representation of the effects of a blocker additive which was not previously possible.

Running HABIT95 in "ADDT" mode allows calculation of the modified slice and attachment energy terms associated with the additive adsorption process detailed above. There are two options: blocker mode accounts for additives larger than the host molecule using the vacancy method; disruptor mode is used for additives of the same size or smaller than the host, here no vacancies are created. In both cases the program alternates between two further modes, with first an additive, and then a host, as the central molecule.

To allow the position of the vacancies (and the additive) to be seen graphically using molecular graphics systems the program outputs a data file in CSSR format (CSE Applications Group, Daresbury Laboratory, UK) containing the atomic coordinates. A visual representation of important interatomic contacts can also be obtained by selecting a single intermolecular bond (an example is given in Table 2).

Blocker Additive Calculation: Naphthalene with Biphenyl as Additive. The vacancy approach to modelling blocker additives has been demonstrated through an examination of the model system of naphthalene/biphenyl (Clydesdale et al., 1994c). Fig. 2(c) shows the predicted morphology of pure naphthalene, Fig. 2(d) shows the predicted effect of biphenyl additive to be an increase in the morphological importance of the {110} forms giving good agreement with experiment (Jetten, 1983). The additive was assumed to bind only at sites with low Δb values i.e., the (1-1-1), (1-10), (1-11) and (020) faces (see Table 2). Here we used $E_{att''}$ values; host E_{att} values were used for all other faces.

The vacancies created in the crystal for the (1-1-1) face can be seen from Fig. 3 which was constructed by selecting the option to write data files and comparing the file with the additive at the centre (which creates vacancies) with the file for the pure host (which has no

Table 2. Changes in binding energy (Δb) and attachment energies (in kcal mol^{-1}) for biphenyl additive in naphthalene lattice. Z=1 (x, y, z); 2 (x+1/2, 1/2-y, z). NV is the number of vacancies created; other symbols are defined in Table 1.

Face (hkl)	Z	Δb	NV	$E_{att''}$	E_{att}
(0 0 1)	1	311.91	2	-5.27	
	2	311.91	2	-5.27	
	(av)	311.91	2	-5.27	-5.97
(1 -1 -1)	1	312.29	2	-10.08	
	2	0.86	4	-8.01	
	(av)	156.57	3	-9.04	-12.24
(1 -1 0)	1	312.71	2	-10.02	
	2	1.43	3	-7.86	
	(av)	157.07	2.5	-8.94	-11.77
(2 0 -1)	1	6.34	3	-10.43	
	2	6.34	3	-10.43	
	(av)	6.34	3	-10.43	-13.08

vacancies). The blocking nature of biphenyl is clarified by indicating the sterically-hindered naphthalene molecules which are replaced (for (1-1-1), site Z=2 there 4 vacancies are created). The intermolecular bonding for this system is discussed below.

Modelling Polar Morphologies: Urea

Classical attachment energy models are incapable of predicting a polar morphology, i.e. crystals which exhibit different growth rates along (hkl) from (-h-k-l) as, by definition, the energy of such pairs of faces is equal. A method to overcome this was proposed (Docherty et al., 1993) by assigning a different set of atomic charges to molecules in the bulk of the crystal and the molecules on the surface of the crystal (these are the same as those for an isolated molecule). The coulombic contribution to each intermolecular bond is thus altered giving rise to a polar morphology. The charges are obtained using *ab initio* quantum-mechanical calculations via programs such as CRYSTAL (Dovesi et al., 1992).

Docherty et al. (1993) outline a successful application of this type of approach to urea which allows the electrostatic nature of the hydrogen-bond network to be accurately modelled. Urea crystals grown from the vapour phase exhibit a well-defined prismatic morphology (aspect ratio 1.5) dominated by {110},

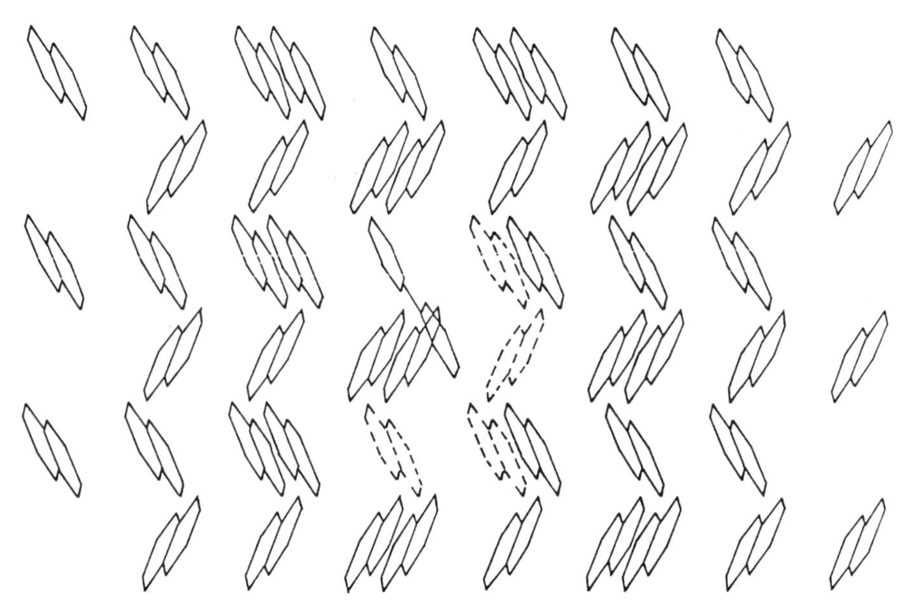

Fig. 3. Example showing the blocking nature of a biphenyl additive molecule in a naphthalene crystal where for steric reasons molecular vacancies (- - -) are created computationally. Reproduced with permission from Clydesdale et al. (1994c). Copyright 1994 Elsevier/North-Holland.

{001} and {111} forms, with the latter being polar, i.e. the {-1-1-1} form is not observed. Fig. 2(e) shows the morphology of urea calculated using a classical attachment energy approach which predicts the growth rates of the {111} and {-1-1-1} faces to be the same. Fig. 2(f) was calculated using polar mode and correctly predicts the polar morphology. Other workers (George et al., 1995) have predicted similar effects and highlighted the potential role played by surface relaxation in the overall growth process.

Bonding Analysis Calculations

Bonding analysis calculations are used for detailed analyses of the intermolecular and interatomic bonding in the crystal, such as hydrogen-bonds. HABIT95 allows the intermolecular interactions to be identified in terms of crystallographic direction; partitioned onto individual atoms; broken down to particular atom-atom contributions and sorted in terms of either strength or contact distance. The intermolecular bonds are helpful in identifying strong bonds in a crystal during a PBC analysis (Hartman and Perdok, 1955), the atom-atom bonds are useful to discover the source of unusual steric interactions (for example close contacts). Comparison with the host results allows direct assessment of the effect of the additive on the intermolecular bonding.

ß-Succinic Acid. Here we illustrate the usefulness of such investigations via the example of the dicarboxylic acid ß-succinic acid ($HOOC-CH_2-CH_2-COOH$ (Leviel et al., 1981)). The results include a list of all the intermolecular interactions sorted by bond strength (Table 3), which are dominated by two strong bonds per origin molecule of -5.99 kcal mol^{-1}. These are hydrogen-bonded molecules (one on either side of the central molecule) which continue along the [100] direction. Using the "Type" assignments (which indicate the assignment of bonds to either E_{att} or E_{sl}) we see that all the strong bonds for the {020} forms are inside the slice (Type = (SL)), thus E_{sl} for these forms is high and E_{att} is low, making these forms one of the most morphologically important (see the morphologies given by Berkovitch-Yellin (1985)).

Table 4 gives a sorted list of the most important interatomic interactions over all intermolecular interactions. This is in the same format used for the breakdown of individual intermolecular interactions (obtained by selecting a specific bond DM, where DM is a number which uniquely defines each intermolecular interaction). From this we can see that the atomic interactions which contribute the most energy to E are a pair of hydrogen-

Table 3. Intermolecular bonding vectors (energy values in kcal mol^{-1}) for ß-succinic acid crystal (growth direction = (020)). DM is a number which uniquely defines each intermolecular interaction; "Dist." is the distance between the centres of gravity of the central molecule [0001] and the second molecule defined by [UVWZ] where U, V and W are multiples of the unit cell dimensions and Z is the number of the asymmetric unit in the unit cell: Z=1 (x,y,z) 2 (x, -y+1/2, z); "Type" indicates which energy term each intermolecular bond contributes to: (SL), (+) or (-) indicate E_{sl}, $E_{att(+)}$ and $E_{att(-)}$, respectively, where the sum of the latter two terms is E_{att} and they are the components of the attachment energy along the positive and negative growth directions.

DM	U V W Z	Dist.	Energy	Type
577	1 0 -1 1	7.62	-5.99	(SL)
340	-1 0 1 1	7.62	-5.99	(SL)
460	0 0 1 1	5.10	-1.80	(SL)
457	0 0 -1 1	5.10	-1.80	(SL)
579	1 0 0 1	5.52	-1.52	(SL)
338	-1 0 0 1	5.52	-1.52	(SL)
458	0 0 -1 2	5.11	-1.43	(+)
435	0 -1 -1 2	5.11	-1.43	(-)
459	0 0 0 2	5.11	-1.43	(+)
437	0 -1 0 2	5.11	-1.43	(-)

oxygen bonds which contribute to the most important intermolecular interactions (DM=577, 340).

The identification of these atom-atom bonds was facilitated using the interface to molecular graphics option, selecting bond number 577. This gave a CSSR file containing the two interacting molecules; displaying this on a suitable molecular graphics system showed the interacting atoms to be hydrogen-bonds in adjacent carboxylic acid groups (see Fig. 4). Note that the large repulsive interactions at the end of the atom-atom lisiting are also due to these groups, in this case hydrogen-hydrogen and oxygen-oxygen repulsion. In this way we can see that the lattice energy is dominated by energy contributions due to the hydrogen-bonds.

The further breakdown of the bonding given in Table 5 partitions each atom in the central molecule's contribution to the lattice energy. The two hydroxyl hydrogens (H3 and H6, see Fig. 4) and the two carbonyl groups (C2-O1 and C4-O3) contribute the most energy to the attractive term of the Lennard-Jones potential (see Eqn. 2), whilst the two hydroxyl oxygens (O2 and O4) tend to be responsible for repulsive interactions.

Table 4. Strongest atom-atom bonds (in kcal mol^{-1}) for ß-succinic acid crystal. "Atom1" and "Atom2" are the atom identifiers (defined in Fig. 4) for the interacting atoms; other symbols are defined in Table 3.

DM	U V W Z	Atom1	Atom2	Dist.	Energy
340	-1 0 1 1	O3	H3	1.69	-14.20
340	-1 0 1 1	H6	O1	1.69	-14.20
577	1 0 -1 1	O1	H6	1.69	-14.20
577	1 0 -1 1	H3	O3	1.69	-14.20
435	0 -1 -1 2	H3	O2	2.86	-8.39
459	0 0 0 2	H6	O4	2.86	-8.39
437	0 -1 0 2	O2	H3	2.86	-8.39
458	0 0 -1 2	O4	H6	2.86	-8.39
340	-1 0 1 1	O4	H3	2.89	-8.30
577	1 0 -1 1	O2	H6	2.89	-8.30
340	-1 0 1 1	H6	O2	2.89	-8.30
577	1 0 -1 1	H3	O4	2.89	-8.30
577	1 0 -1 1	H3	C4	2.58	9.31
340	-1 0 1 1	C4	H3	2.58	9.31
340	-1 0 1 1	H6	C2	2.58	9.31
577	1 0 -1 1	C2	H6	2.58	9.31
340	-1 0 1 1	O3	O2	2.68	9.44
340	-1 0 1 1	O4	O1	2.68	9.44
577	1 0 -1 1	O2	O3	2.68	9.44
577	1 0 -1 1	O1	O4	2.68	9.44
340	-1 0 1 1	H6	H3	2.36	10.16
577	1 0 -1 1	H3	H6	2.36	10.16

Additive Bonding Analysis: Naphthalene with Biphenyl Additive. Table 6 gives the intermolecular bonding analysis for naphthalene/biphenyl. In ADDT mode both host and additive results are presented together to allow easy comparison. The results show that most of the strong interactions are not greatly affected by the additive, however, there is a very large repulsive interaction between the two molecules in the origin unit cell ([0001]-[0002] in [UVWZ] terms) due to the effect of the blocker part of biphenyl (one of the benzene rings).

The results in Table 7 were obtained by selecting this bond to obtain the atom-atom bonding in both ADDT calculation modes: in ADDT mode 1 the interactions are between a biphenyl additive and a naphthalene molecule; in mode 2 we can compare these to the corresponding interaction for two host molecules. Using the interface to molecular graphics option we obtain data files of both these interactions, as displayed in Fig. 5.

We can see that there are no close contacts for the host-host case but that the upper biphenyl ring is sterically hindering the adjacent naphthalene molecule with close contacts less than 1.84Å listed. These are the major sources of the large repulsive term in the intermolecu-

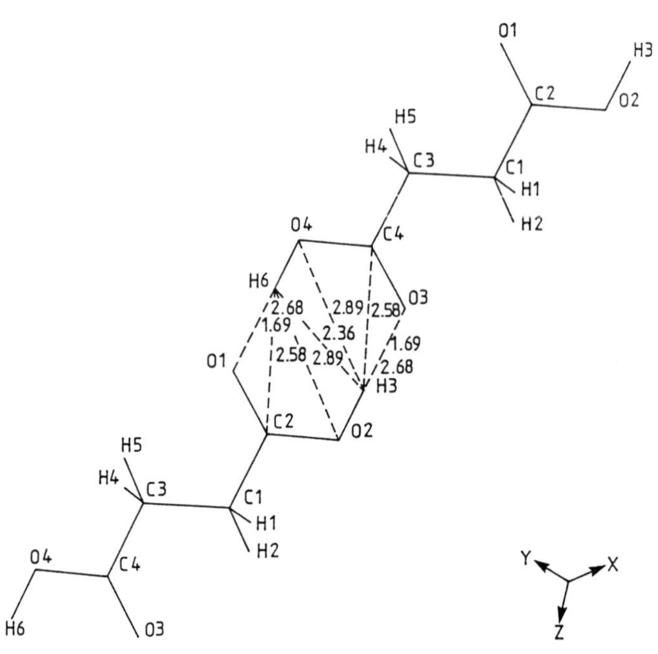

Fig. 4. Dominant atom-atom interactions (interatomic distances in Å) for strongest intermolecular bond in ß-succinic acid crystal. Energy values for these are given in Table 4.

Table 5. Particular atom contributions to the lattice energy for ß-succinic acid, given both in kcal mol^{-1} and as a percentage. Atom labels are defined in Fig. 4.

Atom	Energy	%
C1	-1.14	3.70
C2	-7.70	24.97
O1	-7.77	25.19
O2	2.61	-8.47
H1	-2.57	8.34
H2	-2.10	6.80
H3	-12.17	39.47
C3	-1.14	3.70
C4	-7.70	24.97
H4	-2.57	8.34
H5	-2.10	6.80
O3	-7.77	25.19
O4	2.61	-8.47
H6	-12.17	39.47

Table 6. Intermolecular bonding vectors (in kcal mol^{-1}) for naphthalene with biphenyl additive (growth direction = (1-1-1)). H-H denotes host-host interactions; A-H denotes additive-host interactions. Additive-host bonds with positive bond strengths were replaced by vacancies.

DM	U	V	W	Z	Dist.	H-H	A-H	Type
361	0	0	0	2	5.03	-2.36	303.61	(SL)
348	0	-1	0	2	5.03	-2.36	-2.14	(+)
240	-1	-1	0	2	5.03	-2.36	-1.98	(SL)
253	-1	0	0	2	5.03	-2.36	-2.32	(-)
374	0	1	0	1	5.95	-1.56	4.19	(-)
347	0	-1	0	1	5.95	-1.56	-1.21	(+)
238	-1	-1	-1	2	7.78	-0.75	-0.40	(+)
363	0	0	1	2	7.78	-0.75	-0.86	(-)
251	-1	0	-1	2	7.78	-0.75	-0.64	(SL)
350	0	-1	1	2	7.78	-0.75	-0.70	(SL)

Table 7. Strongest atom-atom bonds (in kcal mol^{-1}) for naphthalene/biphenyl (ADDT mode 1; DM = 361 selected. Symbols are defined in Table 4.

Atom1	Atom2	Dist.	Energy
C6	H7	3.28	-0.195
C6	H8	3.38	-0.190
H3	C7	3.25	-0.177
C4	H4	3.37	-0.175
H4	C10	3.11	-0.173
C5	H6	3.03	-0.171
H3	H7	1.41	8.229
H2	C6	1.61	8.630
C5	C6	2.26	9.075
H3	C9	1.59	9.286
C4	C9	2.17	12.569
H2	H6	1.00	29.979
H2	C8	1.13	40.993
C5	C9	1.83	43.727
C4	C8	1.80	49.366
C4	C6	1.73	62.778

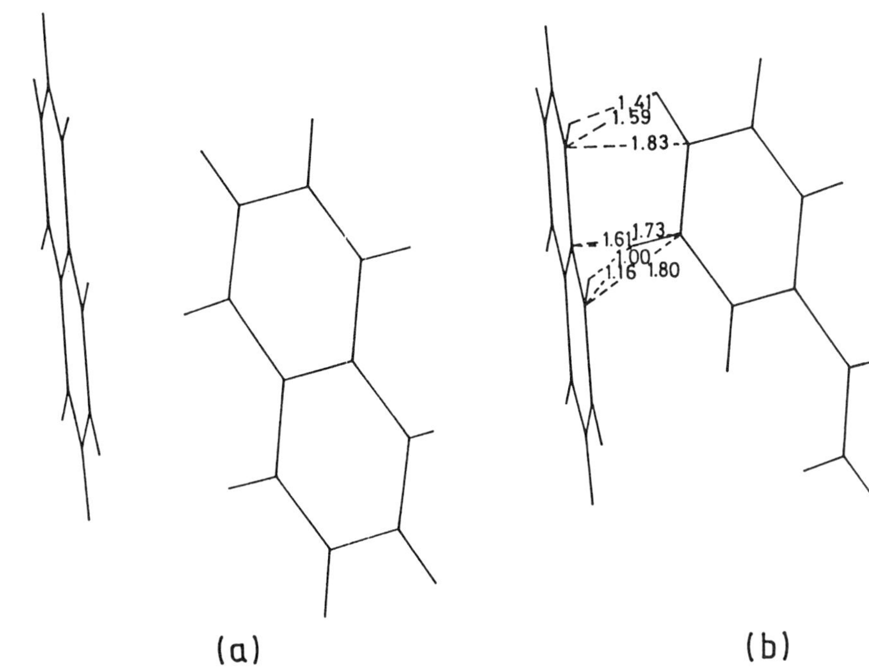

Fig 5. Naphthalene/biphenyl: comparison of interatomic interactions between undoped and doped systems by selecting the intermolecular bond between the two molecules in the origin unit cell. (a) shows the host-host interaction. In (b) the central host molecule has been replaced with biphenyl additive, creating close contacts (those < 1.84Å are shown).

Table 8. Strongest atom-atom bonds (in kcal mol^{-1}) for naphthalene/naphthalene (ADDT mode 2; DM = 361 selected). Symbols are defined in Table 4.

Atom1	Atom2	Dist.	Energy
H3	C5	3.03	-0.168
H2	C1	3.19	-0.163
H3	C1	3.16	-0.161
H5	C10	3.37	-0.160
H2	C5	3.45	-0.159
H5	C7	3.55	-0.157
H3	C3	3.02	-0.152
H5	C6	3.41	-0.151
H5	C3	3.59	-0.141
H3	C6	2.88	-0.140
H5	H8	3.83	0.120
H3	H4	3.61	0.124
H5	H7	3.68	0.124
H3	H6	3.36	0.126
H2	H4	3.58	0.128
H2	H1	3.12	0.147

lar interaction and are detailed in the atom-atom bond listing (Table 8): the strongest repulsion involves atoms 1.83Å apart, within normal bonding distances.

Conclusions

The ability to understand the external shape of crystals has wide and varied applications in both the fundamental and applied aspects of crystal science and engineering. HABIT95 has been developed to predict the morphology of molecular crystals, including the effect of the growth environment, using the attachment energy method. This method has proved to give good agreement between observed and predicted crystals. The prediction of a polar morphology is now possible and the effect of additives of both disruptor and blocker natures can be assessed. The overall schematic presented here in the case of tailor-made additives has also been applied to study growth-induced conformational change (Roberts et al., 1994). The program has also been used to model the case of additives which are not "tailor-made": this work will be presented in future papers.

Current developments, by ourselves and other researchers in the field, centre on a number of areas including the extensions of these techniques to ionic molecular solids, the use of surface modelling approaches to predict the structural nature of the growth interface when mediated by liquid phases, and the prediction of the nature of growth-induced surface roughening processes.

Acknowledgements

Research on morphological modelling at the University of Strathclyde has been supported for a number of years through research grants from the SERC/EPSRC and from a number of industrial sponsors, these we gratefully acknowledge. One of us (KJR) also gratefully acknowledges the EPSRC/SERC for the current support of a senior fellowship.

References

Bennema, P.; Hartman, P. *J. Crystal Growth,* **1980,** *49,* 145

Berkovitch-Yellin, Z. J. Am. Chem. Soc., **1985,** *107,* 8239

Clydesdale, G.; Roberts, K.J.; Docherty, R. *Computer Physics Comm.* **1991,** *64,* 331

Clydesdale, G.; Roberts, K.J.; Docherty, R. In *Controlled Particle, Droplet and Bubble Formation,* Wedlock, D., Ed.; Butterworths-Heinemann: London, 1994a; Ch.4, pp 119-121

Clydesdale, G.; Roberts, K.J.; Docherty, R. *J. Crystal Growth* **1994b,** *135,* 331;*ibid,* **1994c,** *141,* 443

CSSR - Crystal Structure Search Retrieval Database, SERC Chemical Databank System, CSE Applications Group, Daresbury Laboratory, Warrington, England, UK

Dewar, M.J.S.; Theil, W. *J. Amer. Chem. Soc.* **1979,** *99,* 4899

Docherty, R.; Roberts, K.J.; Davey, R.J; Black, S.N.; Saunders, V. *Faraday Discussions* **1993,** *95,* 11

Donnay, J.D.H.; Harker, D. *Amer. Mineralogist* **1937,** *22,* 446

Dovesi, S.; Pisani, C.; Roetti, C.; Causa, M.;
Saunders, V. **1992,** CRYSTAL92, QCPE program no. 577, Quantum Chemistry Program Exchange, Indiana University, Bloomington, Indiana, USA

Dowty, E. Amer. Mineralogist **1980,** *65,* 465

George, A.R; Harris, K.D.M; Rohl, A.L; Gay, D.H. *J. Mater. Chem.* **1995,** *5,* 133

Hartman, P; Perdok, W.G. *Acta Cryst.* **1955,** *8,* 49

Jetten, L.A.M.J. Ph.D Thesis (1983) University of Nijmegan, The Netherlands

Lahav, M.; Leiserowitz, L.; Berkovitch-Yellin, Z.; van Mil, J.; Addadi, L.; Idelson, M. *J. Amer. Chem. Soc.* **1985,** *107,* 3111

Leviel, J.L.; Auvert, G.; Savariault, J.M. *Acta Cryst.* **1981,** *B37,* 2185

Roberts, K.J.; Sherwood, J.N., Yoon, C.S.; Docherty, R. *Chem. Mater.* **1994,** *6,* 1099

Wulff, G. *Z. Kristallogr.* **1901,** *34,* 499

A Comparison of Binding Energy, Metastable Zone Width and Nucleation Induction Time of Succinic Acid With Various Additives

Shyh-ming Jang and **Allan S. Myerson**, Chemical Engineering Department, Polytechnic University, 6 Metrotech Center, Brooklyn, New York 11201

The binding energy of seven alkanoic acids to the major crystal faces of β-succinic acid were calculated. The effect of the alkanoic acids on the metastable zone width and induction time were measured using a Differential Scanning Calorimeter (DSC). The results indicate that alkanoic acids with carbon number more than C_{10} increase the metastable limit. Both the binding energies and metastable zone limit show the alkanoic acids affects on the nucleation of β-succinic acid in ethanol solution increase with increasing carbon number to C_{14} to decrease from C_{14} to C_{16} and then increase again.

Crystallization from the solution is an important process in the manufacture of pharmaceutical and bulk chemicals. Crystallization from the solution is a two-step process - nucleation or the " birth" of a crystal and crystal growth. The rate of nucleation is influenced by factors such as the supersaturation, solution temperature, stirring rate and the presence of impurities. There are hundreds of reports of the effects of impurities and other factors on crystallization in this area which have been summarized in recent reviews. (Myerson, 1993; Mullin 1993)

Surface adsorbed additives can reduce crystal growth rate and alter morphology by binding the movement of steps (Van der Eeerden 1986). In some cases, the additives actually substitute in the binding sequence (Berkovitch-Yellin, 1985). In either case, however, if the additive selectively interacts with a particular crystal face it will slow the growth rate of the face relative to the other faces and alter the morphology of the crystal. The control of crystal morphology by the use of additives is a subject of great interest which has been studied experimentally and through calculation studies employing molecular modeling (Chen, 1994).

Many impurities which effect crystal growth and morphology can also effect nucleation. In many cases these additives and impurities can increase the effective metastable zone width and increase the nucleation induction time making then valuable in applications such as scale inhibition (Bromley,1993). Depending on the amount and strength of adsorption, the effect on crystal growth can be extreme or hardly noticeable. The main purpose of this work is to calculate the binding energy between crystal surfaces and additives and attempt to correlate the additive-crystal interaction (which can be calculated employing molecular modeling techniques) with experimental measurement of the effects of impurities on metastable zone width and induction time.

Binding Energy Calculation

The surface binding energy is the interaction energy between an impurity molecule and a specific crystal surface. If the molecule has significant interaction with the crystal surface the binding energy will be negative. The larger the absolute value of the negative number, the larger the affinity of the additive to the crystal surfaces.

The first step in this calculation involves identification of the major crystal faces of interest for a particular substance. This is done by first obtaining the crystal structure and from the structure calculating the crystal morphology using the Donnay Harker (Donnay 1937)and attachment energy method (Hartman 1980).

The next step cleaves the crystal faces of interest from crystal and attaches a alkanoic acid molecule onto this surfaces. The surface must be large and deep enough so that the value converges and does not increase with increasing size. In this work, a surface with 7x7 unit cells and one unit cell deep was used.

The third step performs a molecular dynamics (MD) calculation to relax initial strain and mechanics minimization (MM) to minimize the inhibitor molecular structure and calculate the binding energy. The energy of the crystal surface, with the impurity molecule is calculated by summing all the interactions between the impurity and all other molecules on the surface including the energy of the impurity itself.

Before performing the MD and MM calculations, a suitable force field needs to be selected and the equilibrium charges of inhibitor needs to be calculated. There are two force field are available in Cerius^{2}1.6 for MD and MM calculations. The calculated lattice energies of β succinic acid are -29.96 Kcal/mol by using Dreiding 2.21 and 85 Kcal/mol by using Universal 1.01. Compared to the experimental value of -28.9 Kcal/mol, the force field used in this study for both inhibitor and crystal was Dreiding 2.21 (May, 1990). The atom partial charges were calculated using the charge equilibrium method of Goddard and Rappe (Rappe, 1991). There are three different parameter sets : Qeq _charged1.0, Qeq_neutral 1.0, and Qeq_charged1.1. They all give us very close results of lattice energy and binding energy calculations (within 5 %).

The final step is to change the inhibitor orientation and position and perform MD and MM calculations repeatedly until the global minimum is obtained.

All calculations in this work were done employing Cerius^2 software obtained from Molecular Simulation Inc. More detail of the procedures can be found in our previous paper (Jang and Myerson,1995).

The binding energy of a given face is obtained by subtracting the energy of the impurity alone from that of the surface with the impurity.

The crystal structure of β-succinic acid was obtained from the literature (Leviel,1981). β-succinic acid is monoclinic with space group $P2_1/c$ and two molecules in the unit cell. The Donnay Harker and attachment energy morphologies for β-succinic acid are shown in Fig. 1.

The binding energies for the low index faces (0 1 0), (1 0 0) of the crystal of β-succinic acid with different alkanoic acids are calculated using the method described above and are given in Table 1. These results show that the binding energies are relatively high while compared to substances such as water, ethanol, and β-succinic acid itself. For example the binding energy of one water, ethanol, and β-succinic acid molecule are -13.2, -15.0, and -21.5 Kcal/mol respectively. That would indicate that the alkanoic acids are possible inhibitors of growing β-succinic acid crystal.

The effect of impurities on crystal growth is ultimately related to the strength of intermolecular bonds that form during the adsorption process (Myerson, 1993). The impurities , which have higher binding energy, will become less mobile molecules on the crystal surface and effectively disrupt the movement of steps, hence the growth rate. Conversely, the

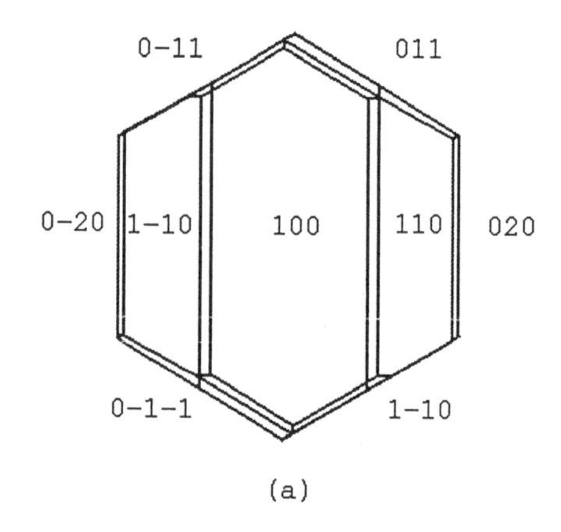

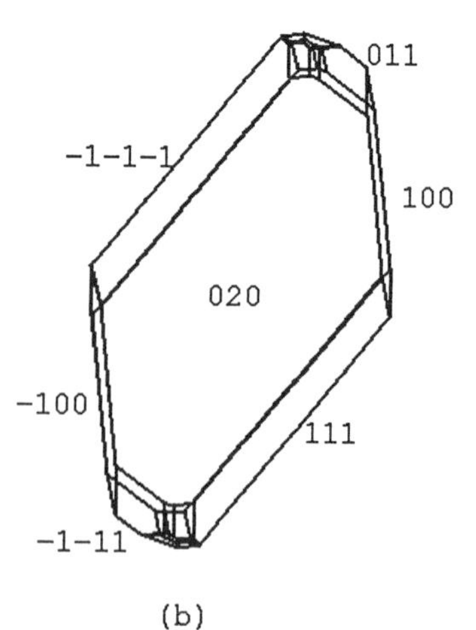

Fig. 1. The theoretical morphology of β-succinic acid crystal using (a) Donnay Harker method, (b) Attachment energy method.

Table 1 Calculated binding energies of different alkanoic acids on the surfaces of β-Succinic acid.

		Binding energies [(a)] Kcal / mole	
Impurity	Formula	(0 1 0)	(1 0 0)
Octanoic acid	$CH_3(CH_2)_6COOH$	-27.76	-25.41
Decanoic acid	$CH_3(CH_2)_8COOH$	-31.17	-35.70
Lauric acid	$CH_3(CH_2)_{10}COOH$	-34.95	-41.69
Myristic acid	$CH_3(CH_2)_{12}COOH$	-41.40	-48.91
Palmitic acid	$CH_3(CH_2)_{14}COOH$	-38.41	-40.39
Stearic acid	$CH_3(CH_2)_{16}COOH$	-41.96	-46.88
Nonadecanoic acid	$CH_3(CH_2)_{17}COOH$	-43.10	-48.08

(a). Calculated based on one impurity molecule on the surface with 7x7 unit cells and 10 Å depth.

more mobile molecules, forming relatively weak bonds, are swept away in the advance of steps and have a much smaller effects on the reduction in growth rate.

Comparing the binding energies with different straight-chain alkanoic acids indicates that the binding energies increase with the number of carbon atoms to a maximum at the point of myristic acid (C_{14}), decrease on the palmitic acid.(C_{16}) then increase again above C_{16} .

The binding forces between impurities and crystal surface basically are electrostatic interactions, hydrogen bonds, and van der waals forces. Generally, adsorption by the electrostatic and van der waals forces increases with the molecular weight. However, from the computer simulations we can see that the stable conformation of alkanoic acids with less than 14 carbon atoms is like a long straight-chain on the crystal surfaces. On the other hand, the stable molecular structure of alkanoic acids with more than 16 carbon atoms are more or less like a coil or helix.

Because of the conformation change, palmitic acid has greater intramolecular interactions and less intermolecular interactions with the crystal surface. Therefore, the binding energy drops from myristic acid to palmitic acid , then increases again with the molecular weight.

Metastable Zone Width and Induction Time Measurement

The effective metastable zone width and induction time can be measured experimentally to evaluate the effect of additives on nucleation processes. The effective metastable zone width is the maximum amount of supersaturation which can be obtained in a system under well defined condition. The induction time is the time required for crystallization to occur of a given supersaturation. The effective metastable zone width and induction time of a supersaturation solution are affected by a number of factors , including the temperature, cooling rate, agitation rate and even the vessel used (Nyvlt, 1985). However, if a well defined experimental system and procedure is employed this measurement can be made reproducibly and used to evaluate the effect of inhibitors.

The metastable zone width and induction time of succinic acid in ethanol solution with and without alkanoic additive were measured in this study.

Succinic and alkanoic acids (> 99.0%) were purchased from Aldrich Chemical and were used without further purification. HPLC grade ethanol was used as the solvent.

The polythermal method (Kern, 1992) was

used to determine the solubility of succinic acid in ethanol. The solubility results are given in Fig. 2.

Thirteen weight % succinic acid solution in ethanol was prepared for metastable limit and induction time measurements. The corresponds to a saturated solution at 39 °C. Alkanoic acid impurities (0.7 mole % to solute) were added to the solution.

The metastable zone width and induction time of succinic acid in ethanol solution without impurity and with different alkanoic acids were measured using differential scanning calorimeter (Perkin - Elmer DSC - 7) . Each sample was maintained isothermally at 50 °C for 20 minutes, then, if the metastable zone width is to be measured, the solution is cooled at a constant cooling rate of 1 °C / min and if induction time is to be measured the solution is cooled to 5 °C quickly (200 °C / min), until the first crystallization peak is observed. The metastable zone width is the difference between the temperature at which crystallization begin and the saturation temperature. The induction time is the time required for crystallization occur after the solution was cooled to 5 °C.

These results are shown in Table 2 & 3 and show that alkanoic acids with carbon number larger than C_{10} increase the metastable zone width and induction time of the succinic acid in ethanol solution. The inhibiting effect increase with the number of carbon atoms in the alkanoic acids until C_{14} (myristic acid) then decrease up to C_{16} (palmitic acid), and increase again with increasing carbon number.

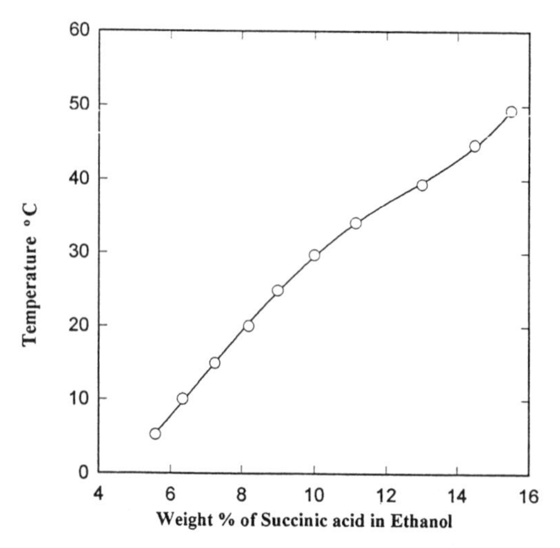

Fig. 2. Solubility of β-succinic acid in ethanol.

Table 2 Metastable zone width of succinic acid in ethanol solution in the presence of alkanoic acids

Impurity	Formula	ΔT_{max} (a) (°C)
Pure solution		35.6 ±2.3
Octanoic acid	$CH_3(CH_2)_6COOH$	35.5 ±2.1
Decanoic acid	$CH_3(CH_2)_8COOH$	40.78 ±2.5
Lauric acid	$CH_3(CH_2)_{10}COOH$	42.96 ±2.9
Myristic acid	$CH_3(CH_2)_{12}COOH$	44.46 ±2.0
Palmitic acid	$CH_3(CH_2)_{14}COOH$	42.55 ±3.1
Stearic acid	$CH_3(CH_2)_{16}COOH$	43.25 ±3.0
Nonadecanoic acid	$CH_3(CH_2)_{17}COOH$	45.28 ±3.0

(a). ΔT_{max} were measured in 13 wt% of succinic acid in ethanol and the impurity concentration 0.7 mole % to the solute. Each sample is maintained isothermally at 50 °C for 20 minutes, then the solution is cooled at a constant cooling rate 1 °C / min until the first crystallization peak is observed.

Table 3 Induction time of succinic acid in ethanol solution in the presence of alkanoic acids.

Impurity	Formula	Induction Time [a] (minute)
Pure solution		3.05 ±1.5
Octanoic acid	$CH_3(CH_2)_6COOH$	4.81 ±2.0
Decanoic acid	$CH_3(CH_2)_8COOH$	33.68 ±3.1
Lauric acid	$CH_3(CH_2)_{10}COOH$	43.18 ±3.8
Myristic acid	$CH_3(CH_2)_{12}COOH$	62.64 ±4.2
Palmitic acid	$CH_3(CH_2)_{14}COOH$	41.80 ±4.0
Stearic acid	$CH_3(CH_2)_{16}COOH$	56.73 ±4.5
Nonadecanoic acid	$CH_3(CH_2)_{17}COOH$	67.04 ±5.0

(a) Induction times were measured in 13 wt% of succinic acid in ethanol with impurity concentration of 0.7 mole % to the solute. Each sample is maintained isothermally at 50 °C for 20 minutes, then the solution is cooled to 5 °C quickly (200 °C/min) until the first crystallization peak is observed.

The effects of impurities on the nucleation rate are believed to be related to the strength of the intermolecular bonds that form during the adsorption process. The impurities recognized as forming strong bonds on the surface will effectively lessen the nucleation rate, and hence increase the metastable zone width. On the other hand, the impurities, forming relative weak bonds, have a much smaller effects on the crystallization process. Fig. 3 & 4. are plotted using the absolute value of the binding energies versus the measured effective metastable zone

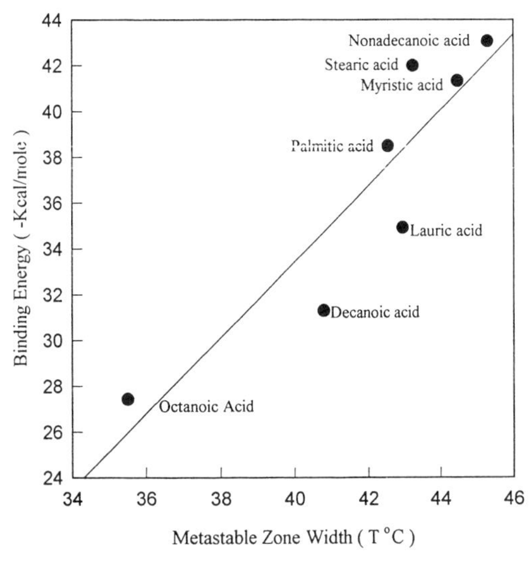

Fig. 3. The binding energy versus metastable zone width on the (0 1 0) face.

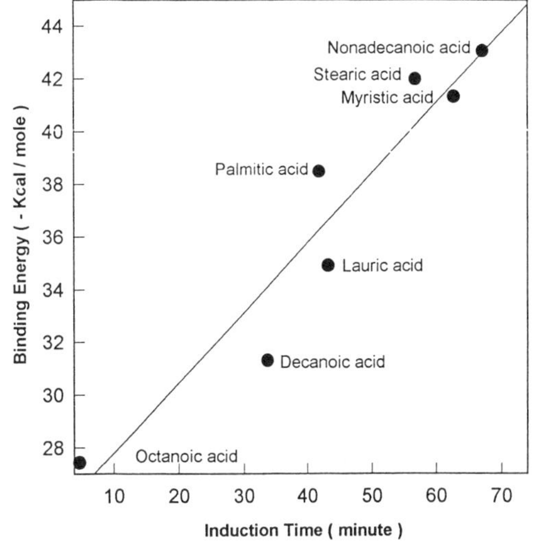

Fig. 4. The binding energy versus induction time on the (0 1 0) face.

width and induction time. The results show the higher the magnitude of the binding energy the larger metastable limit and the longer the induction time.

References

Berkovitch-Yellin, Z. *J. Am. Chem. Soc.*, **1985**,107,8329.

Bromley, L.A.; Cottier, D.; Davey, R.J.; Dobbs, B.; Smith, S. *Langmuir* **1993**, 9, 3215.

Chen, B.D.; Garside J.; Davey, R.J.;Maginn, S.J.; Matsuoka, M. *J. Am. Chem. Soc.*,**1994**,98,3215.

Donnay, J.D.H.; Harker D. *Am. Mineral.* **1937**,22 463.

Ginde, R.; Myerson, A.S. *J. Crystal Growth* **1993**,126 ,216.

Hartman P.; Bennema P. *J. Crystal Growth* **1980**,49,525.

Jang, S.M ; Myerson, A.S. *J. Crystal Growth* **1995,** in press.

Kern ,R.; Dassonville, R., *J. Crystal Growth* **1992**,116 ,191-203.

Leviel, J.L.; Auvert, G.; Savariawalt, J.M. *Acta. Cryst.* **1981**, 37, 2185.

May, S.L.; Olafson, B.D.; Goddard, W.A., *J.Phys. Chem.*, **1990**,94,8897.

Mullin, J. W. *Crystallization*, 3rd ed., Butterworths-Heinemann, London, **1993** P.189-194.

Myerson, A. S. *Handbook of industrial Crystallization,* Butterworth-Heinemann, London. **1993** P.79-82.

Nyvlt, J. *The Kinetics of Industrial Crystallization*, Elsevier, New York, **1985**.

Rappe, A.K.; Goddard, W.A., *J.Phys. Chem.*, **1991**,95,3358

Van der Eeerden,J.P.; Muller-Krumbhaar, H. *Electrochimica Acta* , **1986,**31, 1007

Crystal Habit Modification of Caprolactam in the Presence of Carboxylic Acids - Modelling and Verification -

Silke Niehörster, Sabine Henning, Joachim Ulrich, Universität Bremen, Verfahrenstechnik/FB4, Postfach 330 440, D-28334 Bremen, Germany

A method is presented to calculate the habit of an organic crystal influenced by additives. It is described how the modified attachment energy values are obtained. The results are compared to the face growth rates coming from the experiment. In the experimental section it is examined whether it is possible to change the habit with concentrations far below 10 wt% as it is known for tailor-made additives. The results are described for the system caprolactam/benzoic acid, 2-chlorobenzoic acid and acetic acid.

The crystal habit can be changed significantly in the presence of additives in a crystallization process. Many papers are published in this field of research, but there is no complete theory to explain the mechanisms. Moreover, another problem occurs, namely the additive concentration in the crystallizing system. So-called tailor-made additives (Addadi, 1982, Berkovitch-Yellin, 1982) have to be added usually in concentrations above 10 wt% (van der Leeden, 1989). From an industrial point of view those amounts can often not be accepted. On the one hand the product quality (purity) should not be influenced by the additive. On the other hand the cost of the added additive has to be low.

The detailed understanding of the chemical interactions at the crystalline interfaces is necessary to figure out the effects of additives on the crystal growth process. Additives are bound at preselected crystal faces. The result is a growth rate reduction of the effected faces and a relative enlargement of its surface areas, since the slowest growing faces always dominate the crystal habit.

Generally the molecular packing in an organic crystal is dependent on the possibility of hydrogen bond formation. Hydrogen bonds will be formed whenever suitable hydroxyl, amine and imino hydrogens are available unless prevented by steric hindrances. Hydrogen bonding is often the key to the actions of additives on the growth process. This van der Waals interaction type is stronger and directionally more specific than others. Thus, they can be quite selective to certain crystal faces.

In this study modelling and experiments were carried out using the organic substance caprolactam. All lactames, like the examined one, contain two different groups able to form hydrogen bonds (carbonyl and amino). Caprolactam therefore forms dimers, i.e. two caprolactam molecules are always paired.

It has been shown for caprolactam that it is possible to change its habit as predicted by using a so-called hb-additive (Niehörster, 1995). The prefix stands for hydrogen bond, because the chosen additive ethanol contains a hydroxyl group that can build a hydrogen bond with caprolactam's carbonyl group. Analysis of the [11$\bar{1}$] caprolactam surface structure shows emerging carbonyl groups. Consequently, at these faces the accessibility for the hb-additive is better than at the other faces, resulting in a growth rate reduction.

To investigate the influence of different hb-additives on caprolactam's habit in more detail we used three carboxylic acids, benzoic acid, 2-chlorobenzoic acid and acetic acid. The two benzoic acids were chosen to analyse the influence of the hydrogen bond strength. The molecular structures of both benzoic acids differ, due to the chlorine substitute of 2-chlorobenzoic acid near the COOH-group. This difference must lead to stronger hydrogen bonds between caprolactam and 2-chlorobenzoic acid than with benzoic acid. Acetic acid was chosen to analyse if structurally different additives change the habit, too. Also if it is possible to reduce the additive concentration far below the barrier of 10 wt% that is known for tailor-made additives.

Modelling

Caprolactam's crystal data needed as input for the computation are taken from literature (Winkler, 1975). All calculations were done, using software package CERIUS3.2, distributed by Molecular Simulations Inc.. Analyses of the interaction energies and evaluation of optimum geometries force field DREIDINGII (Mayo, 1990) was used, because of its good applicability for a wide range of organic materials. First the theoretical habit of caprolactam has to be calculated using the attachment energy model to identify the morphological most important faces. In Fig. 1 the resulting habit of pure caprolactam with faces of the form [200], [110] and [11$\bar{1}$] is shown.

The effect of the additives (benzoic acid, 2-chlorobenzoic acid and acetic acid) was studied theoretically in the following way. Starting-point is the Donnay-Harker model of pure caprolactam with its most likely growth faces. Caprolactam has got four general symmetry positions (x y z, -x y -z+1/2, -x -y -z, x -y z+1/2). Always two host molecules belonging to a symmetry position have to be replaced by additive molecules. Energetic stable additive positions inside the unit cell can be found by energy minimization with fixed caprolactam molecules. Then the attachment energy calculation has to be carried out. The generated

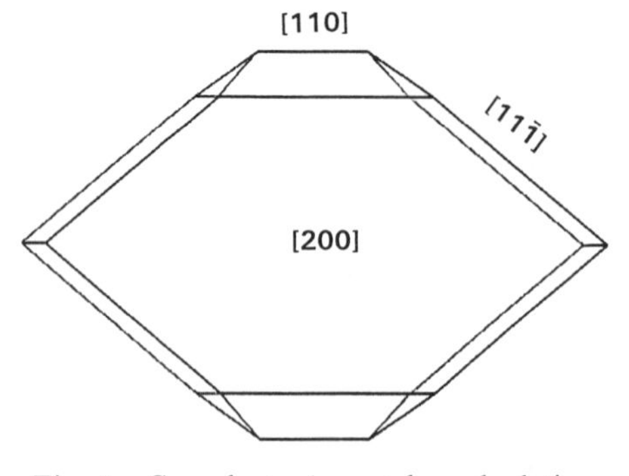

Fig. 1.: Caprolactam crystal - calculation

attachment energy values are multiplied by the ratio lattice energy of pure caprolactam to lattice energy in the presence of the additive. The binding energy is the sum of the slice energy and half the attachment energy (Berkovitch-Yellin, 1985). Differences in binding energies resulting from the incorporation of the additive into the crystal structure are calculated as follows (Clydesdale, 1994)

$$\Delta Eb = Eb' - Eb$$

Binding energy in the presence of the additive is Eb' and without it is Eb. In pure caprolactam a binding energy reduction of one kcal/mol reduces the attachment energy of about 7.17 %. Accordingly the modified attachment energy Eatt' of each face can be calculated with the relation

$$Eatt' = Eatt \cdot (1 - \Delta Eb \cdot 0.0717)$$

In Table 1 the calculated attachment energies of pure caprolactam and those influenced by the additives are shown in dependence of the general symmetry positions. The attachment energies of all faces have changed, caused by the additives. Also presented in Table 1 are the ratios of the attachment energies Eatt'[11$\bar{1}$]/Eatt'[110] averaged over the general symmetry positions. Since the growth rate is proportional to the attachment energy (Hartman, 1980) the results can be explained in the following way. A ratio of 1.0 stands for equal face growth rates. The ratio for pure caprolactam is 1.047, i.e. the faces [111] are growing a bit faster than the faces [110]. In the presence of all chosen additives this ratio decreased below one (benzoic acid 0.808, 2-chlorobenzoic acid 0.703 and acetic acid 0.603). This means that influenced by the additives the faces [11$\bar{1}$] are "growing slower" than faces [110]. The result is an enlargement of faces [11$\bar{1}$] and in the end another habit is achieved. For example in Fig. 2 the theoretical habit of caprolactam in the presence of benzoic acid is shown. Attachment energy values used for this plot are the average attachment energies of Table 1. As discussed before the morphological importance of faces [110] is almost lost in comparison to the pure caprolactam crystal in Fig. 1. The same resulting habits are found for the other additives.

Table 1: Calculated attachment energies

	Sym. Pos.	Eatt[200] [kcal/mol]	Eatt[110] [kcal/mol]	Eatt[11-1] [kcal/mol]	Eatt[11-1]/Eatt[110] [-]	Average Eatt[11-1]/Eatt[110] [-]
Caprolactam		-27,89	-47,94	-50,19	1,047	1,047
CAP+Benzoic Acid	1	-9,17	-20,69	-14,91	0,721	
	2	-9,29	-36,30	-37,51	1,033	
	3	-10,90	-29,42	-13,19	0,448	
	4	-9,29	-36,23	-37,28	1,029	
						0,808
CAP+2-Chloro-benzoic Acid	1	-18,12	-28,04	-9,74	0,347	
	2	-9,11	-33,85	-32,78	0,968	
	3	-9,34	-24,15	-12,74	0,527	
	4	-9,12	-33,82	-32,78	0,969	
						0,703
CAP+Acetic Acid	1	-17,05	-48,13	-31,42	0,653	
	2	-5,63	-32,14	-13,07	0,407	
	3	-2,55	-18,82	-6,75	0,358	
	4	-40,16	-80,00	-79,52	0,994	
						0,603

Experiment

Experiments were carried out in a growth cell placed under an optical microscope. Crystals were observed while growing and either photographed at certain intervals or recorded by a CCD video camera (Kruse, 1990). To avoid influences of moisture due to the hygroscopic properties of caprolactam the growth cell is completely shielded from the environment. A constant melt temperature level was guaranteed by a thermostatic bath connected to the jacket build around the cell. Crystals were grown from a pure melt and in the presence of three carboxylic acids. Concentration of benzoic acid and 2-chlorobenzoic acid was 16.7 wt% and of acetic acid 0.26 wt%. The equilibrium temperature was found experimentally. Undercooling up to 6 K was used as driving force. The experimentally found growth rates of the faces [111] are plotted versus those of faces [110] in Fig. 3. Also, presented in the plot are the results of the regression analysis. The slope of the straight lines is m. In pure caprolactam the slope resp. the ratio of the face growth rates G[11$\bar{1}$]/G[110] is 1.117. This value is the same as faces [11$\bar{1}$] are growing a bit faster than faces [110]. In the presence of the carboxylic acids the ratio changed to 0.803 for benzoic acid, to 0.664 for 2-chlorobenzoic acid and to 0.568 for acetic acid. So all additives decreased the ratio of the face growth rates G[11$\bar{1}$]/G[110]. Thus the morphological importance of faces [110] is reduced significantly, i.e. faces [110] are nearly grown out of existence. In Fig. 4 the growth habit of a single caprolactam crystal grown in a melt containing benzoic acid is shown. The crystal is only bound by [200], [11$\bar{1}$] and very small [110] faces. Same results were achieved for crystals grown in the presence of 2-chlorobenzoic acid and acetic acid.

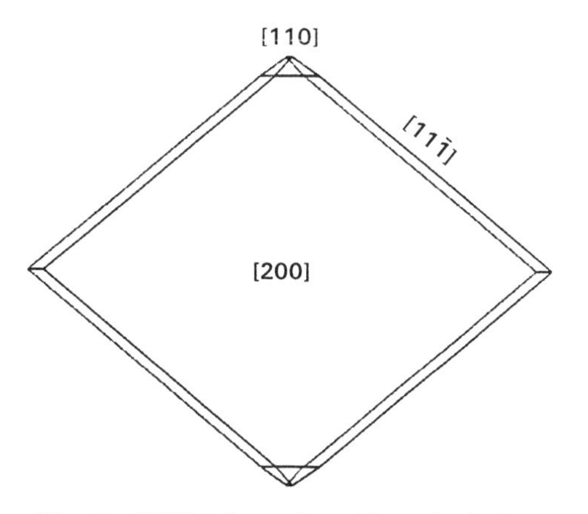

Fig. 2.: CAP + benzoic acid - calculation

Discussion

The agreement of the calculated caprolactam habit (Fig. 1) to crystals grown from pure melt (Chen, 1992, Geertman, 1993) is quite good. In the presence of the additives, the habit of caprolactam has changed significantly. For pure caprolactam the morphological most important

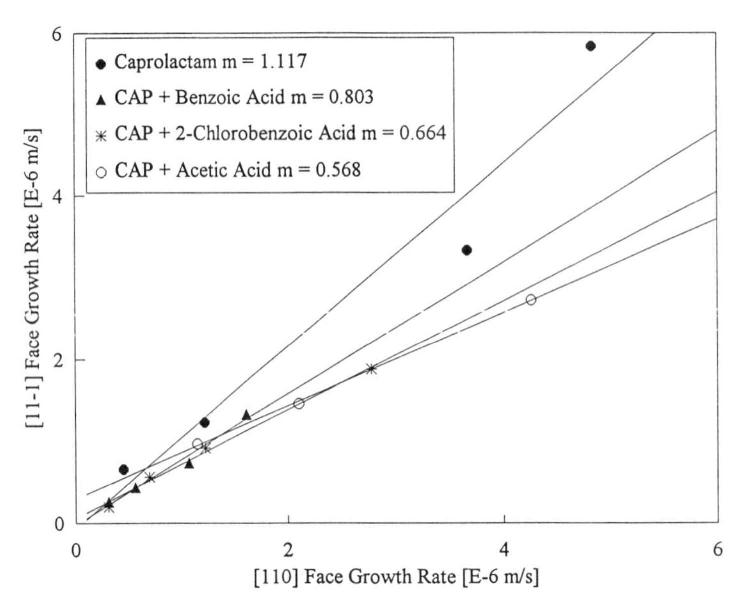

Fig. 3.: Face growth rates - experiment

faces are decreasing in the order [200], [110] to $[11\bar{1}]$. In contrast, the order is [200], $[11\bar{1}]$, [110] in the presence of the additives. The edges built by [110] faces disappeared (Fig. 2, Fig. 4) caused by the additives. Because, the ratio of the growth rates and the attachment energies of faces $[11\bar{1}]$ to [110] decreased.

Question is, how the experimental results can be compared to the calculated ones. The calculated averaged ratios of the attachment energies Eatt'$[11\bar{1}]$/Eatt'[110] and the ratios of the faces growth rates G$[11\bar{1}]$/G[110] are shown in Fig. 5. Agreement between the experiment and calculation is satisfying with a deviation of less than 6.25 %. The ratios reflect also the strength of interaction energies between the host and the additive molecules. As mentioned the additives were selected due to their ability to form hydrogen bonds of different intensities with caprolactam. Since 2-chlorobenzoic acid should form stronger bonds compared to benzoic acid, caused by the chlorine substitute, the reduction of the ratios is bigger. Consequently, the influence of 2-chlorobenzoic acid is shown in a lower value of the growth rate ratio of 0.664 compared to 0.803 for benzoic acid.

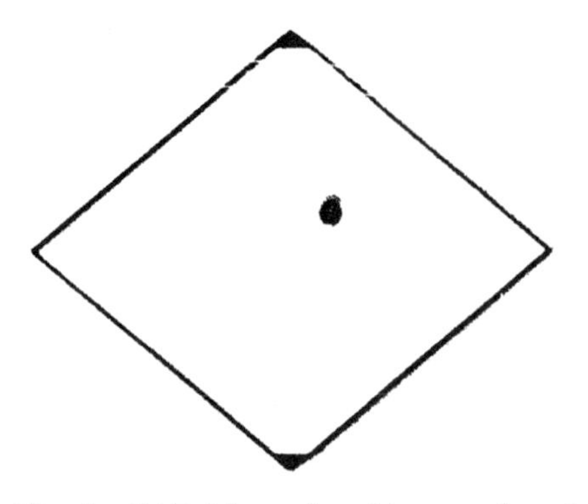

Fig. 4.: CAP + benzoic acid - experiment

A comparison of the molecular weights shows that caprolactam and benzoic acid are nearly equal. The 2-chlorobenzoic acid is heavier depending on the chlorine atom. Therefore both additives are belonging rather to the group of tailor-made additives, particularly regarding the chosen concentrations of 16.7 wt%. The third additive acetic acid with its molecular weight of 60.05 g/mol is structurally different from the host. Therefore, the concentration of 0.26 wt% is very low compared to the usually known 10 wt% for tailor-made additives. Such a small molecule like acetic acid has no problems of steric hindrance to reach the carbonyl groups of caprolactam. Thus forming a hydrogen bond it disturbs the growth process of the effected faces. Also the small sized additive acetic acid reduces the ratio of the face growth rates much more than benzoic acid (0.568 to 0.803). The result is that it is possible to reduce the ratio of the face growth rates and to change

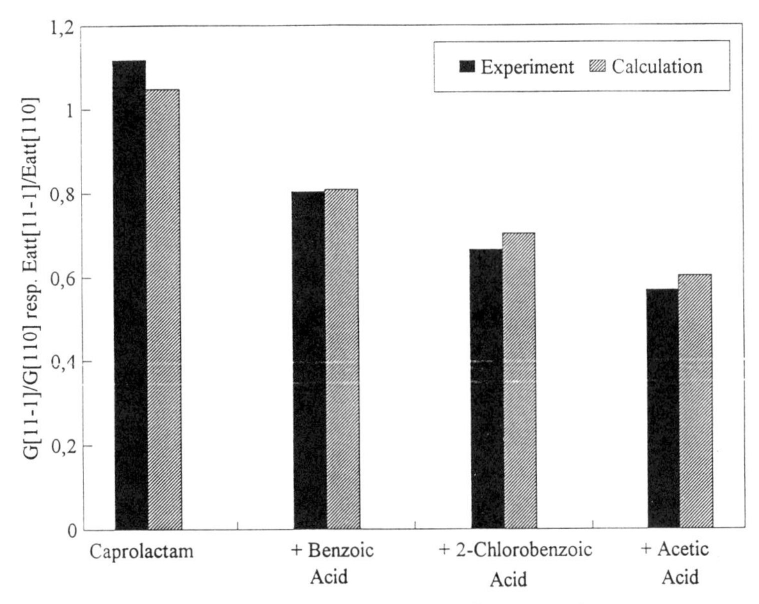

Fig. 5.: Comparison theor. and exp. results

the habit extremely by using structurally totally different additives.

Conclusion

A basic approach to predict the crystal habit modification of the organic crystal caprolactam in the presence of three carboxylic acids is presented. It has been shown that the calculations of the modified attachment energies resulting in the new habit agree quite good with the experimental results. Furthermore the calculations showed according to the experiments how the ratio of the face growth rates will change. Another result is that the habit of an organic crystal can be changed with a structurally different additive. Therefore the necessary concentration to achieve the habit modification can be reduced significantly. In the future this idea has to be checked with other systems to proof its validity as a more general rule.

Acknowledgement

The authors would like to acknowledge their gratefulness for the support of the "Senator für Bildung und Wissenschaft der Freien Hansestadt Bremen", granted for research work included in this paper.

References

Addadi, L.; Berkovitch-Yellin, Z.; Domb, N.; Gati, E.; Lahav, M.; Leiserowitz, L.; *Nature* **1982** 296, 4, 21-26

Berkovitch-Yellin, Z.; *J. Am. Chem. Soc.* **1985** 107, 8239-8253

Berkovitch-Yellin, Z.; Addadi, L; Idelson, M.; Leiserowitz, L.; Lahav, M.; *Nature* **1982** 296, 4, 27-34

Chen, B.D.; *PhD Thesis*, UMIST Manchester **1992**

Clydesdale, G.; Roberts, K.J.; Docherty, R.; *J. Cryst. Growth* **1994** 135, 331-340

Geertman, R.M.; *PhD Thesis*, University Nijmegen **1993**

Hartman, P.; Bennema, P.; *J. Cryst. Growth* **1980** 49, 145-156

Kruse, M.; Stepanski, M.; Ulrich, J.; *Proceedings of Bremer International Workshop for Industrial Crystallization*, Ed. J. Ulrich, Verlag Mainz, Aachen **1990**

Mayo, S.L.; Olafson, B.D.; Goddard III, W.A.; *J. Phys. Chem.* **1990** 94, 8897-8909

Niehörster, S.; Ulrich, J.; *Cryst. Res. Technol.* **1995** 30, 3, 391-397

van der Leeden, M.; van Rosmalen, G.; de Vreugd, K.; Witkamp, G.; *Chem.-Ing.-Tech.* **1989** 61, 5, 385-395

Winkler, F.K.; Dunitz, J.D.; *Acta Cryst.* **1975** B31, 268-269

Crystallization in the Pharmaceutical Industry

REGULATORY CONSIDERATIONS IN CRYSTALLIZATION PROCESSES FOR BULK PHARMACEUTICAL CHEMICALS: A REVIEWER'S PERSPECTIVE.

Wilson H. DeCamp, Center for Drug Evaluation and Research, Food and Drug Administration, Rockville, MD 20857

Abstract
Crystallization procedures are almost universally used as a final purification step in the manufacture of a bulk pharmaceutical chemical. Consequently, they are subject to examination for compliance with FDA's Good Manufacturing Practice regulations. Particular attention should be given to the effect of crystallization conditions on physical characteristics, such as particle size and crystal form. A particular case of potential regulatory concern is the crystallization of chiral drugs. A chiral seed may influence the crystallization of a racemic drug by inducing spontaneous resolution. Potential analytical control methods include optical rotation, melting range, and X-ray powder diffraction. A New Drug Application (NDA) is expected to include a discussion of the solid state forms of the drug substance in its chemistry section. This provides a rationale for the controls on the manufacturing process. The amount and detail of experimental evidence that should be considered for inclusion in the NDA is discussed.

For the most part, a chemical that is potentially useful as a drug for human use is first obtained in a relatively impure state. Whether the purpose is characterization, product development, investigation, or manufacturing, the next challenge of the chemist is to purify the substance. The description of this task is deceptive in its simplicity.

Crystallization procedures are almost universally used as the final purification step in the manufacture of a bulk pharmaceutical chemical (BPC). Steps such as dissolution, crystallization, evaporation, sublimation, distillation, sorption, filtration or centrifugation have as their objective the accomplishment of a physical separation of a desired product from the mixture of other chemicals remaining from the synthesis.

Crystallization and Good Manufacturing Practice

Regardless of the method, the final purification step is critical to the purity of the drug substance, as well as the finished drug product. Consequently, it is subject to examination for compliance with FDA's Good Manufacturing Practice regulations. From a strict point of view, the GMP regulations that FDA has published apply only to finished dosage form drugs. However, the Federal Food, Drug, and Cosmetic Act requires that all drugs be manufactured, processed, packed, and held in accordance with current good manufacturing practice (CGMP). The Act makes no distinction between BPCs and finished pharmaceuticals. Therefore, a failure of either the BPC or the drug product to comply with CGMP is a failure to comply with the requirements of the Act. Thus GMPs for dosage form drugs and BPCs are parallel concepts. (*Guide to Inspection of Bulk Pharmaceutical Chemicals*, 1991)

The essence of GMPs is the documentation of all significant processing steps. These include those steps wherein the material is treated by physical means and/or the transfer and change of energy, but without causing a chemical change of the molecule, as well as those steps wherein the molecule undergoes a chemical change.

A chemical reaction is usually carried out only once on a given lot of a BPC. In contrast, a crystallization step may be, and often is, repeated, perhaps several times. Mother liquors containing recoverable amounts of BPCs are frequently re-used. Such re-use may employ the mother liquor to dissolve the reactants in the next run of that step in the synthesis. It may also allow the crystallization to continue for a long period after the first crop is removed. These secondary recovery procedures should be documented in batch production records.

The fact that a crystallization can be repeated also means that it can involve deliberate in-process blending of synthesis lots. This does not present a regulatory problem, provided it is adequately documented in batch production records. It should be emphasized that in-process blending should still be consistent with the process intent of achieving homogeneity of the batch of finished BPC to the maximum extent feasible. However, blending of batches or lots that individually do not meet specifications with other lots that do is not an acceptable practice.

Recrystallization procedures may also involve the recovery and re-use of the solvent for economic reasons. When fresh and recovered solvents are commingled, a precise lot identity cannot be established. This may be satisfactory if incoming solvents are identified and tested prior to being mixed with recovered solvents and if the latter are tested for contaminants from the process in which they were used previously. The quality of the solvent mixture should also be monitored at suitable intervals. In addition, it would seem to be both prudent and technically sound to avoid using solvent that has been recovered from the manufacture of a different BPC.

Process validation is especially important to the development or modification of a crystallization step. While the concept of validation can be applied to any process, increased emphasis should be given to validation of the BPC production at the crystallization step. This stage of the synthesis is usually the final purification step used for the bulk substance and/or the removal of impurities. Therefore, how this step is controlled may significantly affect the specifications for the bulk pharmaceutical chemical.

Assurance of product quality depends upon the adequate design and control of the manufacturing process. End product testing to specifications is a necessary part of such assurance. It is not necessarily sufficient, because such testing is limited in its sensitivity to reveal all variations that may occur and affect the chemical, physical, and microbiological characteristics of the product. For example, a change in the rate of cooling in crystallization may affect the size of the crystals that are formed. This may, in turn, affect the time that it takes to accomplish a succeeding filtration step. Even if particle size is not a specification for the BPC, it may need to be examined as a part of the validation of a manufacturing process change.

The data necessary to evaluate a process and to demonstrate that it works consistently are accumulated in the development phase. For example, this may include limitations on a purification step, such as ranges of solvent composition and heating or cooling rates. At the time of an inspection, such data may demonstrate that analytical control specifications and procedures are "scientifically sound", as required by 21 CFR 211.160(b). The review of these data is a key responsibility of the field inspector, and is accomplished in a pre-approval inspection. However, the reports describing the development of the parameters for the process, the identification of impurities, and the methods used in the key tests needed for process control also provide perspective for the FDA review chemist. Thus, when the process is scaled up to production batch sizes, or moved to a new site, a comparison can be made with prior batches, and the review can have an improved technical focus.

Crystallization of Chiral Drugs

The crystallization of chiral drugs presents a particularly challenging question. Most chiral compounds that are prepared as racemates crystallize as *racemic compounds*. Specifically, this means that the crystal unit cell contains either a mirror or inversion center as a symmetry element. Thus, the bulk material is constrained to contain exactly the same number of molecules of each enantiomer. In some cases, however, *conglomerate crystallization* may occur. In this case, each individual crystal is a pure enantiomer, even if the bulk composition is that of the racemate. The best known example of this is Pasteur's crystallization of sodium ammonium tartrate, and its subsequent resolution by microscopic examination.

For a racemic chiral drug, therefore, it is essential to consider the enantiomer as a potential different solid state form. Factors that may favor conglomerate crystallization should be evaluated. A solubility for the enantiomer that is lower than that for the racemic compound may suggest the possibility, though not the certainty, that it will crystallize as a conglomerate. It may be that comparisons of the solid state structures of both the racemate and the enantiomer will prove critical to complete understanding of the crystallization process.

There is potential for a chiral seed to influence the crystallization of a racemic drug when conglomerate crystallization occurs. If an enantiomerically pure seed crystal is used, it will favor the formation of crystals of the same handedness. In this event, a first crop of crystals is formed that is enantiomerically enriched, leaving the mother liquor enriched in the opposite enantiomer. The second crop will then be enriched in this enantiomer.

Even when there is overall equality between the enantiomers in solution, there is no reason to believe that any given crystallization of a racemic mixture that is capable of conglomerate crystallization will yield a racemic product. Examples of such random variability, though rare, have been published. In some cases, the variability may be quite large, leading to ratios in excess of 70:30 in a single crystallization. On the average, however, the ratio of enantiomeric crystals in a conglomerate will approach unity. (Bernal, 1992; Kipping and Pope, 1897)

This situation appears to call for special analytical studies to show that the controls on the crystallization process are scientifically sound. These could include tracking the identity of each crystallization crop by an analytical method that is sensitive to the crystalline form, probably X-ray powder diffraction or differential scanning calorimetry. Alternatively, the enantiomeric excess of both the crystallized product and the mother liquor could be similarly monitored in-process by optical rotation or chiral HPLC.

Chemistry Review and Crystallization

Having focused initially on issues related to Good Manufacturing Practice, let us now turn to how this affects the way that the review chemist looks at a crystallization step. Again, to keep the proper perspective, we must emphasize that issues of GMP compliance are the responsibility of the FDA field inspector, as are the audit of the original data for the process validation studies. On the other hand, the CDER review chemist is responsible for the review of the analytical control specifications and procedures to determine that they are scientifically sound in concept and appropriate to regulate the identity, strength, quality, and purity of the drug product and the BPC from which it is manufactured. The fact that these responsibilities are divided does not mean that they are independent of each other. The review chemist has an overall regulatory perspective, but must rely on the observations of the field inspector as to the implementation of the manufacturing procedure. In contrast, the field inspector depends on the headquarters review chemist for the technical evaluation of the manufacturing processes and controls.

How the process of crystallization proceeds is poorly understood, whether from a solution or the melt. Although the thermodynamic principles are well known, the randomness of nucleation makes this a phenomenon that is often invoked to explain unexpected events. The anecdotal evidence for "disappearing polymorphs" seems to far exceed the experimental evidence. Nevertheless, some thorough studies have been reported, and were reviewed recently by Dunitz and Bernstein (1995).

The critical factor in a crystallization process is not limited to which crystalline form has the lowest solubility. Thermodynamics tells us that the form obtained in the liquid-to-solid phase change will be the one with the lowest free energy under the conditions of the process. Kitaigorodskii (1961) has suggested that the density of packing is a fundamental factor affecting free energy. Since the heat of sublimation and, to a lesser degree, the melting point, increase with density, it follows that studies of these fundamental physical chemical characteristics of a BPC may contribute in a significant way to the understanding of the dynamics of its crystallization.

Polymorphs are not always random and unpredictable in their occurrence. The more common case seems to be that multiple crystalline forms can be reproducibly produced in a manufacturing environment. Unless seeding is carefully controlled, the results still may be unpredictable. Even then, it is impossible to control the presence of other microscopic particles that may act as crystallization nuclei, promoting the formation of unexpected forms. For this reason, exhaustive efforts should be undertaken to identify all crystalline forms or solvates that may be stable under the range of conditions that may reasonably be anticipated during synthetic or manufacturing processes. While experimental evidence is preferred at this time, future developments in the dynamic modelling of crystal structures may lead to more reliable predictions of crystal structures and their phase transformation behavior. (Gavezzotti, 1994)

Let us consider one recent event as an example of the complex way in which these responsibilities interact. Ranitidine hydrochloride is marketed by Glaxo Pharmaceuticals (now Glaxo Wellcome Inc.) under the trade name Zantac. The drug was approved in 1983 for the treatment of duodenal ulcers. The original patent application was filed in 1977. In 1980, a new crystalline form was discovered. A patent for its pharmaceutical use was filed, and was granted in 1985. When an Abbreviated NDA (ANDA) was filed in 1991, infringement of this patent was alleged, and was upheld by a court decision in 1993. (*Glaxo, Inc. v. Novopharm*, 1993)

This much is historical fact, obtainable from public records. The question here is not further elaboration of the technical details of the patent litigation, but rather to explore what we can learn <u>from</u> the situation about regulatory review.

In brief, in the original crystallization procedure, ranitidine base is dissolved in a solvent containing HCl gas, then a cosolvent is added to aid in crystallization. As a result, crystals of Form 1 ranitidine hydrochloride precipitated from the solution. A change in the solvent system led to the discovery of Form 2. Differences in both the infrared spectrum and the X-ray powder diffraction patterns were claimed in the patent. The court found that only Form 2 ranitidine HCl has ever been marketed.

Given the knowledge that the manufacture of a bulk drug substance was based on a process such as described above, let us consider how an FDA review chemist might approach the task of examining the information submitted about the recrystallization process. The objective of the CMC review is to establish whether the control specifications and methods described in the NDA are adequate to insure the batch-to-batch quality of the BPC. In reaching this decision, the reviewer will often go beyond the specifications for end-product testing.

In circumstances such as those described for the recrystallization of ranitidine HCl, the process parameters that might alter the quality of the final BPC include the choice of solvent and cosolvent, the amount of cosolvent added, whether the pH of the solution is adjusted, the rate at which the solution is cooled, and whether the mixture is allowed to age before filtering. A change in any of these process parameters could result in the formation of a different crystalline form, and are reasonably within the scope of the chemistry, manufacturing, and controls (CMC) review.

Whether any of these variables could, in fact, alter the quality of the BPC can be determined only by the results of an appropriately designed study. In particular, a claim that a variation in a particular parameter has no effect on the quality of the product may be especially subject to question if evidence to support the claim is missing. The reviewer may, of course, conclude on the basis of general chemical knowledge that such a claim is technically reasonable. A conclusion that the lack of supporting data constitutes a deficiency in the application may be equally reasonable.

Using contemporary standards, process validation studies should include at least one, if not more, checks to verify that the process yields the desired polymorph. To support the scientific soundness of the specifications, data should also be included in the NDA to show how the measurements would distinguish between the crystalline forms.

Lest my purpose be misinterpreted, it should be noted that the original submission of the Zantac NDA was well before the comprehensive "NDA rewrite" of 1985, as well as the issuance of our guidelines. My mention of this case as an illustrative example does not mean either that such data were or were not submitted. Furthermore, it does not mean that there either have or have not been inspection observations related to polymorphism. *(Guideline for Submitting Supporting Documentation in Drug Applications for the Manufacture of Drug Substances*, 1987)

Crystallization in the New Drug Application

A New Drug Application (NDA) is expected to include a discussion of the solid state forms of the drug substance and their relation to bioavailability in its chemistry section. The guideline requires, where appropriate, specifications characterizing the drug substance so as to assure the bioavailability of the drug product. It is well established that polymorphic form, degree of solvation or hydration, and particle size profoundly affect dissolution and bioavailability from solid dosage forms or suspension drug products. Although these properties are less important for solution dosage forms and for drug substances that are highly water soluble, this does not extend to semi-solid dosage forms. For creams and ointments, the active drug is often a micronized solid dispersed in either the oil or water phase. Its solid state properties may have a major effect on the partitioning of the drug between the phases.

At the time that an NDA is submitted, we expect that the applicant will have established whether or not the drug substance can exist in multiple solid-state forms, whether these affect the dissolution and bioavailability of the drug product, and whether particle size is also important for dissolution and bioavailability. It is not necessary to create additional solid-state forms by techniques or conditions unrelated to the synthetic process for the purpose of clinical trials. However, submission of a thorough study of the effects of solvent, temperature, and possibly pressure on the stability of solid state forms should be considered. For example, if the unsolvated form is more dense than the solvated form, milling or micronization procedures may induce phase changes. It should not be necessary to routinely include detailed investigations of the thermodynamic values mentioned earlier. However, for those drugs that are particularly sensitive to the conditions of crystallization, such studies may be useful in interpreting the observations obtained during drug development.

The information submitted in support of a conclusion that a change in solid-state form does not occur when the drug substance is manufactured or that different forms occur but do not result in a bioavailability problem should be supported by appropriate analytical data. Historical data alone are insufficient on these points, unless a variety of solvents have been used in the synthesis lots during development. In particular, a conclusion that polymorphism does not occur should be supported by crystallization studies using a range of solvents. This should include not only the solvent or solvents used in the final recrystallization, but also any solvents encountered during the manufacture of the drug product. For aqueous solvents, such as those used in a wet granulation, pH may also be a factor. If the drug product is a dispersion of the solid drug in a liquid phase, the potential for a change in crystalline form over the shelf life of the drug should be investigated as a part of the stability studies. In drug products where the active ingredient is dissolved, if it is present at a concentration that is near its solubility at room temperature, studies should be conducted to establish whether crystalline material forms on cooling or freezing, and whether it redissolves when the product is allowed to return to room temperature.

This discussion began with a comment about the deceptive simplicity of the process of purification of a drug. What happens, in fact, is far more complex than merely adding an impure solid to a solvent, heating it, and then allowing it to cool to form crystals. To identify the critical parameters of the process, to establish a scientifically sound basis for their control, and to develop appropriate analytical methods that can be routinely implemented - these are the challenges that face the chemist who is responsible for developing and implementing a crystallization process.

References

Bernal, I. *J. Chem. Educ.* **1992**, *69*, 468-469.

Crookes, D.L. U.S. Patent 4,521,431, **1981**; cf. *Chem. Abs.* 97:61014g.

Dunitz, J.D. and Bernstein, J. *Acc. Chem. Res.* **1995**, 28, 193-200.

Gavezzotti, A. *Acc. Chem. Res.* **1994**, *27,* 309-314.

Glaxo, Inc. and Glaxo Group Limited, Plaintiffs, v. Novopharm Ltd., Defendant. (1993) 830 F.Supp. 871, No. 91-759-CIV-5-BO. United States District Court, E.D. North Carolina, Raleigh Division. Sept. 17, 1993.

Guide to Inspection of Bulk Pharmaceutical Chemicals. U.S. Food and Drug Administration:Rockville, MD, Revised September 1991.

Guideline for Submitting Supporting Documentation in Drug Applications for the Manufacture of Drug Substances. U.S. Food and Drug Administration:Rockville, MD, February 1987.

Kipping, F.S. and Pope, W.J. *J. Chem. Soc. Trans.* **1897**, *71*, 611.

Kitaigorodskii, A.I. *Organic Chemical Crystallography*; Consultants Bureau:New York, **1991**.

Crystal Growth of L-SCMC seeds in a DL-SCMC Solution of pH 0.5

K. Toyokura, K. Mizukawa and M. Kurotani, Dept. of Applied Chemistry, Waseda Univ., 3-4-1 Okubo, Shinjuku-ku, Tokyo 169 Japan

The effect of pH on the solubility of aqueous solution of DL-SCMC (S-Carboxymetyl-DL-cysteine) was studied by addition of HCl. When the pH of the solution was decreased to 0.5, the solubility increased rapidly. Crystal growth rate of L-SCMC in low pH solution was much faster than that in the solution of pH 2.8 which corresponds to the isoelectric point. The shape of crystal grown in low pH solution was also observed to be little changed in the same growth rate condition. Crystallization phenomena were observed in the stationary solution. The differences between pH 0.5 and 2.8 solution were supposed to be from the difference in dimensionless supersaturation, and some merits on operations at pH 0.5 solution for optical resolution was discussed on industrial application.

On crystallization processes, decision of a stable operational condition for production of crystal under high yield is important in chemical industry, and a smaller saturated concentration of solute in a solution is considered to be effective for high yield of feed material. But when sparklingly soluble solution is selected for operation, too many nuclei is apt to be produced and relatively large crystal separated easily from the solution by filtlation, is difficult to be produced without some special devices. Changes of operational conditions as pH and others, increase saturated concentration of a solute sometimes, and are considered to be effective to produce better crystals easier. Crystallization of L-SCMC (S-Carboxymetyl-L-cysteine) was studied in supersaturated DL-SCMC solution of pH 2.8 and some fundamental research results for optical resolution were reported [1) 2)]. In this study, the solubility of DL-SCMC solution was observed to increase with decrease in pH and larger crystal growth rate of L-SCMC in an aqueous solution of pH 0.5 was obtained in comparison to one of pH 2.8. Crystals obtained in these solutions are also compared and industrial optical resolution in these solutions is discussed.

1. Experiments

1.1.Effect of pH on the solubility of SCMC

In this study, an aqueous DL-SCMC solution of 5wt% NaCl was fed into an agitated vessel and enough amount of DL-, and D-, or L-SCMC crystal were added and suspended in it for two days to make saturated solution. The solution was then separated from suspended SCMC crystal through a 0.45 μm membrane filter and the concentration was analyzed by HPLC(High Performance Liquid Chromatography). PH of the solution was adjusted by addition of HCl and operational conditions are shown on **Table 1**.

1.2.Crystal growth rate of a L-SCMC crystal fixed on the end of stainless needle, in

Table.1 Operational conditions for observation of solubility of SCMC

test' run	range of pH	temperature[K]	moleratio D/L
A-1	0.5,1.0,1.5,2.0,2.8	299	1
A-2	0.5	283~308	1
	2.8	283~343	1
A-3	0.5	299	0~1

1054–7487/96/0072$12.00/0

supersaturated solution of DL-SCMC with 5wt% NaCl

Aqueous solution of DL-SCMC with 5wt% NaCl was prepared to be saturation at a desired temperature and set in an agitated vessel of 500ml. The solution was agitated for a period of time at a temperature of 5K higher than the saturated one and then cooled down to the operational temperature shown in **Table 2,** by the cooling rate of 0.3[K/min.]. When the temperature of the solution reached the operational supercooling, the solution was kept at the supercooling level under agitation for 30min., and then a L-SCMC seed crystal fixed on the end of a stainless needle was set in the solution. The shape of seed crystal is easily modeled as shown in **Fig.1** and the length of **La** in this figure was observed every hour since the seed was set in it.

Other tests for increasing rate of crystal surface area were also carried out in a supersaturated DL-SCMC of pH 2.8 and 0.5 in which linear growth rate were 0.04[mm/hr] by adjustment of supersaturation in both cases. In these tests, the crystal was taken out from the solution every one hour and taken photograph. From crystal photos, crystal shape was observed and crystal surface area was estimated.

1.3.Crystal shape of SCMC

When DL-SCMC solution saturated at 308K was kept in the stationary condition at 298K for a week, until SCMC crystals appeared and grew up to some size. The crystals were sampled and kept in the saturated solution of L-SCMC. Change of the crystal shape was observed by the microscope and taking photograph.

2.Experimental Results and Discussions

2.1.Effect of pH on solubility of SCMC

Experimental results obtained under the conditions of A-1 in **Table 1** are plotted in **Fig.2**.

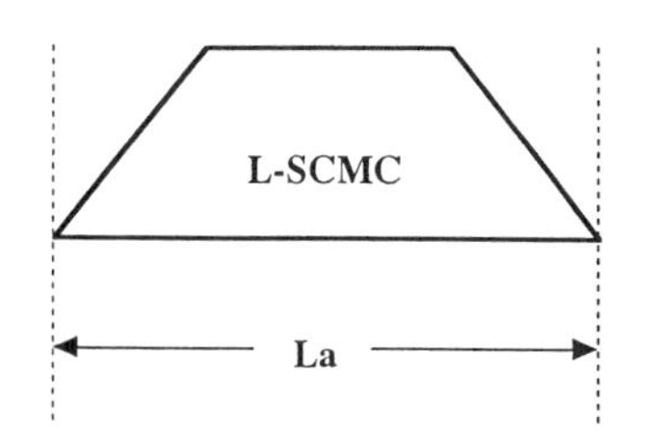

Fig.1 Modeling shape of L-SCMC crystal

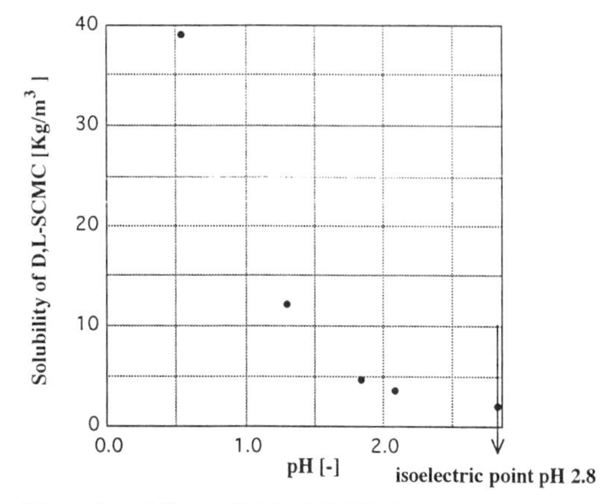

Fig.2 Solubility of DL-SCMC for pH 0.5 to 2.8 at 299K (test run A-1, Table1)

When DL-SCMC crystals were dissolved in water, the pH in the DL-SCMC solution was 2.8 and the saturated concentration was 2.0[kg/m^3]. When HCl was added into aqueous solution of DL-SCMC, the saturated concentration of DL-SCMC increased to 41.2[kg/m^3] for pH 0.5 which was about 20 times of that of pH 2.8. Saturated concentrations of aqueous solution of DL-SCMC obtained on test run A-2 in Table 1 are shown in **Fig.3**. From these plots, **Eq.(1)** was obtained.

$$y=17.4\times 10^{0.015(t-273)} \quad (1)$$

Here, y and t are concentration of DL-SCMC [kg/m^3] and temperature, respectively. The

Table.2 Operational conditions for tests of crystal growth

test' run	Operational temperature[K]	revolution rate of agitator[min^{-1}]	pH
B-1	283	0,50,150,250	0.5,2.8
B-2	299	50,150,250	0.5,2.8

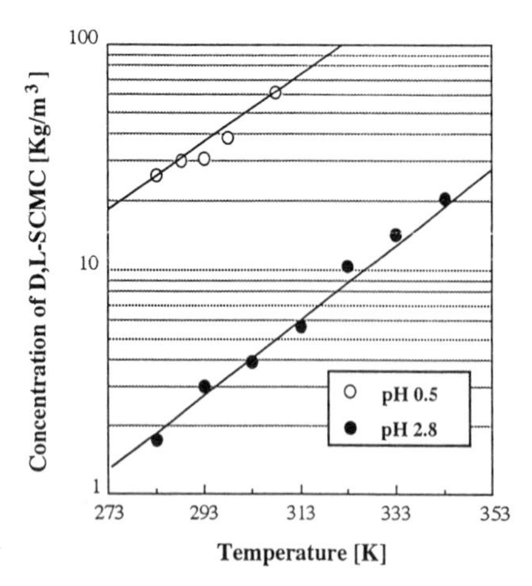

Fig.3 Solubility of DL-SCMC solution of pH 0.5 and 2.8 (test run A-2,Table1)

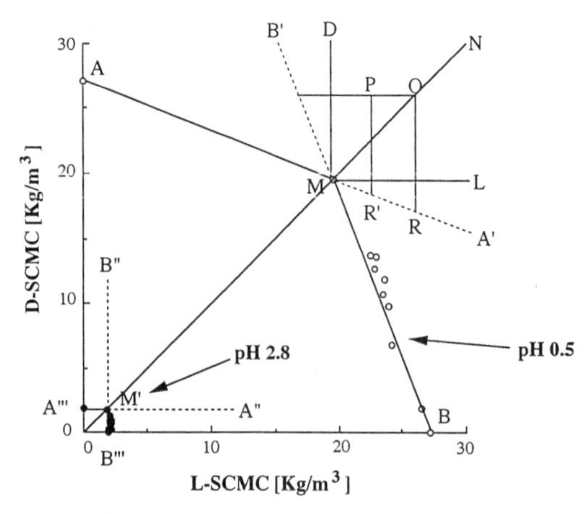

Fig.5 Phase diagram of L-SCMC and D-SCMC at 299K

solubility of SCMC was also observed at 299K for several compositions of L-SCMC against D-SCMC (A-3 in Table1), and the plots in **Fig.4** were obtained. From these plots, the solubilities of L-SCMC and D-SCMC against mole ratio of L-SCMC to D-SCMC are shown in **Fig. 5** for pH 0.5 and 2.8. M and M' in the figure were the concentration saturated at 299K for aqueous solution of pH 0.5 and 2.8, respectively. When saturated concentration of L-SCMC was considered for the solution of pH 0.5 at 299K, the line of MB was obtained for the range that L-SCMC concentration was more than D-

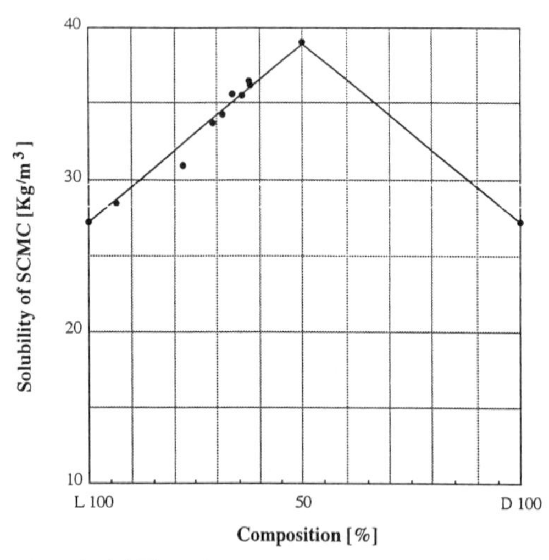

Fig.4 Solubility diagram of SCMC (Operational Condition 299K, pH0.5, 5wt% NaCl, test run A-3,Table1)

SCMC. When the supersaturated solution of DL-SCMC at 299K was considered, point O in Fig.5 is plotted for the composition of the solution, and supersaturation is shown as the distance OM. When L-SCMC seed crystals are assumed to be added into this supersaturated solution and assumed to grow without decrease of concentration of D-SCMC, the composition of the solution changes to point P from point O. When supersaturation of D-SCMC is considered in this figure, supersaturation, the driving force for crystallization of D-SCMC, is expressed by OR or PR' in these solutions. Therefore supersaturation for D-SCMC is considered to decrease with the growth of L-SCMC crystals without any crystallization of D-SCMC. But when the crystallization of L-SCMC was carried out in the DL-SCMC solution at pH 2.8, saturated composition of L-SCMC or D-SCMC is expressed by the line of B'' B''' or A'' A''' and supersaturation for D-SCMC was considered to be constant in spite of L-SCMC crystal growth. From these discussions, optical resolution of DL-SCMC by crystal growth of L-SCMC seed is considered to be favorable on operation at pH 0.5, compared with one at pH 2.8.

2.2.Crystal growth rate of L-SCMC seed

Typical examples of change in L-SCMC crystal size are plotted against the growth time in **Fig.6**. Supersaturation was generally supposed to decrease with growth of L-SCMC crystal seeds. But on this test, as a few tiny seeds were put in

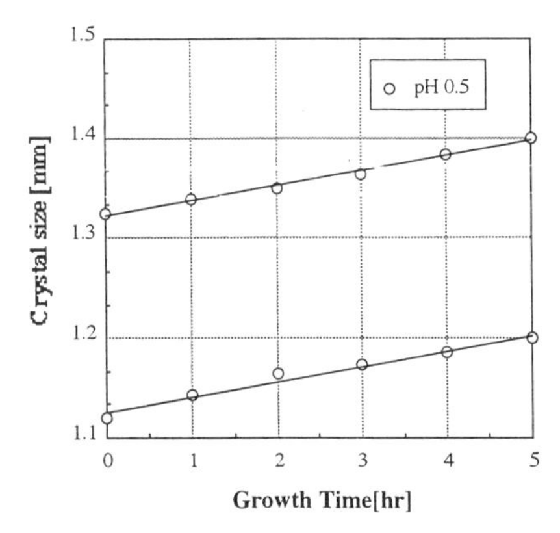

Fig.6 Change of the size of L-SCMC seed expressed by La in Fig.1 against growth time (supersaturation 11.7[kg/m^3], pH0.5, 150r.p.m.])

about 500ml of DL-SCMC supersaturated solution, the decrease of the supersaturation by growth of seeds was supposed to be neglected. From these considerations, plots of crystal size change against growth time became straight lines as shown in Fig.6. Therefore, the linear growth rates of L-SCMC crystals expressed La in Fig.1 were estimated from the slopes of the lines obtained from tests, and are plotted against supersaturation in **Fig.7**. Plots in Fig.7 were deviated in some range, but were independent of the revolution rate of an impeller which was less than 250 rpm. From these tests, the surface reaction was supposed to be dominant on crystal growth mechanism, as in the crystal growth in the DL-SCMC supersaturated solution of pH 2.8 [1]. Data obtained at 299K and pH 2.8 were also plotted in it. The operational dimensionless supersaturation, S, at pH 0.5, expressed by the ratio of actual concentration to saturated concentration, is much less than that at pH 2.8, though supersaturation expressed by difference ΔC at pH 0.5 is about three times and the maximum growth rate observed on these tests at pH 0.5 was about 50% more than that at 2.8 pH. Growth rate data in Fig.7 are plotted against dimensionless supersaturation, S, in **Fig.8**. On the other hand, the critical size, γ_c for nucleation was correlated by **Eq.(2)** [3],.

$$\gamma_c = \frac{2\sigma V^*}{RT \ln S} \qquad (2)$$

Here σ, V*, R are surface energy, molar volume, gas constant, respectively. From Equation (2), dimensionless supersaturation is considered to affect nucleation rate. Operational dimensionless supersaturation at pH 2.8 is much higher than that at pH 0.5, and nucleation rate at pH 2.8 is considered to be higher than that at pH 0.5. When industrial crystallization is considered, operational conditions for large crystal growth rate and small nucleation rate, are convenient to produce reasonable size of crystal for industrial operations, and operation at pH 0.5 is supposed

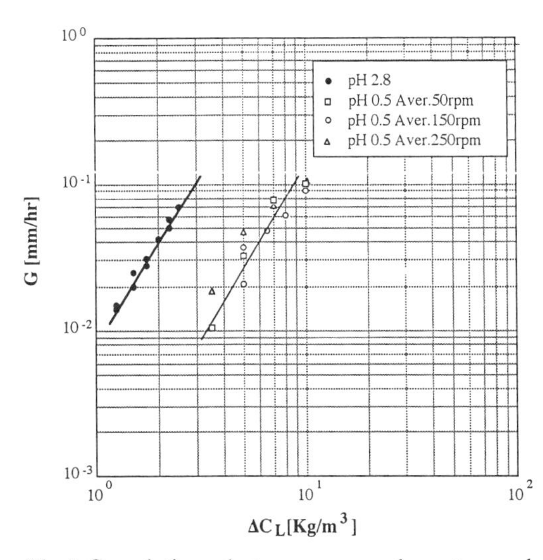

Fig.7 Correlation between growth rate and supersaturation (Operational temp.299K)

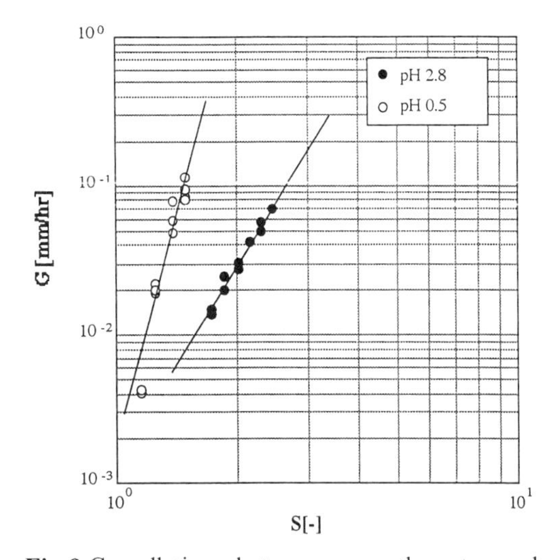

Fig.8 Correllation between growth rate and dimensionless supersaturation

to be more available. Increasing rate of surface area of fixed seed crystal in either solution was observed as in **Fig.9** and schematic pictures of growing seed are shown in **Fig.10**. From these comparisons, crystallization at pH 0.5 is slightly different from one at pH 2.8.

2.3.Shape of crystals produced in a supersaturated DL-SCMC solution at pH 0.5

The solution of pH 0.5 and 5wt% NaCl staturated by DL-SCMC at 308K settled at 299K for about a week. A few fines appeared after about two days and small crystals whose size were from 0.7 to 1.0mm on L_a and from 0.15 to 0.2mm on L_b were obtained. A typical photogragh of the crystals is shown in **Photo.1**. Composition of this crystal was 50% of L- and D-SCMC. When this crystal was put in a saturated solution of L-SCMC, parts of the crystal dissolved gradually as in **Photo.2** and after 7 hours' dipping, the shape did not change. From these tests' results, the shape and structure of crystal generated in DL-SCMC solution of pH 0.5 expressed by the model shown in **Fig.11**. On the other hand, crystal appeared in the solution of pH 2.8 were needle and when the crystals were dipped in a saturated solution of L-SCMC, a transparent needle crystal changed to a muddied white and then finally fragmental. From these tests' results, DL-SCMC crystals precipitated in the solution of pH 2.8 is supposed to be composed of tiny fines of L- and D-SCMC crystals. The detailed mechanism has not been studied yet, but these differences were considered to be caused by the difference in the nucleation rates. Consequently, operation at pH 0.5 is supposed to have some excellent merits for optical resolution on an industrial processes.

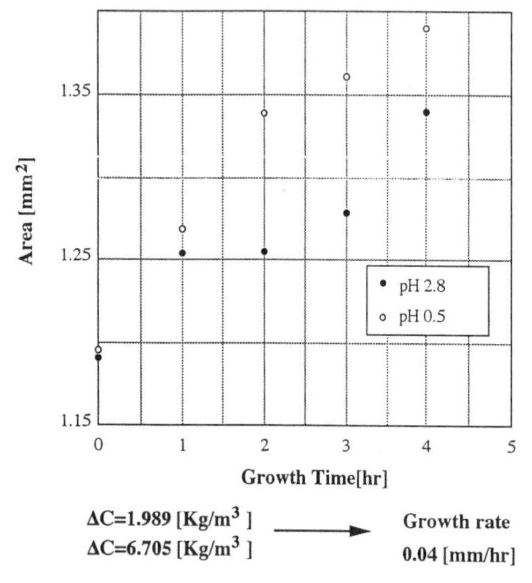

Fig.9 Increase of crystal surface area against growth time

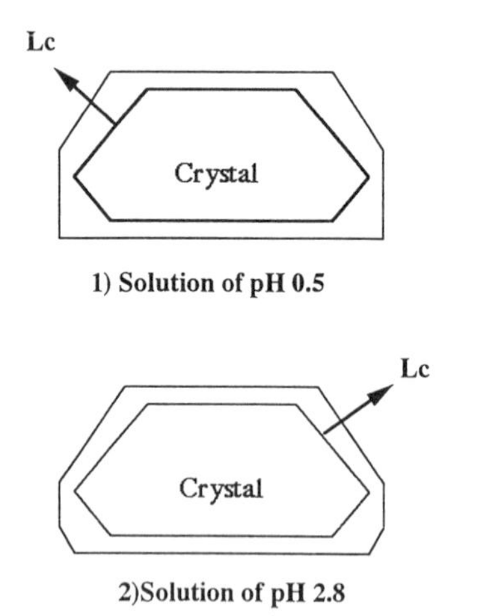

Fig.10 Change of crystal shape by growth

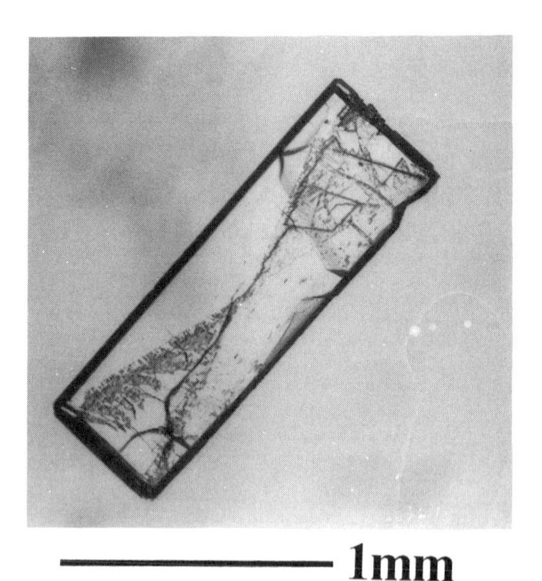

Photo.1 DL-SCMC Crystal from stationary DL-SCMC solution of pH0.5, 5wt%NaCl

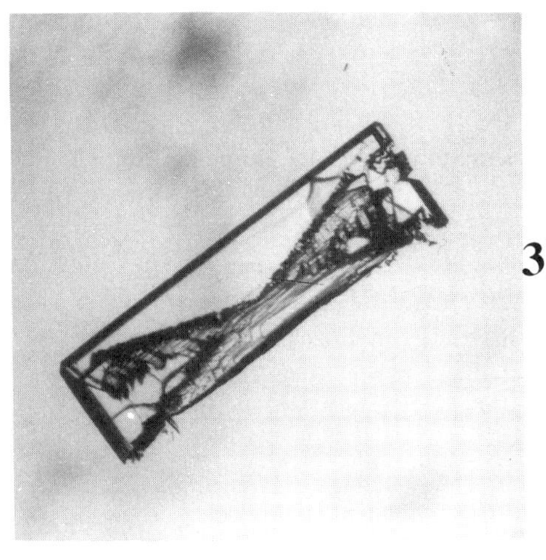

3hr

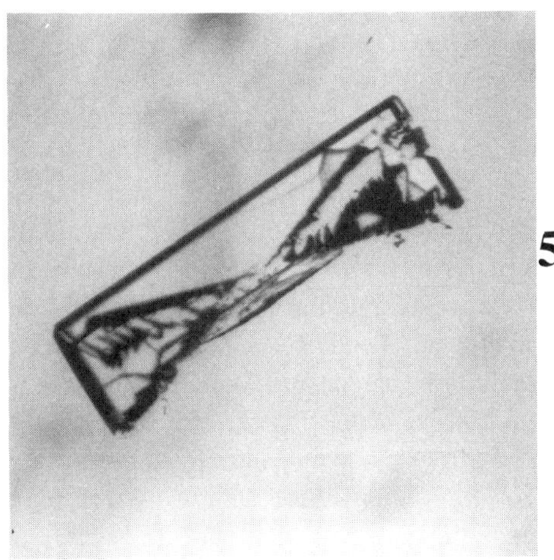

5hr

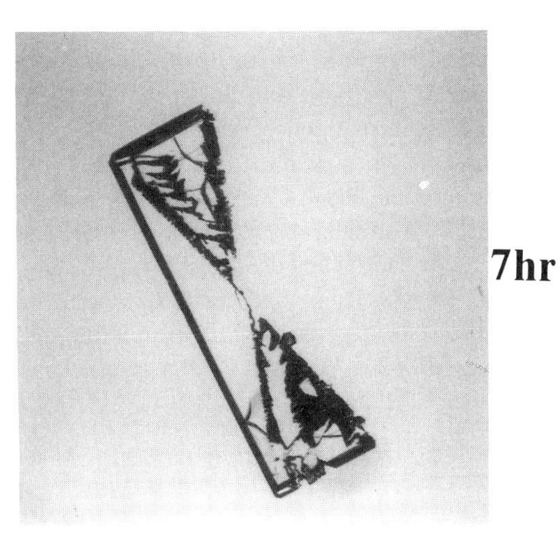

7hr

1mm

Photo.2 Change of shape of DL-SCMC crystal

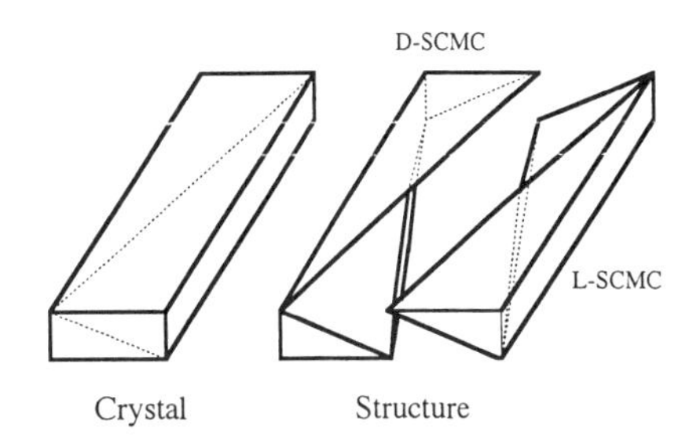

Fig.11 Modeling shape and structure of DL-SCMC crystal precipitated in a stationary solution of pH0.5,5wt%NaCl

References

1)Yokota,M.; Toyokura, K.;
Ind. Crystallization'93 vol.2 **1993**, 203-208
2)Yokota,M.; Toyokura, K.;
Chemical Engineering vol.39 **1994**, 69-76
3)Lamer,V.K.; *Ind. Eng. Chem.* 44 **1952**, 1270

Acetaminophen Crystal Habit: Solvent Effects

Daniel A. Green and **Paul Meenan,** E. I. du Pont de Nemours & Co., Experimental Station, E304, Wilmington, DE 19880-0304

Crystal morphology has been predicted for the analgesic acetaminophen (p-hydroxyacetanilide, paracetemol) using attachment energy methods. However, there are significant differences between the predicted and observed morphology when acetaminophen is grown from water solution (Meenan and Green, 1994). These differences can be attributed to the effect solvent has on the growth rate of the various crystal habit faces. The present work was undertaken to elucidate this effect, particularly the role of hydrogen bonding between solvent and crystal surface. To this end, molecular modeling is combined with the results of simple experiments charting the habit shift of acetaminophen as a function of the hydrogen bonding tendency of the solvent or solvent mixture.

Acetaminophen, a popular over-the-counter analgesic is of course a very important pharmaceutic crystallized in large quantities. Beyond its immediate value, it also serves as a model compound for the development of techniques such as those discussed here for the resolution of differences between experimental and predicted morphologies.

Crystal habit has great impact on solids/liquid separations, processability, and crystal surface/environment chemistry of a material; and therefore purity, cost of manufacture, and end use properties are affected. It is therefore very important to be able to predict and control crystal habit.

Crystal habit modeling has advanced to the state where habit prediction for organic materials is relatively straightforward. However, significant differences between predicted and actual habit are frequently encountered for solution grown materials. This is because standard crystal habit modeling does not take interactions between the growing crystal and the surrounding solution into account.

For acetaminophen crystallized from common solvents, solvent/crystal interactions significantly affect crystal habit. In particular, hydrogen bonding between acetaminophen and the solvent plays a large role in determining crystal growth habit, In our previous work, we crystallized acetaminophen from water, finding that the growth of the {001} and {110} faces was significantly slowed; therefore, these faces were relatively larger than predicted as a result of hydrogen bonding with the solvent. Evidently, the attachment of solute molecules is slowed because of the interference of relatively strong hydrogen bonding between crystal face and solvent (in this case water). To further elucidate this phenomenon and to begin to quantify the effect of hydrogen bonding with the solvent, in the current work we crystallize acetaminophen from 3 solvents which represent a range of hydrogen bonding potential from water (highest) through ethanol to acetone (lowest). In addition, we crystallized acetaminophen from 3 acetone/water solvent mixtures. We found that the hydrogen bonding potential of the solvent does in fact significantly affect crystal habit and that this can be accounted for by qualitatively examining the hydrogen bonding potential for the crystal habit faces involved.

Experiments

Crystals were grown at ambient temperature by slow solvent evaporation for the

1054–7487/96/0078$12.00/0

three solvents and three solvent mixtures. Solvent evaporation rate was controlled by partially covering the top of a small beaker containing acetaminophen solution and controlling the amount of open area. Open area was adjusted to give approximately the same growth rate for each of the solvents studied, adjusting for the relative volatility of the solvent used. The beakers were placed in a hood overnight. The best samples were obtained when the solution was not evaporated to dryness. Since crystals, once nucleated, settle to the bottom of the beaker, their habit may be affected by the glass surface on which they grow. To determine the extent of this effect, crystals were also grown by slow cooling from seeds hanging in solutions. The differences noted between techniques are small, there was some thinning of the crystals noted in the direction normal to the {001} face. The habit of crystals grown by evaporation are discussed below unless otherwise noted.

We chose three solvents representing a range of polarity and therefore hydrogen bonding potential. These are acetone, ethanol and water. We attempted to grow crystals from less polar solvents such as heptane, but found the solubility of acetaminophen to be too low for effective crystal growth. Fig. 1 shows growth morphologies of acetaminophen grown in these three solvents. For the water grown crystals, {110} are the dominant faces, with {001}faces nearly as large, while the {001} faces are dominant for the crystals grown from acetone and ethanol. Acetone grown crystals have both {020} and {200} faces, while ethanol grown crystals have the {200} but not the {020} faces , and the water grown crystals have neither. In addition, we find the following trends with increasing solvent polarity from acetone to water: the relative area of the {011} faces increases, {$\bar{2}$01} faces decrease and {110} faces increase.

To further probe the effect of solvent polarity, we grew crystals from 3 mixtures of acetone and water. Fig. 2 shows growth morphologies for crystals grown from these mixtures as well as pure acetone and pure water. Again, {001} are the dominant faces except for the water grown crystals, whose dominant faces are {110}. The {200} faces are present for all except the water grown crystals. There is a slight increase in the relative area of the {110} faces as the solvent polarity increases from pure acetone to pure water. In addition, the {011} faces are present only for pure water and pure acetone grown crystals, and the {$\bar{2}$01} faces are approximately constant, except for the crystals grown from 75% acetone/25% water where these faces are comparatively slightly smaller.

Crystal Habit Modeling

Acetaminophen crystallizes in a $P2_1/a$ monoclinic space group, with unit cell parameters a=12.93Å, b=9.40Å, c=7.10Å; β=115.9° (Haisa, et al. 1976).

In our previous work (Meenan and Green, 1994), we reported habit predictions based on lattice geometry (Bravais, 1913; Friedel, 1907; Donnay et al., 1937) and on attachment energy (Hartman and Perdok, 1955; Berkovitch-Yellin, 1985) using the force fields of Lifson, et al. (Lifson, et al. 1979), Momany, et al. (Momany, et al. 1974), and Scheraga, et al. (Nemethy, et al., 1983). These force fields are parameterized for different classes of systems. They primarily use Coulomb's Law to sum electrostatic interactions and a Lennard-Jones 6-12 force field to sum the van der Waals interactions. Some (Momany, Scheraga) also include a 10-12 potential to model hydrogen bonding (within the crystal lattice). Partial charges may be taken from the literature (Lifson, et al. 1979) and used with the Lifson force field or may be calculated, e.g. (Stewart, 1990) for use with the other force fields. A commercial molecular modeling package was used to obtain these predictions (CERIUS^2, 1994). There are slight differences among the habit predictions using lattice geometry and the three force fields studied (Fig. 3), with the most obvious perhaps being the presence or absence of the {020} and the {200} faces. The lattice geometry and Lifson predictions have the {200} faces and not the {020} faces, while the Momany and Scheraga predictions have the {020} and not the

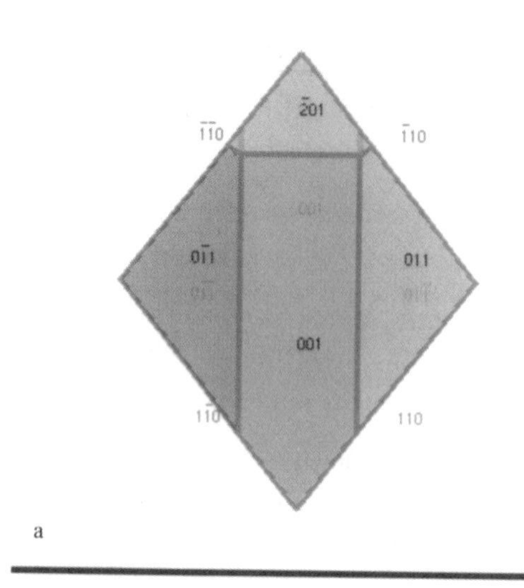

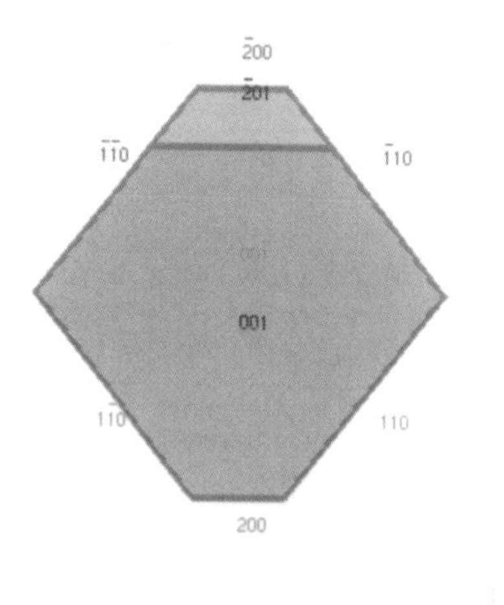

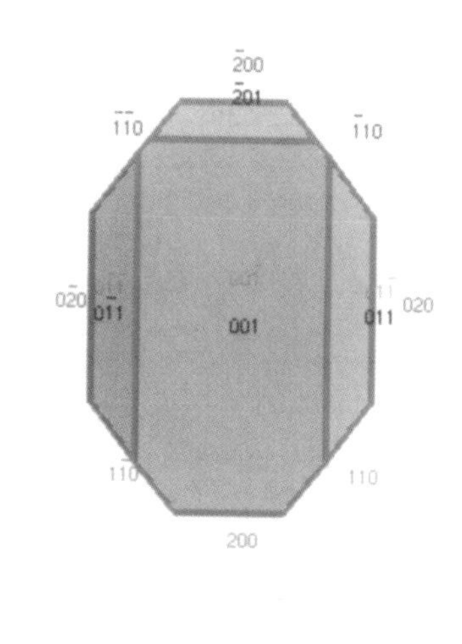

Fig. 1. Observed morphologies for acetaminophen crystals grown by slow evaporation from: a) water, b) ethanol, and c) acetone.

{200} faces. Previously, we found the attachment energy prediction using the force field of Momany et al. was the best fit to the water grown crystal habit, and now find that the attachment energy prediction using the Lifson force field is a better fit to the acetone grown crystal habit, although the {020} faces present in the growth morphology are not present in the prediction.

Discussion

Generally, the agreement between predicted and grown crystal morphologies is good. The overall prediction of a fairly blocky crystal, and the identification of the major faces is rather good, particularly for crystals grown from hanging seeds by slow cooling, which do not in general exhibit the thinning described

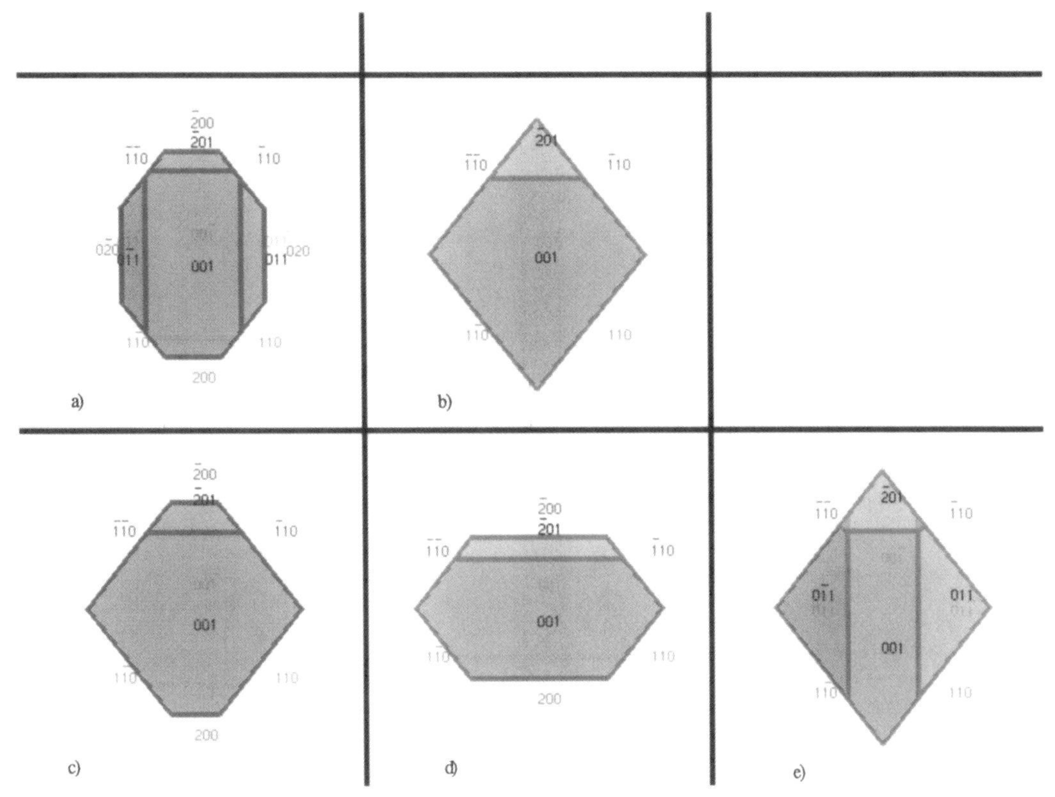

Fig. 2. Observed morphologies for acetaminophen crystals grown by slow evaporation from a) acetone, b) 75% acetone/25% water, c) 50% acetone/50% water, d) 25% acetone/75% water, and e) water.

above which is likely due to the presence of the glass surface in the evaporative experiments. However, there are significant deviations, and these deviations are found to depend on the solvent used in the crystallization, particularly on the polarity of the solvent and therefore its hydrogen bonding potential.

This effect of the solvent polarity can be understood by examining the hydrogen bonding potential of the various crystal faces present in these morphologies. Faces with greater hydrogen bonding potential can be expected to more strongly bind to solvent molecules. For growth of the face to occur, solute molecules must displace hydrogen bound solvent molecules. Therefore, hydrogen bonding of the solvent to the face must slow the growth rate of that face and the stronger the hydrogen bonding the greater the retardation of the growth rate of the face. Both the hydrogen bonding of the face itself and the solvent molecule determine the strength of the hydrogen bonding and rate retardation.

The retardation of face growth will be superimposed on whatever growth rate is imposed by the energetics of solute molecule incorporation into the crystal lattice. Since the slowest growing faces have the largest relative area in the growth morphology, the effect of slowing the growth rate of a particular face is to increase its relative size in the crystal.

The faces predicted by both lattice geometry and attachment energy are, approximately in order of decreasing relative size, $\{\bar{2}01\}$, {110}, {020} or {200}, {211}, {111}, {001}, and {011}. Faces observed in the growth morphologies are: {001}, {110}, {011}, and $\{\bar{2}01\}$ for all the crystals, plus {200} and {020} for some of the crystals grown. Therefore, the predicted {211} and {111} faces

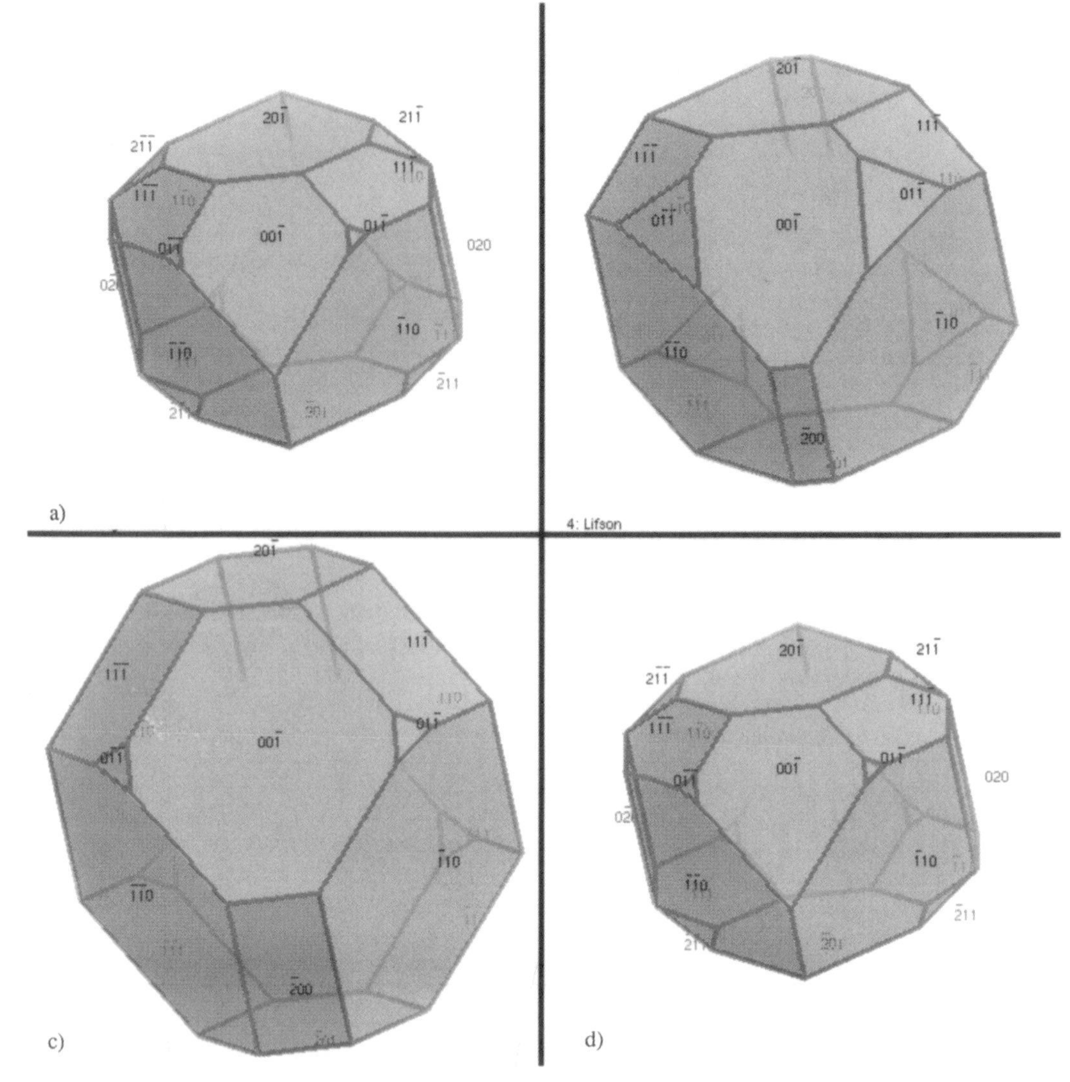

Fig. 3. Predicted acetaminophen crystal morphologies based on a) attachment energy, force field of Scheraga, et al., b) attachment energy , force field of Lifson, et al., c) lattice geometry, and d) attachment energy, force field of Momany, et al.

were not observed. Differences between all of the growth morphologies and the predictions suggest that the solvent is affecting the crystal habit for each of the solvents studied, of course to varying degrees.

Examining the position of the molecules in the crystal lattice relative to the crystal faces, particularly the positions of the carbonyl (C=O), hydroxyl (OH), and the NH of the amide group of the acetaminophen molecule which are the preferred sites for hydrogen bonding, we can qualitatively asses the hydrogen bonding potential of the various crystal faces and perhaps rank them from highest to lowest hydrogen bonding potential. Doing this for the major crystal faces, we get: {110} (Fig. 4) with OH groups projecting above the surface and NH and C=O groups at the surface, {011} with OH groups projecting from the surface and carbonyl groups at the surface, {200} with OH groups projecting from and C=O groups near the surface, {001} with OH groups projecting from

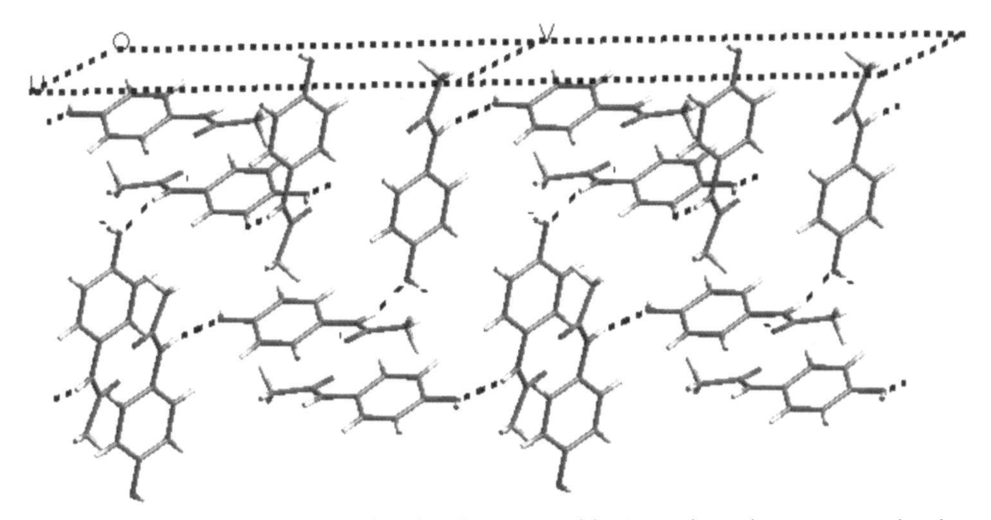

Fig. 4. Molecular arrangement at the {110} surface showing hydrogen bonding within the crystal lattice and to solvent water molecules.

the surface, {020} with the OH group partially obstructed and the N atom adjacent with H-bonding within the lattice inclined only slightly from parallel to the surface, and $\{\bar{2}01\}$ with the OH group turned away from the surface (essentially embedded in the lattice) and the N atom at the surface but partially obstructed by a methyl group (Fig. 5). We expect that the stronger the hydrogen bonding potential of a face the more its growth will be affected by the stronger hydrogen bonding solvents. That is, the growth rate of a strongly hydrogen bonding face will be slowed more by a strongly hydrogen bound solvent that a more weakly hydrogen bound solvent. Conversely, the faces showing only weak hydrogen bonding will be slowed less by variations in hydrogen bonding potential of the solvent. It should be kept in mind that the final growth morphology is the result of the *relative* growth rates of all the faces.

Considering the habit shift associated with increasing solvent polarity found for the solvent series: acetone-ethanol-water, the {011} and {110} faces increase in relative area with increasing solvent polarity, which is consistent with their strong hydrogen bonding potential. {200} faces are observed for ethanol and acetone but not for water, and {020}faces are observed only for acetone. This indicates that these faces grow more rapidly *relative to the others on the crystal* for more polar solvents, which is consistent with the comparatively weak

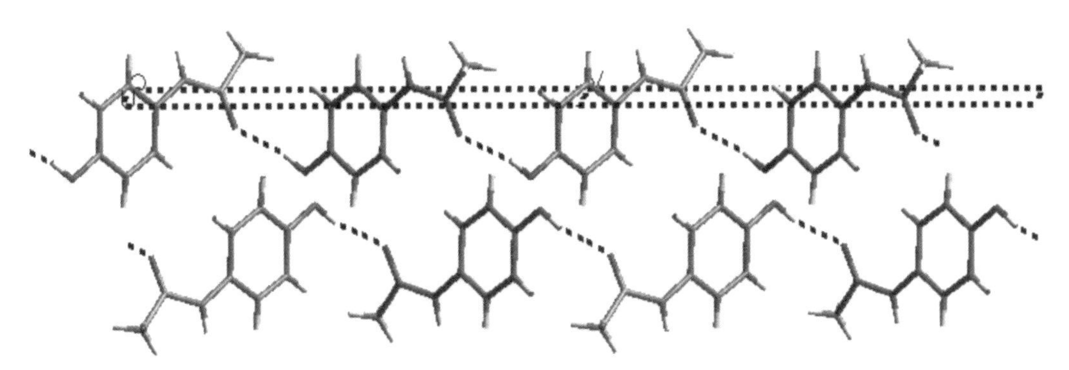

Fig. 5. Molecular arrangement at the {201} surface showing hydrogen bonding within the crystal lattice.

hydrogen bonding of the {020} faces and the intermediate hydrogen bonding of the {200} faces. The change in the relative area of {001} faces is hard to asses from our samples; it appears to be relatively constant. Our molecular modeling suggests intermediate hydrogen bonding for this face.

For the acetone/water solvent mixture experiments, the absence of the {020} faces except for the least polar solvent and the presence of the {200} faces for all but the most polar solvent are consistent with our hypothesis since they weakly and intermediately hydrogen bond respectively and are expected to be larger for less polar solvents. The slight increase in {110} with increase in solvent polarity is also consistent. {011} is a strongly hydrogen bonding face and is present for the most polar solvent, pure water, which is consistent, but it is also present for pure acetone, which we can not now explain. We also can not explain why the relative area of the {$\bar{2}01$} face is maximum for the intermediate concentration, unless the crystal analyzed was not representative.

Conclusions

Both lattice geometry and attachment energy methods predict acetaminophen crystal habit that approximates growth morphologies. There are, however, significant differences between predicted and observed morphologies which are likely caused by hydrogen bonding between solvent and crystal faces. The changes in morphologies of crystals caused by solvents of varying polarity are generally consistent with our hypothesis that hydrogen bonding between solvent and crystal faces causes these changes and the departure from predicted morphologies.

References

Berkovitch-Yellin, Z., *J. A. C. S.*, 107, 8234, **1985**

Bravais, A., *Etudes Crystallographiques*, Paris, **1913**

CERIUS^2 molecular modeling suite, Molecular Simulations, Inc., Burlington, MA., **1994**

Donnay, J. D. H. & Harker, D., *Am. Mineral.*, 22, 463, **1937**

Friedel, G., *Bull. Soc. Fr. Mineral.*, 30, 326, **1907**

Haisa, M., Kashino, S., Kawai, R., and Maeda, H., *Acta Cryst.*, B32, 1283, **1976**

Hartman, P. & Perdok, W. G., *Acta Cryst.*, 8, 49, **1955**

Lifson, S., Hagler, A. T. & Dauber, P., *J. A. C. S.*, 101, 5111, **1979**

Meenan, P. and Green, D. A. "Prediction and Control of Crystal Morphology", *Proceedings, First International Particle Technology Forum*, American Institute of Chemical Engineers, New York, August, I, 316, **1994**

Momany, F. A., Carruther, L. M., McGuire, R. F. & Scheraga, H. A., *J. Phys. Chem.*, 78, 1579, **1974**

Nemethy, G., Pottle, M. S., & Scheraga, H. A., *J. Phys. Chem.*, 87, 1883, **1983**

Stewart, J. J. P., *Journal of Computer-Aided Molecular Design*, 4, 1, **1990**

Use of Single Crystal X-Ray Structural Data in Prediction of Solid State Properties of Alprazolam and Diltiazem Hydrochloride

Esa Muttonen, Postgraduate Studies in Pharmaceutical Technology, The School of Pharmacy, University of Bradford, West Yorkshire, U.K. and Orion Corporation, Orion-Farmos R&D, Espoo, Finland.
Veli Pekka Tanninen, Orion Corporation, Orion-Farmos R&D, Espoo, Finland.
Len Shields, Department of Chemistry and Chemical Technology, University of Bradford, Bradford, West Yorkshire, U.K.
Peter York, Postgraduate Studies in Pharmaceutical Technology, The School of Pharmacy, University of Bradford, Bradford, West Yorkshire, U.K.

Prediction of some solid state properties of crystalline organic pharmaceutical materials is possible from their molecular conformation and packing model. The predicted properties can then be compared with those determined experimentally. This approach has been used in this study for crystal habit determination alprazolam and for the crystallographic stability of diltiazem hydrochloride.

Physical properties of active ingredient in a drug formulation can affect the manufacturing process, stability of the dosage form and finally, the bioavailability of the preparation (Florence, 1988). Pharmaceutical drug substances are almost without exception manufactured and processed further in solid state as powders. The primary methods in production of pharmaceutical materials are batch crystallisation from supersaturated solutions by cooling or by precipitation. Scaling-up difficulties and unsatisfactory control of the large vessels used in batch crystallisation often lead to variation in the characteristics of the end product, such as particle size and/or crystal properties. This batch to batch inconsistency may or may not cause problems in the manufacture of pharmaceutical preparations depending largely on the scale and quality of preformulation studies, formulation development and manufacturing process validation.

A concerning factor in the solid state properties of drugs is polymorphism. Polymorphs often appear in different crystal habits and particularly they exhibit different rates of dissolution. Polymorphism is a common feature for pharmaceutical materials. Numerous samples have been reported to appear in at least two different molecular arrangements with some even in 12 different crystal modifications, such as phenobarbitone (Wells, 1988). Typical methods for probing polymorphic and crystal properties of a drug entity are X-ray powder diffractometry (XRPD), FT-IR spectroscopy and calorimetry.

In the development of a pharmaceutical dosage form extensive research work is necessary to scrutinise the physical properties of the chosen drug entity. Purity, solubility, salt selection studies, compatibility study with possible excipients, stability and shelf life studies in different conditions must be carried out prior to pharmaceutical formulation

1054–7487/96/0085$12.00/0

development. Particle engineering forms a vital part of such preformulation work. The aim of the particle engineering is to obtain optimal physico-chemical properties for the definitive drug agent, such as purity, stability and aqueous dissolution.

Until recently, the solid state properties of pharmaceutical raw materials have only been measured experimentally and the empirical particulate data, such as surface properties, flowability and brittleness, have been used to model the manufacturing processes. However, a more fundamental approach, molecular pharmaceutics, is needed to define and explain the crucial causes and reasons for particle and powder properties and behaviour, molecular pharmacy. The aim of molecular pharmacy is to provide predictive opportunity based on molecular conformation and/or crystal structure to describe important characteristics of a solid material .

Such prediction of solid state properties is of value in the preformulation phase of a solid dosage form. This is possible by calculating the theoretical conformation of the molecule or by using from the established conformation and packing characteristics of the molecule to calculate various properties. For example, IR- and Raman vibrational characteristics (Hehre, 1986), crystal habit (Berkowitch-Yellin, 1985), X-ray powder diffraction pattern (Smith, 1991), density (Schultze-Rhonhof, 1974), heat of fusion (Stewart, 1990) and some mechanical properties (Roberts, 1991) of the solid crystalline material may be compared to theoretically derived parameters. Predictive molecular pharmaceutics can also provide data for estimating physical stability, such as the activation energy for phase transitions.

Materials

Alprazolam, a potent 1,4-benzodiazepine derivative sedative and diltiazem hydrochloride, a common cardiac failure drug (batches 91E327 and 91H2259 respectively, manufactured by Orion Corporation, Orion Pharma International FERMION, Espoo, Finland) and re-crystallised alprazolam form 2 sample (U.S. Pharmacopoeia form) were used in the study. As summary of the two crystal structures available for alprazolam form 1 and diltiazem hydrochloride are tabulated in Table 1 and the molecular formulae in Fig. 1.

Table 1. Crystal structures of alprazolam (Muttonen, 1995) and diltiazem hydrochloride (Kojic-Prodic, 1984).

	Alprazolam form 1	Diltiazem hydrochloride
Formula	$C_{17}H_{13}ClN_4$	$C_{22}H_{26}N_2O_4S$ x HCl
M_w	308.77	450.97
Lattice system	Triclinic	Orthorhombic
Space group	$P\bar{1}$	$P2_12_12_1$
Z	2	4
a	7.318(5) Å	42.18(1) Å
b	9.829(13) Å	9.079(1) Å
c	11.103(11) Å	6.035(2) Å
α	90.84(1) °	90 °
β	95.36(2) °	90 °
γ	109.58(4) °	90 °
V	748.3 $Å^3$	2311.1 $Å^3$
$D_{x\text{-}ray}$	1.371 g/cm^3	1.296 g/cm^3
$D_{measured}$	1.33	-
R_w	0.035	0.061

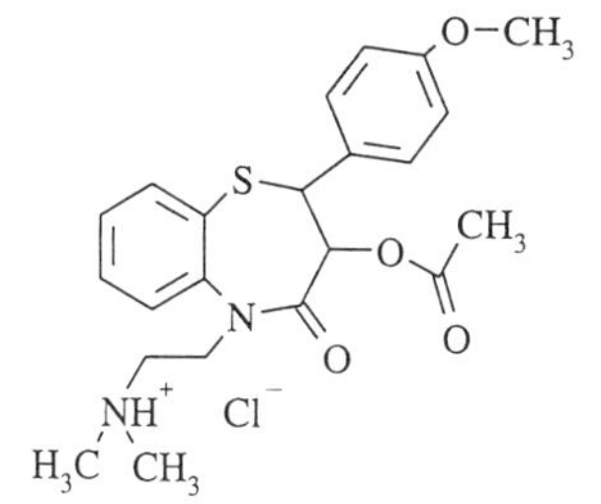

Fig. 1. Molecular formulae of a) alprazolam and b) diltiazem hydrochloride.

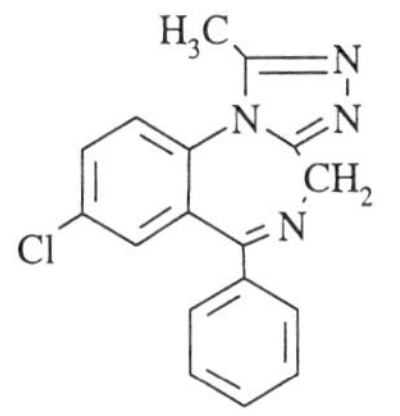

Caution - Alprazolam may be habit forming. Alprazolam is a controlled substance (depressant) listed in the U.S. Code of Federal Regulations, Title 21 Part 1308.14 (1985).

Experimental

The atomic electrostatic potential charges (ESP) for interpretation of the molecular interaction of alprazolam were calculated with MOPAC v.6.00 (Stewart, 1990) using the MNDO Hamiltonian.

Cerius^2 v1.5 software (Molecular Simulations Inc., MA, U.S.A.) operating with the charges calculated by MOPAC was applied for the morphology prediction of alprazolam form 1 using the attachment energy method (Berkowitch-Yellin, 1985) and the Dreiding 2.21 force field (Mayo, 1990).

The prediction of X-ray powder patterns for alprazolam and diltiazem hydrochloride was performed with Lazy v2.0 for Macintosh (Lazy Pulverix program (Yvon, 1977)) using the co-ordinates and cell parameters obtained from the structural studies.

X-ray powder diffraction (XRPD) patterns of alprazolam and diltiazem hydrochloride were obtained by X-ray diffractometers (Siemens D 5000 and D 500, Karlsruhe, Germany). A copper target X-ray (Cu K_α $\lambda = 0.15418$ nm) tube was operated at 45 kV x 40 mA power. The measuring range of the pattern was 3 - 53 ° 2θ with a step size of 0.020 ° 2θ, continuous scan, and a measuring time of 1.2 °/min. Powder sample was placed in a shallow-depth cavity of a sample holder and smoothed with a glass slide.

Scanning electron micrographs were obtained with JEOL-840A Scanning Microscope (Jeol Ltd, Tokyo, Japan). The samples were ion sputter coated in an argon atmosphere with approximately 50 nm of gold before the SEM study. An acceleration voltage of 2.5 kV and an 1×10^{-11} A electron beam current was used in the study.

Alprazolam crystal structure and habit

The molecular packing revealed a π-π bonding and a polar interaction between the methyl-triazolo groups of the two molecules in the unit cell. The packing mechanism was confirmed with molecular orbital charge calculations by MOPAC using MNDO Hamiltonian. The X-ray density was 1.37 g/cm^3 being slightly higher than the measured apparent true density value by He-pycnometry 1.33 g/cm^3, and this comparison provides useful information in estimating the crystallinity of samples.

The predicted XRPD pattern by Lazy v2.0 was similar to that obtained experimentally from the supplied raw material (form 1) used in the study. This confirms the quality of the structure refinement and packing model. The absence of additional intensity maxima in the experimental pattern also verifies the crystalline phase purity of the starting material up to the detection limit of this method at approximately 1% w/w. The minor differences noticed in the peak intensities may be due to preferred particle orientation in XRPD sample preparation. However, the calculated XRPD pattern from single crystal data of alprazolam is clearly different from the measured pattern of the pharmacopoeia standard (form 2). The measured XRPD diffraction patterns of the alprazolam polymorphic forms 1 and 2 and that calculated from the structural data are illustrated in Fig. 2.

Crystal morphology prediction was attempted using Cerius v1.5 software (Molecular Simulations Inc., MA, USA) operating with the charges calculated by MOPAC using the attachment energy method and the Dreiding 2.21 force field (Mayo, 1990). The predicted habit resembles alprazolam form 1 particles rather than particles of form 2 as seen in Fig. 3. Alprazolam polymorphic form 1 consists of angular particles similar to the predicted habit, whilst the acicular morphology of form 2 is clearly different from the predicted morphology. Particles of form 2 are much more elongated and the relative surface area of lattice faces is different from the simulated crystal morphology. Exact match for form 1 could probably not be achieved due to the approximations used in the morphology prediction method, whilst the attachment energy method has been shown to give good results with simple organic molecules and packing arrangements. The reasons for the lack of exact match in this case might cover crystallisation of sample from a liquid (compared with the theoretical morphology model calculated in vacuum conditions) and some molecular interactions not being included in the calculations, such as the possible weak Cl···H (3 Å distance) interaction in the alprazolam unit cell.

Thus, the conformation and the packing model of alprazolam used in this study can only be used for prediction of solid state properties form 1 and not for powders presenting the polymorphic form 2, as the USP reference form sample studied. Inconsiderate use of single crystal X-ray structural data in property prediction may lead to wrong prediction of solid state properties as shown in the case of alprazolam particle morphology.

Stability of diltiazem hydrochloride

The crystallographic solid state stability of diltiazem hydrochloride was probed by comparing the experimentally obtained initial X-ray powder diffraction pattern to that obtained after 2 years of controlled storage. The acquired patterns were also compared to JCPDS International Centre for Diffraction Data (Swathmore, PA, U.S.A.) data cards for diltiazem hydrochloride. However, the data cards give different details, Powder Diffraction File 42-1857 includes a peak at 4.2 ° 2θ position whereas PDF 38-1755 and 40-1986 do not. The X-ray powder diffraction pattern for crystal structure comparison was calculated with Lazy v2.0 using the published structural data (Kojic-Prodic, 1984). The hydrogen

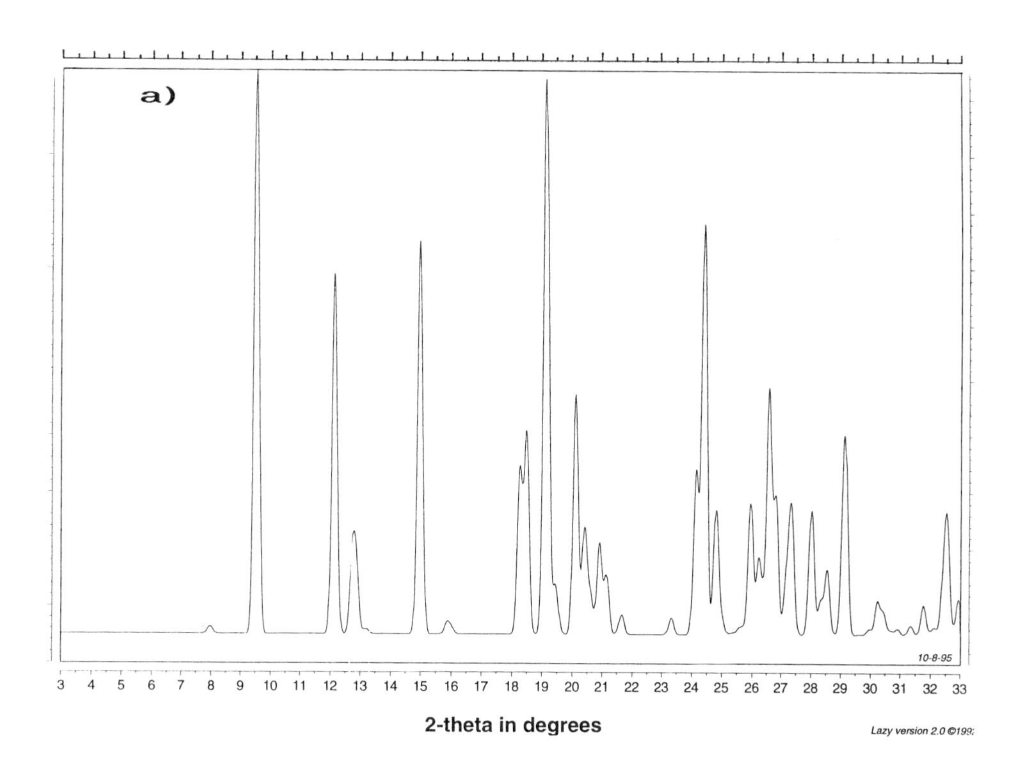

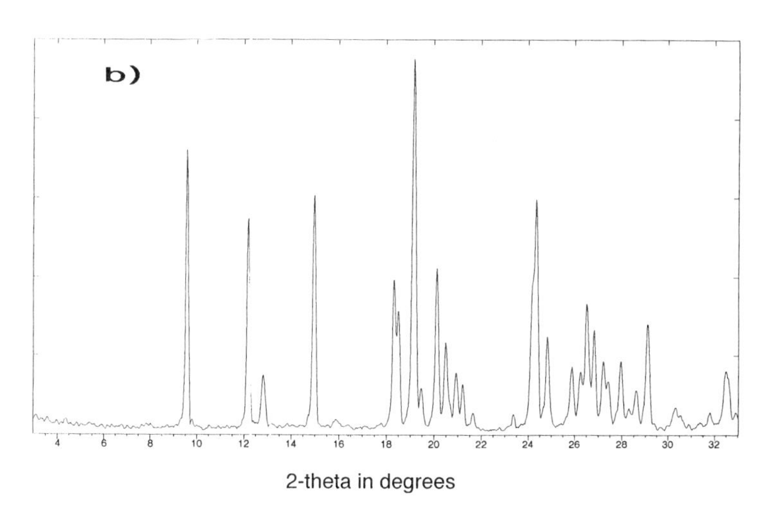

Fig. 2. Powder X-ray diffraction patterns calculated from a) the structural data of alprazolam and b) that of polymorphic form 1 and c) polymorphic form 2.

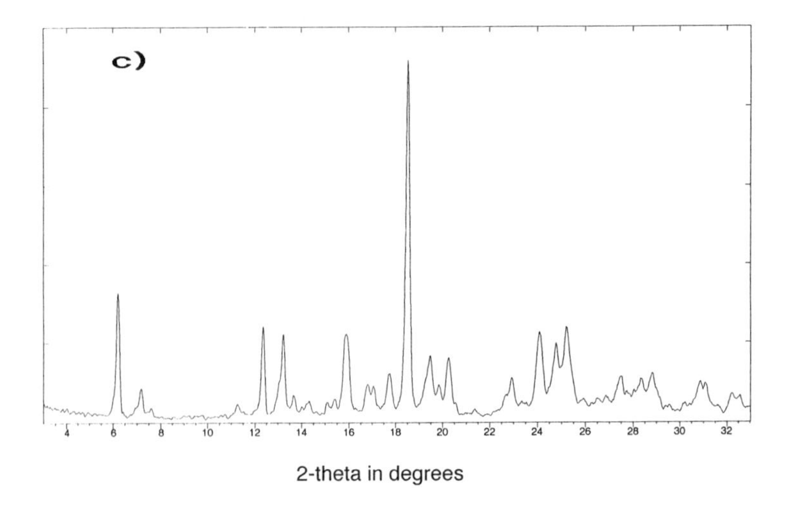

Fig. 2. *(Continued).*

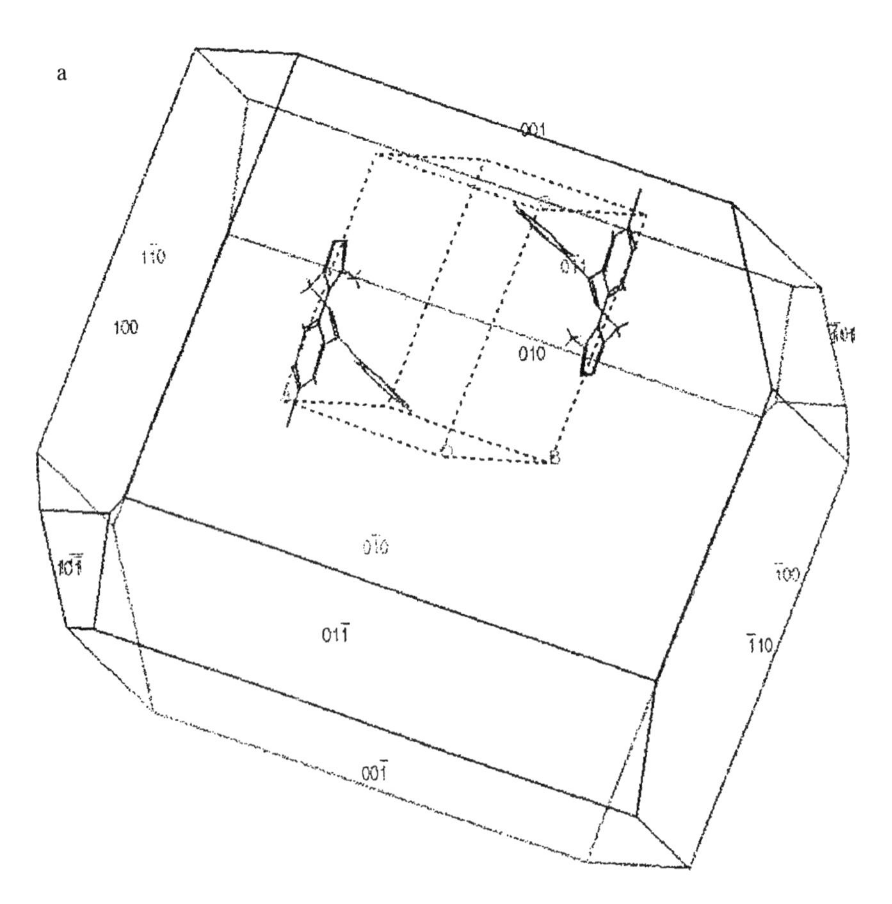

Fig. 3. The predicted crystal habit of alprazolam a) and SEM micrographs of polymorphic form b) 1 and c) 2 powders.

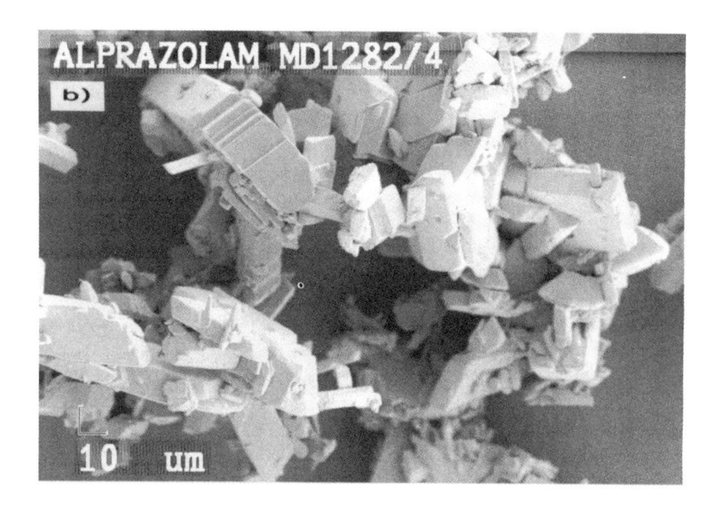

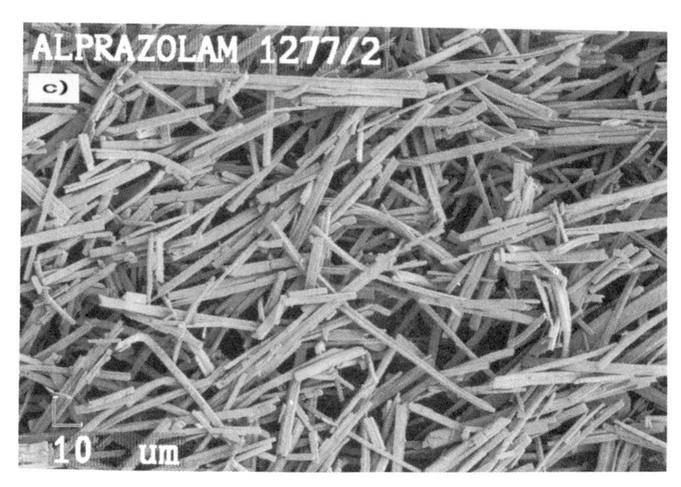

Fig. 3. *(Continued).*

atoms were not positioned before the pattern simulation presented in Fig. 4. The calculated diffraction maxima positions were similar to the measured patterns of the diltiazem hydrochloride commercial batches studied before and after the shelf life study. However, the peak intensity of the first diffraction maximum, positioned at 4.2 ° 2θ, dramatically increased during three years storing as illustrated in Fig. 5.

The observed increase in peak intensity could be due to an unknown instability in the crystal lattice or, alternatively, the peak intensity differences noticed may be partly due to preferred orientation caused by the sample preparation. Preferred orientation effect is strongly dependent on particle morphology, with flaky or needle shaped particles are the most likely particle habits to cause this orientation. Texture and particle size affect the relative peak intensities of XRPD pattern. However, these do not fully explain the intensity differences reported.

An indication of amorphous features in the crystal structure is seen as a broad baseline of the diffraction maxima at 4 - 5 ° 2θ. This

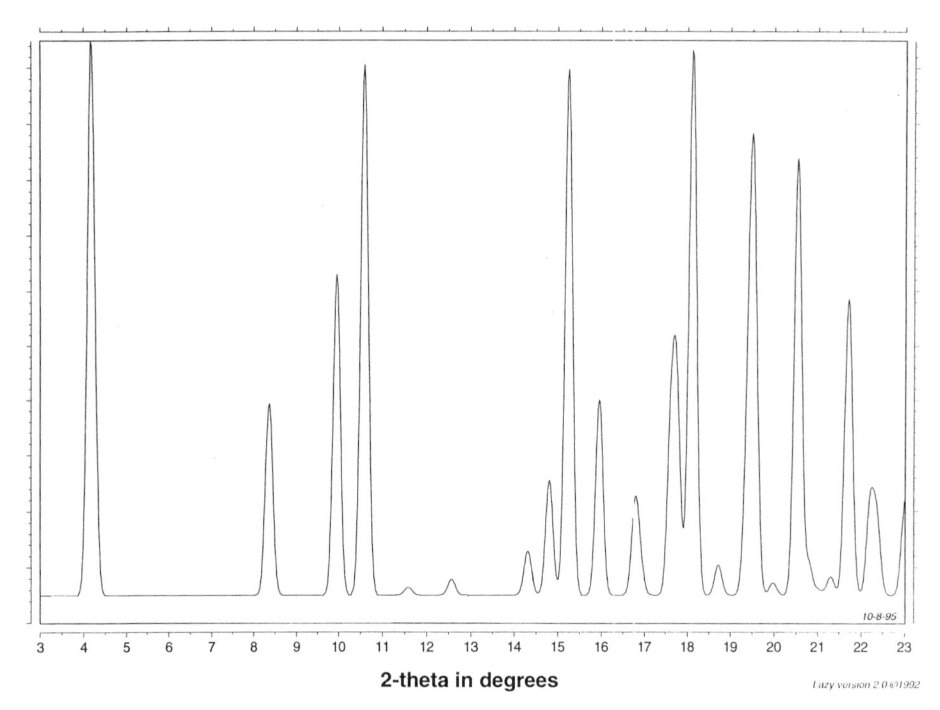

Fig. 4. Calculated X-ray powder diffraction pattern of diltiazem hydrochloride using the reported crystal structure at range 3 - 23 2θ.

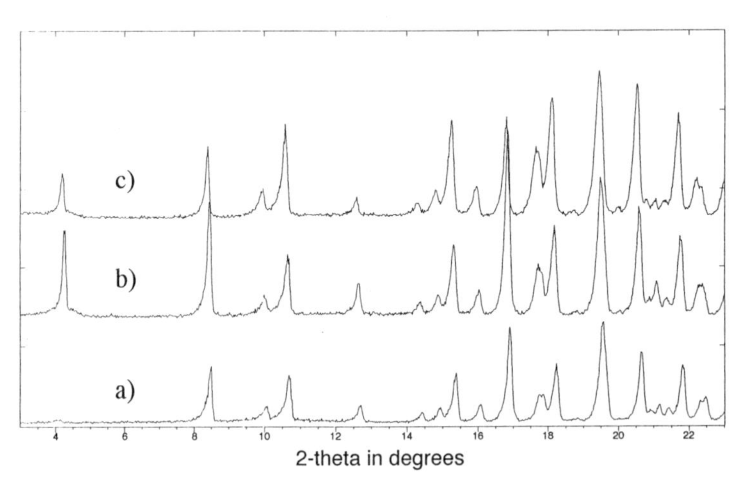

Fig. 5. X-ray powder diffraction patterns of diltiazem hydrochloride measured a) as original in 1992, b) 1995 and c) after milling.

observation, together with the observed peak intensity differences between the simulated pattern and the measured patterns might be a sign of instability in the chlorine atom position in the crystal lattice.

According to Bragg's law, the measured diffraction angle 4.2 ° 2θ corresponds a lattice spacing of 21 Å. Lattice constant *a* was calculated to be approximately 42 angstrom by single crystal X-ray study, being exactly

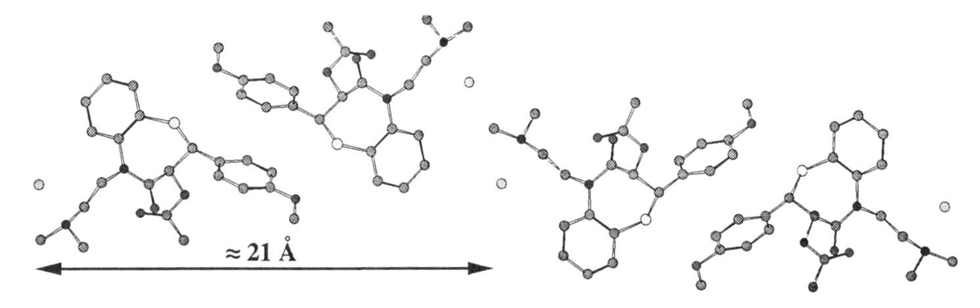

Fig. 6. Molecular packing of diltiazem hydrochloride in the unit cell without hydrogen atoms and the approximate distance between the chloride atoms in the crystal lattice drawn by Chem3D™ Plus v2.0.1.

double the experimental d-spacing value obtained by XRPD. Comparison of the predicted diffraction patterns with the experimental batches, exhibition of both patterns in Fig. 4 and 5, simulated from the known crystal structure of diltiazem hydrochloride, suggests a doubling of the crystal lattice. Twinning takes place along the *a*-axis of the unit cell subsequently forming a superlattice. The differences noticed in patterns reported in JCPDS data cards can be assigned to the presence or absence of the superlattice. The distance between the chlorine atoms in the unit cell is approximately 21 Å. The molecular packing in the unit cell and the Cl - Cl distance marked for diltiazem hydrochloride are shown in Fig. 6. The short range hydrogen bond between N(6)···Cl is 3.001(5) Å determining the overall lattice bonding van der Waals and Coulomb force along *c*-axis. Thus, the formation of the superlattice increases the ordering of the crystal, establishing a more stable crystal lattice.

Conclusions

A bonding interaction not commonly found in molecular crystals of pharmaceutical materials was identified for the polymorphic form of alprazolam in the study. The lattice bonding mechanism was explained using computational chemistry with molecular orbital calculations. Since molecular packing interactions determine the physical stability of molecular crystals, knowledge of the intermolecular bonding energetics and mechanism is of great value in estimating and predicting solid state stability of the crystalline form.

The ability to predict several solid state and particulate properties and to explain changes in crystal properties from molecular detail of drug substances has been shown in this study for alprazolam and diltiazem hydrochloride. Such research provides key information in directing the synthesis, preformulation, development and processing in pharmaceutical product design.

References

Berkowitch-Yellin, Z.; van Mil, L.; Addadi, L.; Idelson, M.; Lahav, M. and Leiserowitz, L. *J.Am.Chem.Soc.* **1985**, *107*, 3111-3122.

Florence, A.T. and Attwood, D. *Physicochemical Principles of Pharmacy*; The Macmillan Press Ltd.: London, 1988.

Hehre, W.J.; Radom, L.; Schleyer, P.v.R. and Pople, J.A. *Ab Initio Molecular Orbital Theory*; John Wiley & Sons: New York, 1986.

Kojic-Prodic, B.; Ruzic-Toros, Z. and Sunjic, V. *Helv.Chim.Acta* **1984**, *67*, 916-926.

Mayo, S.L.; Olafson, B.D. and Goddard III, W.A. *J.Phys.Chem.* **1990**, *94*, 8897-8909.

Muttonen, E.; Shields, L. and York, P. Crystal Structure of Alprazolam form 1., *Unpublished results* **1995**.

Roberts, R.J.; Rowe, R.C. and York, P. *Powder Technology* **1991**, *65*, 139-146.

Schultze-Rhonhof, E. *Method Chim.* **1974**, *1*, 514-546.

Smith, D.K. and Gorter, S. *J.Appl.Cryst.* **1991**, *24*, 369-402.

Stewart, J.J.P. *J.Computer Aided Mol.Design* **1990**, *4*, 1-105.

Wells, J.I. *Pharmaceutical Preformulation*; Ellis Horwood Ltd.: Chichester, 1988.

Yvon, K.; Jeitchko, W. and Parthé, E. *J.Appl.Cryst.* **1977**, *10*, 73-74.

Incorporation of Structurally Related Substances into Paracetamol (Acetaminophen) Crystals.

Barry A. Hendriksen, Eli Lilly and Co, Lilly Research Centre, Windlesham, Surrey, GU20 6PH, UK
David J.W.Grant, Department of Pharmaceutics, College of Pharmacy, University of Minnesota, Health Sciences Unit F, 308 Harvard St. S.E., Minneapolis, MN 55455-0343.
Paul Meenan and Daniel A. Green, Central Science and Engineering, E.I.DuPont de Nemours and Company, Experimental Station, Wilmington, DE 19880-0304.

Substances structurally related to paracetamol have been incorporated into paracetamol crystals by evaporation, by cooling and by gel crystallization , and their morphologies studied by image analysis. The levels of uptake have been determined by HPLC analysis. Some additives (PABA, metacetamol) have relatively high segregation coefficients (0.30) while orthocetamol's is only 0.025. Molecular modelling of the paracetamol crystal surface is shown to offer the basis of explanations for differences between additives in their segregation coefficients, their influence on the inhibition of nucleation and upon overall crystal morphology, on the basis of the additive molecules' ability to disrupt or block the growing surfaces.

A previous paper (Hendriksen and Grant, 1994) has shown that some structurally related substances, present as additives, inhibit the primary nucleation of paracetamol from aqueous solutions. It was suggested, from general analogy with known crystal growth modification by structurally related additives, that the growth of the paracetamol nucleus was modified by a mechanism similar to that previously reported. In the present work the growth-rate modifications of paracetamol crystals by the same related substances, as additives, are studied to elucidate their mechanisms, and that of the inhibition of the nucleation process. The crystal structure data of paracetamol are used as the basis for constructing a computerised molecular model of the crystal, which is subsequently used to predict morphology according to models described below, and to cleave hypothetically and to view individual faces. The molecular structures of the additives are individually considered for their ability to disrupt or block the surface or lattice, and thereby explain the observed effects upon inhibition of nucleation and upon crystal morphology.

Experimental

Materials

The materials used, their sources and the abbreviations in this paper are as follows. *p*-acetamidophenol (acetaminophen, paracetamol) was supplied by Eli Lilly and Co (Basingstoke, UK). HPLC analysis of this material showed no significant peaks attributable to impurities. The additives, acetanilide, (AA, 97%), methyl *p*-hydroxybenzoate (methylparaben, purity not stated), *p*-acetamidobenzoic acid (PABA, 98%), *o*-acetamidophenol (orthocetamol, 99%), and *m*-acetamidophenol (metacetamol, 97%) were supplied by Aldrich (Gillingham, UK). The additive *p*-acetoxyacetanilide (PAA) was synthesised by one of us, and was shown by HPLC analysis to contain 0.18 mole-% paracetamol. Agar (grade 1) was supplied by Oxoid (UK).

Morphological Studies

Crystals of paracetamol were prepared for morphological examination by evaporation, cooling and gel-crystallisation, with and without additives. Supersaturation in this paper means the ratio of the actual mole fraction concentration to the saturated value, and thus always exceeds unity.

Evaporative Experiments.

Additives were added to a 13 g/l solution of paracetamol in a series of plastic Petri dishes at concentrations equivalent to 2 and 4 mole-% with respect to paracetamol, and allowed to evaporate slowly at ambient temperature. The resultant crystals were photographed and their sizes and shapes examined by image analysis (Quantimet 500, Leica).

Cooling Experiments.

Additives were added to a 13.37 g/l solution of paracetamol in a series of vials at concentrations equivalent to 1, 2 and 4 mole-% with respect to paracetamol, and cooled in a refrigerator at 7.5°C, at which the supersaturation was 1.46. The crystals were examined as described above.

Gel Crystallisation.

Some pilot crystallization experiments using paracetamol without additives afforded well-formed crystal at an agar concentration of 0.6% based on the total solution, and at ambient temperature supersaturations of 1.4 to 1.7. Accordingly a series of gel crystallizations were carried out in 15 ml vials under these conditions at additive concentrations equivalent to 1, 2.5 and 4 mole-% with respect to paracetamol. The samples were allowed to crystallise at ambient temperature (≈22°C) and were inspected at intervals.

Analysis of the Mole Fractions of Additives Incorporated into Crystals

Paracetamol samples were crystallised in the presence of known amounts of additives at the conclusion of the nucleation-inhibition experiments described previously (Hendriksen and Grant, 1994) by allowing them to cool to ≈10-15°C. Samples were air-dried for two days. Each crystalline paracetamol sample was dissolved in a solution containing aqueous phoshate buffer, pH 3.8 and 15% v/v methanol (30% in the case of PAA). These solutions were assayed by isocratic HPLC with the same solvent system as the mobile phase. The concentration of each component in each solution was calculated from the peak areas at 220 nm. Standard solutions were run consecutively. Concentrations of additives were expressed as mole-% with respect to paracetamol in the crystals.

Molecular Modelling of Paracetamol Surfaces.

A previous paper (Green and Meenan, 1994) outlined how morphological calculations could be made using (a) the geometrical laws of Bravais-Friedel-Donnay-Harker (BFDH) (Bravais, 1913; Friedel, 1905; Donnay and Harker, 1937) and (b) lattice energy calculations, based on the theories of Hartman and Perdok (1955), which have been shown to be valid for a number of growth theories (e.g. Bennema and Hartman, 1980). Predictions based on both these models have been shown to exhibit excellent correlation with observed habits. A number of computational codes are currently available (Docherty et al, 1988; Clydesdale et al, 1991) for application in the simulation of organic systems. The simulations described in this paper have been obtained using the molecular modelling suite CERIUS^2 (Molecular Simulations Inc, 1994). Predicted morphologies were obtained using the BFDH model and also based on lattice energy predictions, using the force field of Momany et al (1974) and partial atomic charges calculated semi-empirically using the MOPAC code (Stewart, 1990). Using the CERIUS^2 modelling suite, individual crystal surfaces from the morphology can be viewed and hypothetically cleaved so that the crystal chemistry of that particular face may be elucidated.

Results

Morphology

The length to breadth ratios of crystals prepared by evaporation and cooling are given in Table 1 and Figs. 1 and 2. Control samples are tabular crystals with low aspect ratios of 1.23 from cooling or 1.54 from evaporation. Several additives did not change the crystal habit. However, metacetamol, PAA, and PABA changed the morphology dramatically, particularly in the cooling experiments, where aspect ratios of up to 7 were noted. These additives also exerted a strong inhibitory effect upon the nucleation process (Hendriksen and Grant, 1994). To a lesser extent AA caused elongation in the cooling experiments only. The effects were generally concentration-related as observed by Chow et al (1985).

Gel Crystallization

Samples crystallised from an 0.6% agar gel were examined after 8 days (control samples) or after 3 days (those with additives), the crystals were counted, and their overall appearance was noted. The number of crystals, listed in Table 2, reflects the number of nuclei and thus the nucleation rate. As expected, higher supersaturations consistently lead to more crystals, but surprisingly there is a tendency towards more crystals in the presence of metacetamol and PAA, two additives which were effective inhibitors of nucleation in non-gelled crystallisation. The number of nuclei or crystals observed is not entirely consistent with the concentrations of the additives.

The appearance or morphology of the crystals varied considerably. Controls comprised agglomerates of many individual crystals growing out from each other in two directions, always in an overall V shape. The same shape was reported for agar gels by Femi-Oyewo and Spring (1994) in their studies on the growth of paracetamol crystals from aqueous solutions containing various polymeric materials. The less effective nucleation-inhibiting additives (methylparaben, orthocetamol and AA) generally produced the same shape as did the controls, but the more effective additives

Table 1. Aspect Ratios of Crystals prepared by evaporation or by cooling

Related substance (or additive)	Level	Aspect Ratio					
		Evaporation			Cooling		
		Mean	SD	n	Mean	SD	n
Orthocetamol	1				1.4	0.34	25
	2	1.31	0.27	20	1.34	0.39	21
	4	1.49	0.26	20	1.93	0.48	25
Methylparaben	1				1.27	0.21	22
	2	1.49	0.22	20	1.46	0.37	16
	4	1.68	0.39	20	1.75	0.72	2
PABA	1				3.27	1.17	25
	2	2.01	0.56	20	4.72	1.44	25
CONTROL		1.54	0.29	20	1.23	0.24	26
AA	1				1.37	0.39	28
	2	1.54	0.47	20	3.76	0.77	21
	4	1.34	0.31	20	2.84	0.87	26
PAA	1				2.88	0.61	20
	2	3.6	1.27	20	6.84	2.58	22
	4	4.39	1.63	20	5.09	1.88	24
Metacetamol	1				3.5	1.3	22
	2	3.35	0.91	20	5.15	2.05	26
	4	3.49	2.63	20	7.33	2.99	11

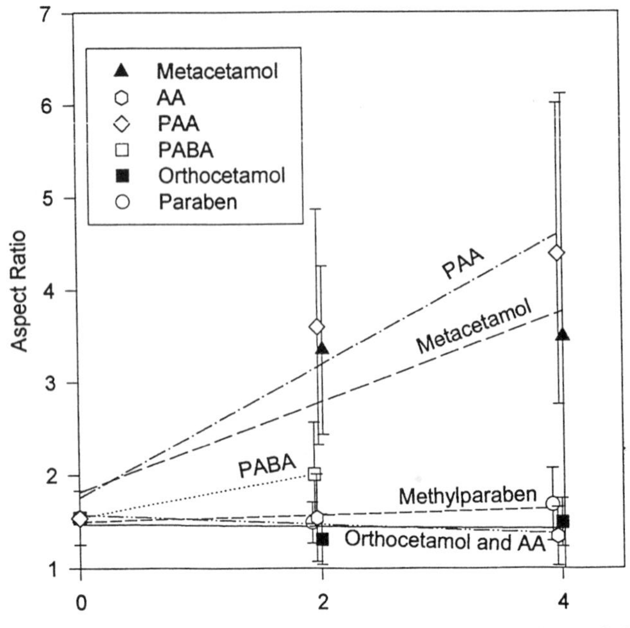

Fig. 1. Aspect Ratios of Paracetamol Crystals Prepared by Evaporation in the Presence of Additives.

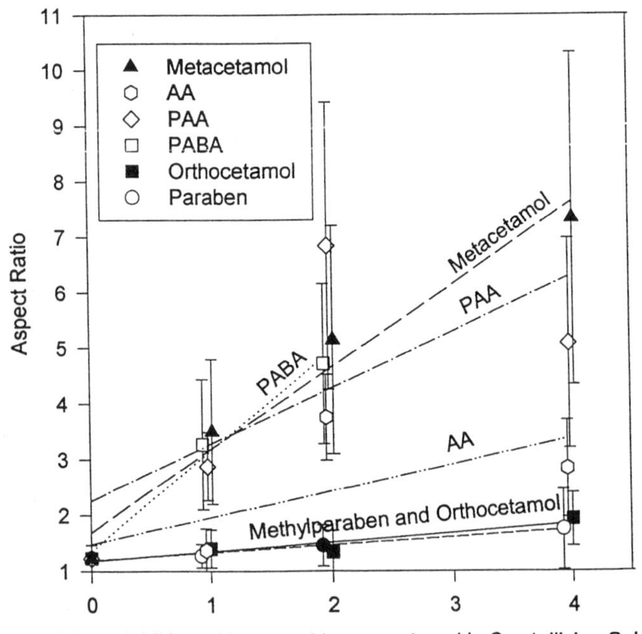

Fig. 2. Aspect Ratios of Paracetamol Crystals Prepared by Cooling in the Presence of Additives.

Table 2. Gel crystallizations; number of crystals seen (m = many)

Related Substance (or additive)	Level	Supersaturation 1.4	1.5	1.6	1.7
Orthocetamol	1	0	2	2	1
	2.5	0	0	3	6
	4	0	0	2	4
Metacetamol	1	1	4	m	7
	2.5	0	15	m	m
	4	0	4	m	m
Acetanilide	1	0	2	7	m
	2.5	0	0	8	11
	4	0	2		6
Me Parabens	1	1	2	6	10
	2.5	0	4	7	11
	4	0	0	7	10
Acetoxyacetanilide	1	0	3	13	25
	2.5	0	2	10	20
	4	0	1	4	12
Control (after 8 days)	-	0	-	0	-

produced different shapes. Metacetamol produced superficially rectangular crystals with a slight but definite curvature, while PAA produced long needle-like crystals which agglomerated together to give a pronounced overall curved shape. PABA was not studied in this series of tests.

Incorporation Of Additives Into The Crystals

The results are summarised in Figure 3. The molar ratio incorporated is approximately proportional to the molar ratio in the crystallization solution. PABA and metacetamol are incorporated much more efficiently than the others. AA, PAA and methylparaben are much less efficiently incorporated, while orthocetamol is hardly incorporated at all. Even in the most efficient case, the segregation coefficient, the mole ratio of additive to paracetamol in the crystals divided by that in the crystallizing solution, was only 0.3, so incorporation was generally disfavoured.

Structural Studies On Individual Crystal Surfaces

Hydrogen bonding plays an important role in the crystal structure of paracetamol. The crystal structure (Fig. 4) shows that the molecules form hydrogen-bonded chains, linked in a zig-zag fashion through the crystal structure. Disruption of this hydrogen bonding will lead to alteration of crystal growth rates and hence to modification of crystal morphology.

A previous study (Green and Meenan, 1994) noted that, theoretically, the {201} forms dominated, with {110}, {020}, {211}, {111}, {001}, and {011} forms also present to a lesser degree. Predicted aspect ratios ranged from 1.45-1.80, depending on the predictive model used. The experimental aspect ratios were slightly lower, 1.23-1.54, whereas the experimental growth morphology (Green and Meenan, in press) was tabular, with {110} forms dominating, with {001}, {011}, {201}, and {110} forms also present, and with an approximately hexagonal main form which exhibited reasonable agreement to the predicted morphology.

Discussion

The causes of morphological modification lie in a combination of the ease with which the additive molecule can be incorporated into the host paracetamol crystal, and the subsequent ease with which a further paracetamol molecule

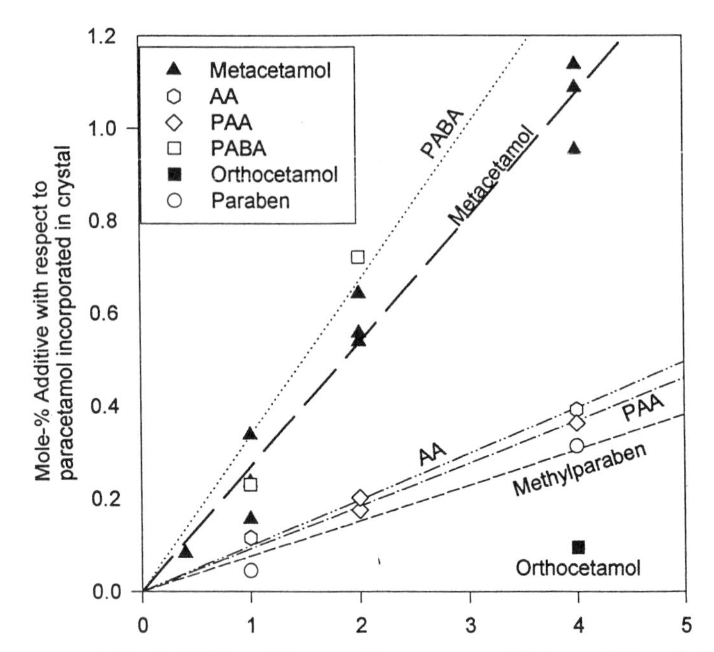

Fig. 3. Uptake of Additives by Paracetamol Crystals.

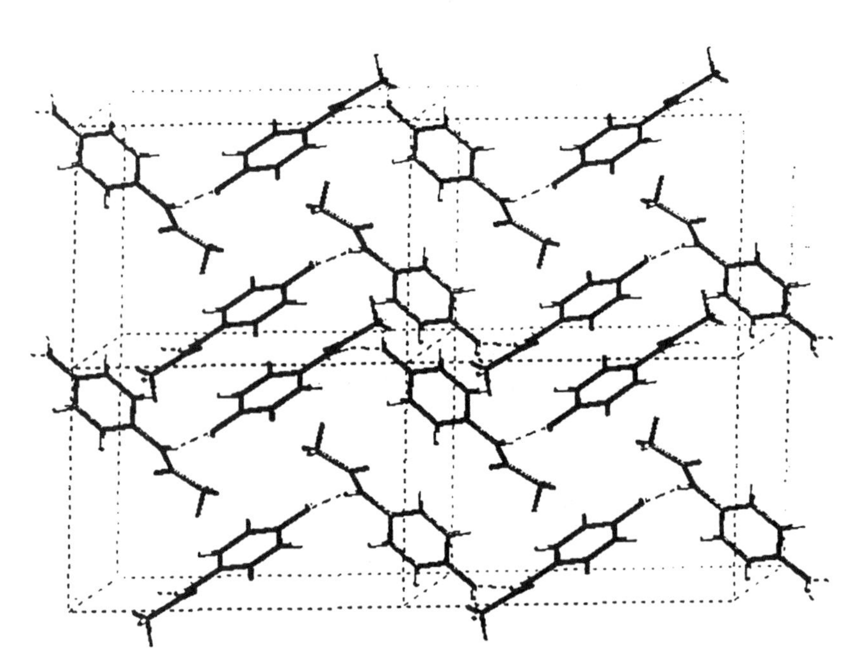

Fig. 4. Unit Cell of Paracetemol, visualized using the CERIUS® software. Hydrogen bonding is illustrated by the dotted lines between molecules.

can be incorporated onto or next to the guest molecule. The ease of incorporation depends upon the molecular similarity (see Fig. 5), sterically and energetically, to the host molecule. Such reasoning has traditionally been applied to the modification of habit with structurally related substances (e.g. He et al, 1994; Weissbuch et al, 1986).

The modification of the crystal habit by a structurally related molecule can be rationalised as the incorporation of either a disruptor or a blocker molecule into the host lattice. A disruptor molecule acts by disrupting crystal packing, for example, by the lack of a functional group. A blocker molecule, due to steric hindrances, physically inhibits the path of incoming host molecules to the growth face and typically interferes with crystal packing. Each of these molecular types will influence crystal growth rates and hence nucleation and crystal habit.

PABA and Metacetamol

Two additives, PABA and metacetamol, with a segregation coefficient of 0.3, are more readily incorporated into the paracetamol lattice than the others. Both have been shown in the earlier work (Hendriksen and Grant, 1994) to be very powerful inhibitors of the nucleation process, and both are shown here to be powerful modifiers of crystal habit, particularly in cooling experiments.

The differences in molecular structure with changes in conformation obviously impose steric constraints on the packing of molecules in the crystal structure and hence will disrupt it's hydrogen bonding patterns. In the case of PABA, the presence of a carboxylate "COOH" group rather than the hydroxyl "OH" group of paracetamol imposes a steric effect on further crystal packing, acting as a blocker to further molecular deposition. However, the "COOH"

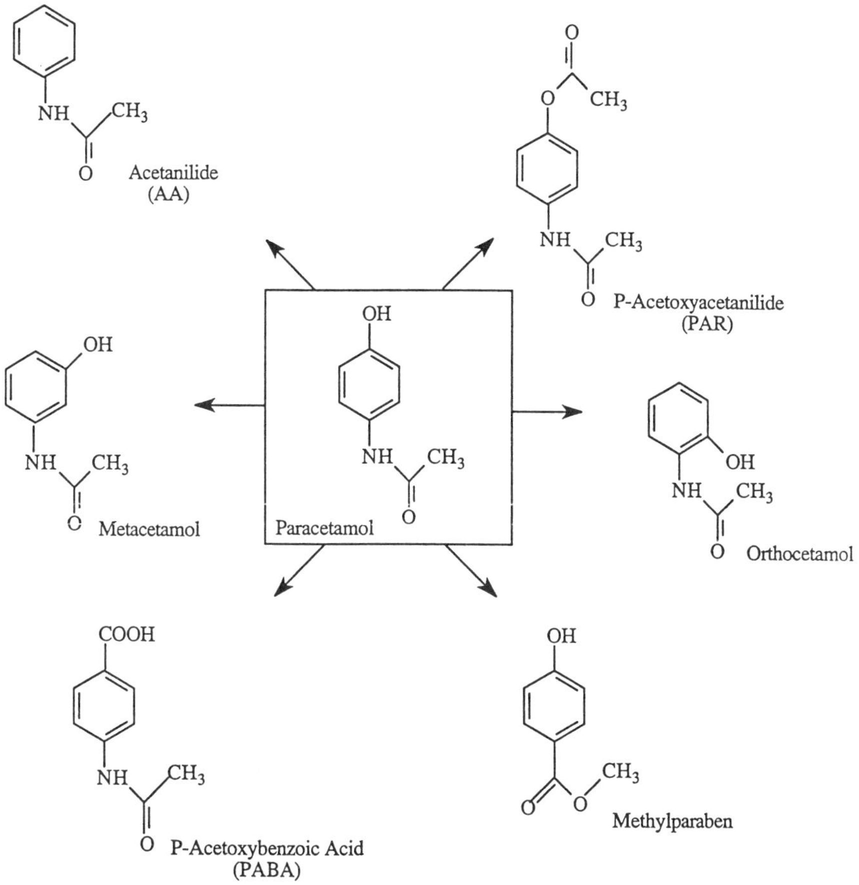

Fig. 5. Molecular Structures of Paracetamol and Molecularly Related Additives.

group will allow the hydrogen bonding pattern to be retained. Thus, this molecule will probably be the most powerful habit modifier of the systems examined.

PABA and metacetamol, with hydrogen bond donors ("COOH" or "OH") in the para or meta position from the acetamido-group, can compete with paracetamol molecules to a limited extent for "docking" onto the growing faces of the paracetamol crystal and onto the growing embryos. Hence, the segregation coefficient is appreciable (~0.3), nucleation is inhibited and growth of the preferred face is slowed down. Since growth of the other faces is less inhibited, the crystal habit is changed significantly.

Metacetamol acts as a disruptor molecule; the shift to the meta position of the hydroxyl group definitely disrupts the hydrogen bonding patterns within the structure, and hence will alter the growth rates and ultimately the crystal morphology.

Methylparaben, AA and PAA

Three additives, methylparaben, AA and PAA are less readily incorporated, with a segregation coefficient of ~0.09. In spite of the similarity in their uptake, their influence on nucleation and habit varies enormously. For example, PAA very strongly inhibits nucleation and modifies the habit, whereas AA is less effective in both respects, and methylparaben is ineffective.

Methylparaben, AA and PAA each have one hydrogen-bonding group in common with paracetamol, either "OH" in the case of methylparaben, or *p*-acetamino in the cases of AA or PAA. These molecules can, therefore, dock onto the growing faces of paracetamol crystals and onto the growing embryos, but to a limited extent because of limited compatibility or incompatibility of the group in the para position, methylcarboxylate in methylparaben, hydrogen in AA and acetoxy in PAA. Hence the segregation coefficient is low (~0.09). Methylparaben differs from paracetamol only very slightly, with O⟨ instead of NH⟨, so it shows little or no selectivity for any particular faces and therefore does not modify the original habit. It does not inhibit nucleation because of the above similarity. AA, which lacks the OH group of paracetamol, slows down the growth of certain faces and the embryos. AA therefore significantly modifies the crystal habit and significantly inhibits nucleation. PAA, which has a para-acetoxy group instead of the OH group of paracetamol, effectively prevents the growth of certain faces and the embryos. PAA is therefore a powerful modifier of crystal habit and a powerful inhibitor of nucleation.

The question arises whether the impurity molecule act as a blocker or a disruptor during crystallization. For PAA the additional steric constraints imposed by an ester group "$OCOCH_3$" instead of a hydroxyl group will initially inhibit uptake of the molecule. The ester group will also significantly disrupt hydrogen bonding due to steric considerations and will thus act as a habit modifier.

Acetanilide (AA), due to its molecular similarity, will also become incorporated into the crystal structure, but the lack of the hydroxyl group will reduce uptake capacity significantly, since it deleteriously influences the "lock and key" mechanism prevalent in the uptake of tailor made additives. This lack of functionality will also disrupt the hydrogen bonding patterns within the crystal structure and thus the bulk morphology.

The fact that methylparaben is taken up to a certain extent but is ineffective as a nucleation inhibitor or habit modifier indicates that it neither acts as a blocker nor a disruptor to any great extent. The hydrogen bonding within the crystal structure of paracetamol is from the hydroxyl oxygen to the amine hydrogen. Within methylparaben, the lack of an "NH" group is compensated by the fact that the hydrogen bonding still occurs between the oxygen in the hydroxyl group and the carbonyl carbon. Thus the hydrogen bonding pattern within the crystal structure is still retained and hence no effects on nucleation and habit are observed.

Orthocetamol

Orthocetamol with a hydroxyl group in the otho position cannot easily "dock" onto the growing faces of the paracetamol crystal. Therefore very little orthocetamol is incorporated, so the segregation coefficient is only 0.025. In line with its low incorporation, orthocetamol does not differentially inhibit the growth of the crystal faces, so the crystal habit is unchanged. Orthocetamol cannot easily "dock" onto the growing embryos, so nucleation is not inhibited.

Combination of PAA and Metacetamol

When PAA and metacetamol are added in combination, each at 2 mole-% with respect to the paracetamol concentration, their levels of incorporation are in line with that expected from either independently, i.e. they do not interfere with each others' uptake. The effect upon habit was not studied, but in nucleation experiments the combination was highly effective, equal to that of PAA at 4 mole-% used alone. It appears that the effects of PAA and metacetamol are additive, presumably on account of their different modes of action arising from their different molecular structures discussed above.

Agar

In the agar gel crystallization studies, the effects on nucleation and on habit modification were not always consistent with equivalent observations in aqueous solutions. Femi-Oyewo and Spring (1994) have shown that polymeric additives, including agar, when present in the crystallizing solution, led to crystals with different properties, such as residual moisture content and crystal size, yield, and strength. To various degrees the polymers inhibited crystal growth and therefore presumably inhibited nucleation. The unpredictable combination of additive effects with agar-related effects probably led to this inconsistency.

Segregation

In the present systems, incorporation is not thermodynamically favoured, since the crystallization process tends to concentrate the additives in the solution phase as opposed to the crystalline phase (segregation coefficient <1.0), indicating that the crystallization process is also a purification process. There are examples for which the additives concentrate in the crystalline phase at the expense of the solution phase (segregation coefficient >1.0). Walton (1967) quoted examples of lanthanum oxalate at 0∞ which took up the europium, americium and yttrium trivalent cations with distribution coefficients (segregation coefficients uncorrected for ionic charge) of 4.8, 3.8 and 3.7 respectively. It should be mentioned that the uptake of an additive by a crystallizing or a precipitating solid may be controlled by kinetic factors as well as by thermodynamic conditions. In such cases additives cannot simply be removed and will remain to influence crystal nucleation and growth and thus the physico-chemical properties of the principal solute. It is anticipated that different solvents, solvent blends or even pH will influence the relative solubilities and hence the thermodynamics of uptake, and will thus permit the otherwise difficult purification (separation) from additives.

References

Bennema, P.; Hartman, P., *J Cryst. Growth*, **1980,** 49, 145.

Bravais, A.; *Etudes Crystallographiques,* Paris, **1913**

Chow, A. H. L.; Chow, P. K. K.; Zhongshan, W.; Grant, D. J. W., *Int. J. Pharm.*, **1985,** 24, 239.

Clydesdale, G,.; Docherty, R.; Roberts, K. J., HABIT: A Program for Predicting the Morphology of Molecular Crystals: *Comp. Phys. Comm.* **1991,** 64, 311.

Docherty R, Roberts K J and Dowty E, MORANG - *Comput. Phys. Comm.*, **1988,** 51, 423.

Donnay, J. D. H.; Harker, D., *Am . Mineral.*, **1937,** 22, 463.

Femi-Oyewo, M. N.; Spring, M. S., *Int. J. Pharm.*, **1994,** 112, 17.

Friedel, G., Bull. Soc. Fr. Mineral., **1907,** 30, 326.

Green, D. A.; Meenan, P., published in the proceedings of the 1st Powder Technology Forum, Denver **1994**

Green, D.A.; Meenan, P., to be published in the proceedings of the Third Crystal Growth of Organic Materials (CGOM-3), **1995**

Hartman, P.; Perdok, W. G., *Acta. Cryst.*, **1955** , 8, 49.

He, S.; Oddo, J. E.; Tomson, M. B., *J. Coll. Int. Sci.*, **1994,** 162, 297.

Hendriksen, B. A.; Grant, D. J. W., published in the proceedings of the 1st Powder Technology Forum, Denver **1994**

Molecular Simulations Inc., CERIUS^2: version 1.6, St John's Innovation Centre, Cambridge, U.K. **1994**

Momany, F. A.; Carruthers, L. M.; McGuire, R. F.; Scheraga, H. A., *J. Phys. Chem.*, **1974,** 78, 1579.

Stewart, J. J. P., MOPAC - A Semiempirical Molecular Orbital Program, *J. of Computer-Aided Molecular Design*, **1974,** 4, 1,2

Walton A G, *The Formation and Properties of Precipitates*, Interscience, John Wiley & Sons, New York, NY, **1967**, pp 79-112.

Weissbuch, I.; Shimon, L. J. W.; Landau, E. M.; Popovitz-Biro, R.; Berkovitch-Yellin, Z.; Addadi, L.; Lahav, M.; Leiserowitz, L., *Pure and Appl. Chem.*, **1986,** 58, 947.

Optical Resolution of DL-SCMC in a Cooled type Batch Operation

K. Toyokura, M. Kurotani and I. Hirasawa, Dept. of Applied Chemistry, Waseda Univ., 3-4-1 Okubo, Shinjuku-ku, Tokyo 169 Japan

Crystallization of L-SCMC (S-Carboxymethyl-L-cysteine) seeds was studied in a DL-SCMC supersaturated solution in a stirred batch crystallizer and the change of concentration of D- and L-SCMC was observed against the operation time. When initial supersatuation for D-SCMC was 6.15×10^{-2}[g/100ml], nucleation of D-SCMC was not recognized in the initial three hours' operation, and decreasing rate of the L-SCMC concentration in DL-SCMC solution was also observed. When the initial supersturation for D-SCMC was more than 7.36×10^{-2}[g/100ml], the decrease in the concentration of D-SCMC was observed in the solution. Operating points expressed by concentration of D- and L-SCMC were plotted in the D- and L-SCMC phase diagram, and the operational condition for the optical resolution of L-SCMC from DL-SCMC solution by crystallization was discussed.

Fundamental crystallization phenomena for the development of optical resolution have been studied in the aqueous solution of DL-SCMC(S-Carboxymethyl-DL-cysteine), and surface nucleation on growing seeds and crystal growth of suspended seeds have been studied [1]. And also the operational conditions for optical resolution are needed to be proposed, considering the phase diagram. In this study, the decrease in the concentration of L-SCMC in a cooled type batch crystallizer was studied in aqueous solution of DL-SCMC with 10wt% NaCl, and available operational range expressed by composition of D- and L-SCMC was discussed with operational supersaturation.

1.Experiments

DL-SCMC solution saturated at 308K with 10wt% NaCl was heated up to dissolve small amount of L-SCMC crystals. The solution prepared to have a desired composition, was fed into a 500ml vessel and agitated in the condition of 200rpm. Then the solution was cooled down at the constant rate of 0.2[K/min] to reach the desired supersaturated condition. After 2.5g of L-SCMC seeds of which size was from 300 to 550 μ m in the longitudinal length were fed to the supersaturated solution, and the slurry of suspended crystals was sampled every hour to be separated through a 0.45 μ m membrane filter. The concentration of the filtrate was observed by HPLC (High Performance Liquid Chromatography). Experimental conditions are shown on **table.1**.

2.Experimental Results and Discussions

2.1.Decrease in the concentration of L- and D-SCMC in a supersaturated solution

Observed concentration of L- and D-SCMC in the sampled solution was plotted against the operation time in **Fig.1**. From plots in **Fig.1-(a)** and **Fig.1-(b)**, the concentration of L-SCMC decreased rapidly at first and then fell down in almost constant rate till the concentration of it reached nearly 0.185[g/100ml]. In these operations, decrease of concentration of D-SCMC was hardly observed for initial three hours. On the initial operation period of run1 and run2 under initial supercooling of 10K for D-SCMC, the surface nucleation of D-SCMC on L-SCMC seeds was not considerd to be taken place and then nucleation of D-SCMC crystals occured after induction period of about three hours. Nuclei of D-SCMC on the surface of L-SCMC seeds was supposed to grow and the concentration of D-SCMC in the solution was also considered to decrease. These phenomena were in good accordance with data reported by

Table.1 Operational conditions

Run number	Operational temperature[K]	Excess amount of L-SCMC [g/100ml]	initial supersatuation D-SCMC *10^-2 [g/100ml]	initial supersatuation L-SCMC *10^-2 [g/100ml]	saturated concentration of D- or L-SCMC at operational temp. *10^-1 [g/100ml]
1	298	0.01	6.15	7.15	1.34
2	298	0.02	6.15	8.15	1.34
3	295.5	0.02	7.36	9.36	1.22
4	293	0.01	8.46	9.46	1.11
5	293	0.02	8.46	10.5	1.11

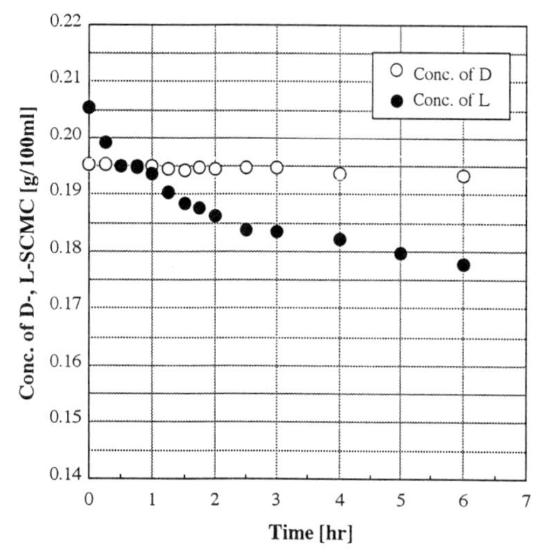

Fig.1-(a) Change of concentration of D- and L-SCMC against operation time (Run.1, ΔC_D=6.15×10^{-2}[g/100ml])

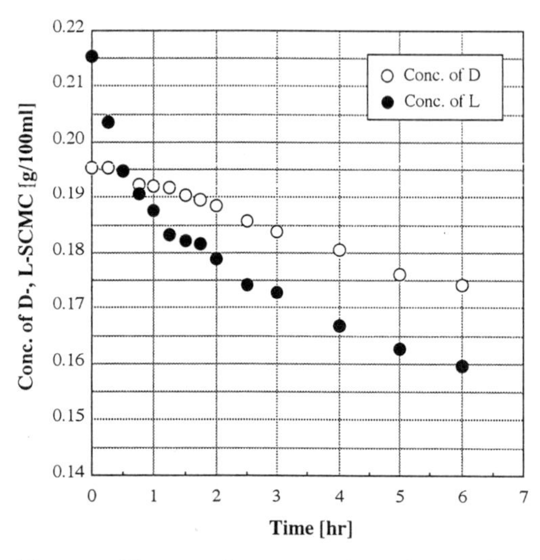

Fig.1-(c) Change of concentration of D- and L-SCMC against operation time (Run.3, ΔC_D=7.36×10^{-2}[g/100ml])

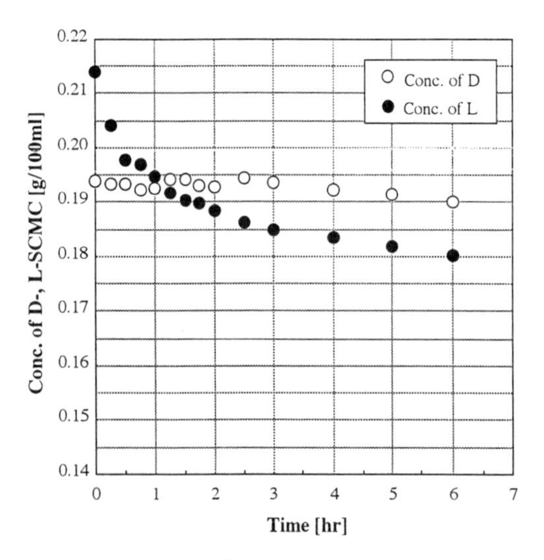

Fig.1-(b) Change of concentration of D- and L-SCMC against operation time (Run.2, ΔC_D=6.15×10^{-2}[g/100ml])

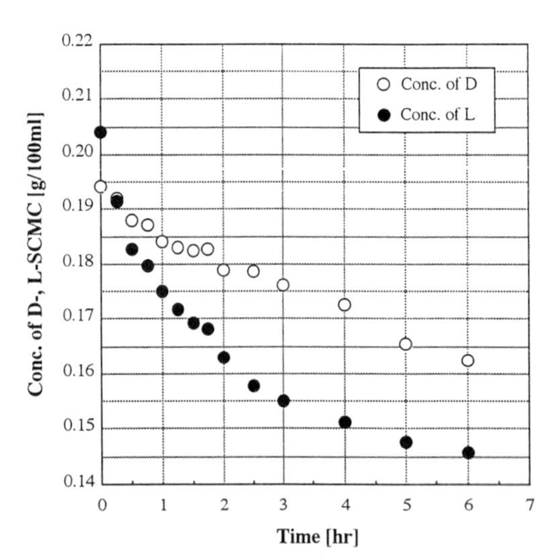

Fig.1-(d) Change of concentration of D- and L-SCMC against operation time (Run.4, ΔC_D=8.46×10^{-2}[g/100ml])

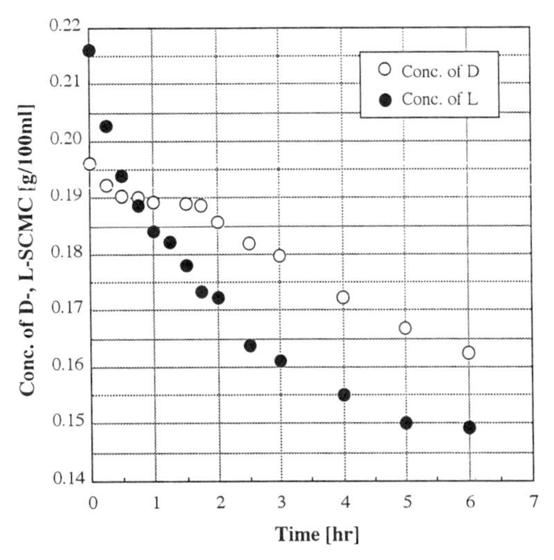

Fig.1-(e).Change of concentration of D- and L-SCMC against operation time
(Run.5, $\Delta C_D=8.46\times10^{-2}$[g/100ml])

Yokota et al.[2]. When the operation time was longer than three hours, the decreasing rate of L-SCMC concentration slowed down. When the phase diagram of D- and L-SCMC is considered, L-SCMC crystals do not grow in the range of the concentration of D-SCMC which is higher than the eutectic point. When the difference in the concentration between D-SCMC and L-SCMC became beyond 0.01[g/100ml], the decreasing rate of the concentration of L-SCMC seemed to be affected by that of D-SCMC. In the test of run3 under initial supercooling of 12.5K for D-SCMC, the concentration of L-SCMC in the solution decreased more rapid than that of run1 and run2 as shown in **Fig.1-(c)**. On the other hand, the concentration of D-SCMC gradually decreased without induction period for the decrease in the concentration of D-SCMC. In tests of run4 and run5 under initial supercooling of 15K for D-SCMC, the decreasing rates of concentration of D- and L-SCMC were much more rapid than those in Fig.1-(a) to (c), and the concentration of L-SCMC reached 0.145 to 0.150[g/100ml] after six hours' operation, as shown in **Fig.1-(d)** and **Fig.1-(e)**.

2.2.Crystallization of D- and L-SCMC

2.2.1.Crystallization of D-SCMC

Plots of the D-SCMC concentration in Fig.1 show that D-SCMC nucleation did not occur for three hours' operation under initial supercooling of 10K, but when supercooling was larger than 12.5K, D-SCMC nucleation occured without observation of induction period. Average decreasing rate of concentration of D-SCMC were approximately estimated from the slope of the concentration decrease in **Fig.2**. The decreasing rate of D-SCMC on initial supersaturation 6.15×10^{-2}[g/100ml] was observed after three hours' operation.

2.2.2.Crystal growth rate of suspended L-SCMC crystals

When concentration of the solution of L-SCMC is expressed by C_L, the rate of change of C_L is correlated with supersaturation ΔC_L in **Eq.1**.

$$\frac{dCL}{d\theta}=k\Delta CL^{q_L} \qquad (1)$$

Here θ, k and q_L are the operation time, crystal growth rate coefficient and power number of ΔC_L. In these tests, amount of seeds was almost same, and then k affected by seeds was considered to be almost same. When the decreasing rate of the solution was obtained from the slope of the plots in Fig.1, the decreasing rate of the solution was classified into three regions. The first region was characterized by that the amount of L-SCMC was more than that of D-SCMC in the solution and the decreasing rate in this region was the largest. It was reasonablly understood by data reported by Yokota et al.[3].

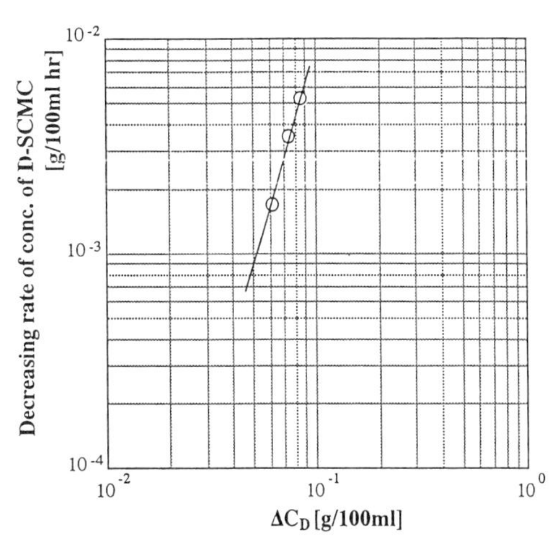

Fig.2.Decreasing rate of D-SCMC concentration

The second region was that the concentration of L-SCMC was nearly same to that of D-SCMC, and the decreasing rate of the concentration was assumed to be almost constant. L-SCMC decreasing rate of the concentration was correlated with supersaturation in **Fig.3**, and k and q_L became $3.47 \times 10^2 [g/100ml]^{1-q_L} [hr]^{-1}$ and 3.7. The third region was that change of concentration of L-SCMC in the solution was nearly same to that of D-SCMC, and in this region, the concentration of L-SCMC decreased, accompanied by the decrease in D-SCMC. The difference of the concentration between L-SCMC and D-SCMC was 0.01[g/100ml] and a little more, and the difference was affected by decreasing rate of the concentration.

2.3.Composition of grown parts of L-SCMC seeds

Composition of grown parts of L-SCMC seeds was estimated from the data shown in Fig.1. Typical examlples of compositon were plotted in **Fig.4-(a)** to **Fig.4-(c)**. When the initial supersaturation for D-SCMC was less than 6.15×10^{-2}[g/100ml], the optical purity of grown parts of L-SCMC was higher than 90%. But when supersaturation was larger, the optical purity of L-SCMC in the grown parts of seeds decreased with growth of seed crystals. They are supposed to be caused by the increase of surface area of D-SCMC against that of L-SCMC.

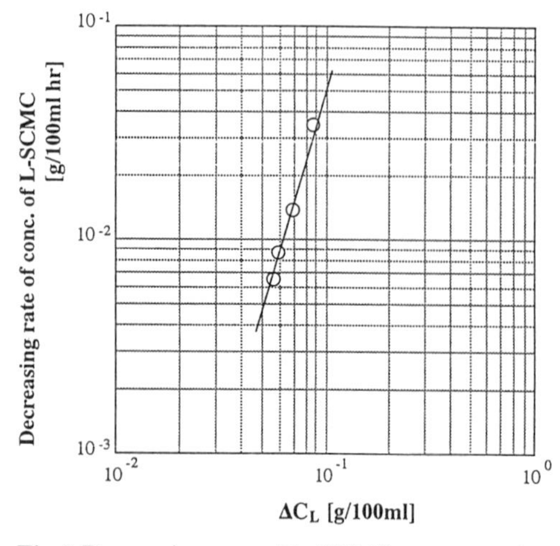

Fig.3 Decreasing rate of L-SCMC concentration

2.4.Operational condition for the optical resolution of DL-SCMC solution

Operating points expressed by combination of concentration of D-SCMC and L-SCMC in Fig.1 are plotted in **Fig.5**. In this Figure, saturated concentration of D- and L-SCMC are also plotted for operational temperture for tests' run1 to 5. From these plots, operating points are

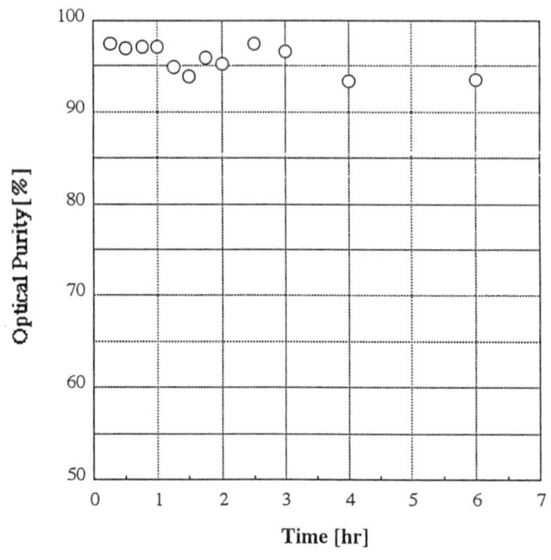

Fig.4-(a) Change of optical purity against operation time
(Run.1, $\Delta C_D = 6.15 \times 10^{-2}$ [100g/ml])

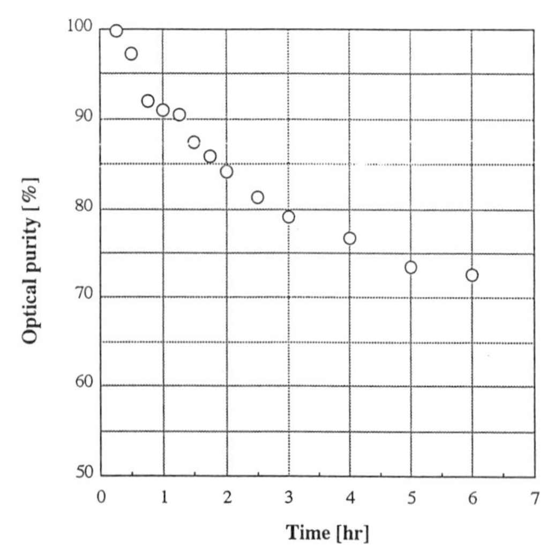

Fig.4-(b) Change of optical purity against operation time
(Run.3, $\Delta C_D = 7.36 \times 10^{-2}$ [g/100ml])

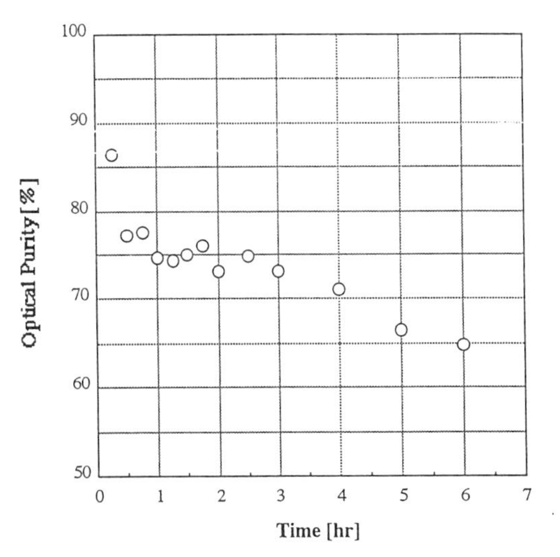

Fig.4-(c) Change of optical purity against operation time
(Run.4, ΔC_D=8.46×10^{-2} [g/100ml])

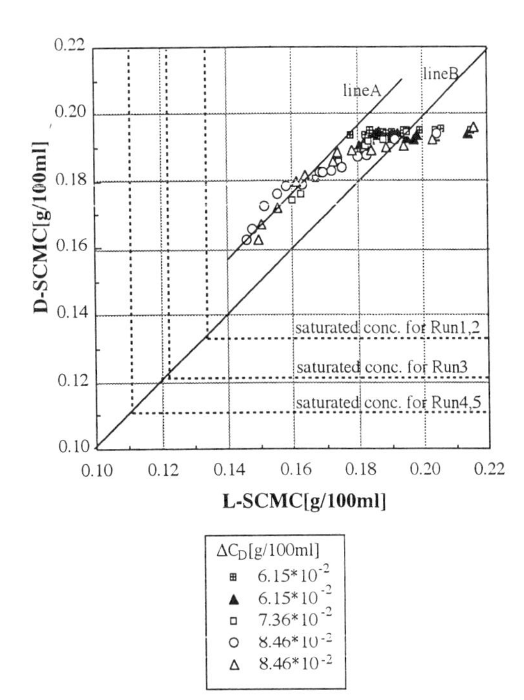

Fig.5 Operating points for growth of L-SCMC in DL-SCMC diagram

supposed to lie on the right side of the lineA. This line is the limit for precipitation zone of L-SCMC crystal and almost parallel to the lineB of the eutectic points decided by the temperature. When the phase diagram of D- and L-SCMC system is considered, the range between lineA and lineB was considered to be pseudo-precipitation zone for L-SCMC crystal and the lineA is supposed to be affected by the operational conditions which influence crystalization rate. From plots of operating points, data obtained under the initial supersaturations of 7.15 × 10^{-2}[g/100ml] for L-SCMC and 6.15 × 10^{-2}[g/100ml] for D-SCMC was parallel to the horizontal axis and almost pure L-SCMC crystals were expected to grow. But crystal growth was apt to stop, when the concentration approached the lineA. But when the initial supersaturations for D-SCMC was larger than 7.36 × 10^{-2}[g/100ml], the concentration of D-SCMC decreased with decrease in that of L-SCMC, and operating points moved from the right upper parts to the left lower parts in Fig.5. From plots of operating points, crystallization for the optical resolution of L-SCMC from DL-SCMC solution is recommened to be operated on right side far from the lineA under the initial supersaturation of D-SCMC less than 6.15×10^{-2}[g/100ml].

References

1)Yokota, M.; Toyokura, K.
Ind.Crystallization'93 vol.2 **1993**, 203-208.
2)Yokota, M.; Toyokura, K.
Chemical Engineering vol.39 **1994**, 69-76.
3)Yokota, M.; Toyokura, K.
unpublished

Industrial Crystallization of Organic Materials

Aspects of Melt Crystallization as a Separation Process

Joachim Ulrich, Universität Bremen, Verfahrenstechnik/ FB 4, Postfach 33 04 40, D-28334 Bremen, Germany

Solid layer melt crystallization processes - if analysed in detail - consist of more than a unidirectional crystallization on a cooled surface. Effects, unavoidable under industrial conditions, such as the inclusion of liquid melt droplets, the migration of these droplets and the occurance of cracks due to temperature gradients arise. These effects demonstrate the limitations for the highest achievable purity of a solid layer melt crystallization process. Consequences resulting from the effects mentioned are discussed.

Compared to other thermal unit operations melt crystallization can have the following advantages (Ulrich, 1993; Ulrich et al., 1995; Matsuoka, 1995):

- high selectivity
- low temperature level
- low energy consumption
- no solvent
- no waste waters.

The awareness of such advantages leads to a more frequent installation of melt crystallization units than in the past. In this contribution not crystallization itself, which can be conducted as solid layer or as suspension technique, but side effects always accompanying melt crystallization are to be considered; here especially those of the layer techniques will be discussed. As side effects could be regarded those that can be influenced by post-crystallization treatments like sweating and washing. Post-crystallization treatments, however, mainly reduce effects called forth by the way the crystallization is executed and in case excess purification beyond the achievements of the crystallization is wanted. But there are other, non-intentional effects, influencing the final purity of the remaining crystalline materials.

Fast crystallization leads to rough solid-liquid interfaces, causing problems with the solid-liquid separation. Here sweating and washing are good tools to improve the product purity. Even a certain amount of impurities integrated in solid layers due to some constitutional supercooling could be reduced by the post-crystallization treatments sweating or diffusion washing - a concentration-driven process purifying by liquid-liquid diffusion out of the pores of the layers -.

One of the unavoidable effects though during solid layer melt crystallization processes is the migration of liquid inclusions within the crystal layers due to the temperature gradient over the layer thickness. Another unavoidable effect is the appearance of cracks in the crystalline layers. The cracks are being filled with highly impure melt.

Structure of Crystalline Layers

As to be seen in Fig. 1 (Scholz et al., 1991), there are three main types of impurity sources in a layer, reducing the overall purity gained by crystallization.

The first is the incorporation of impurities due to nucleation. Nuclei usually spread like "stars" i.e. concentrically around the starting point on the cooled surface. Places for inclusions are the meeting points of those "stars" as well as the remaining space where no crystals cover the surface. Since the crystal layer insulates at the crystal-free places, good heat transfer is possible and, as a consequence, fast crystal growth takes place. Further acceleration leads to dendritic growth which again causes a lot of liquid inclusions. The impurity concentration of the latter is much higher than that of the feed. Crystallization in a dendritic structure only leads to a local separation and not to an overall separation as is aimed at by the melt crystallization process.

The second source of impurities in a layer is

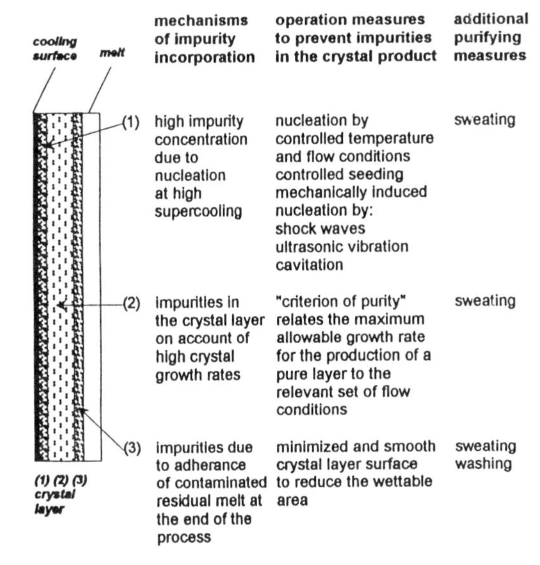

Fig. 1. Mechanism of impurity inclusions and measures for process improvement

a growth process being too fast. Extremely slow growth leads to very pure material. However, high yields are of interest in order to create small equipment dimensions or short retention times. Yet crystal growth rates being too fast evoke constitutional supercooling which in its turn, getting too extensive, leads to dendritic crystal growth dramatically reducing the purification efficiency, as described above. If the criterion of purity - a dimensionless value of the growth rate over the mass transfer coefficient - is taken into account (e.g. Wintermantel, 1972/ 1986; Scholz, 1993), a relatively good purification efficiency can be achieved by the crystallization step.

The third source of impurities is the highly contaminated residue adhering to the layer after the crystallization has been finished and the residue has been drained. This residue is not only effected by being the impurity-enriched boundary layer of the crystallization process, but it is over and above resulting from the most impure melt present in a process cycle since it remains at the end of crystallization when a large amount of pure material has been crystallized from the original feed stream.

In dependence on the growth rate the number of inclusions and the number of pores open to the liquid melt will be larger or smaller. The slower the growth rate is, the smoother is the surface of the layer, the fewer are the inclusions and pores and, naturally, the less contaminated is the liquid residue holdup after the draining.

Sweating

Sweating (e.g. Matsuoka et al., 1986) - a temperature-induced post-crystallization treatment of crystalline layers (Saxer et al., 1993) - reduces most of the adhering residue by making it drain. Furthermore, it drains a certain amount of the impure liquid out of the open pores by reducing its viscosity and by partly remelting its surroundings.

The more pores exist in a layer, the more impure liquid will be drained by a sweating step.

The amount of sweated liquid is about 10 % of that of a crystal layer (e.g. Özoguz, 1991 or Kuszlik, 1990 or Wangnick, 1994). A sweating carried out at a temperature level just below the melting point of the pure compound usually stops after a certain time (approx. 20 min.). Subsequently no more material will drain and a further raising of the temperature would lead to a total remelting or a slipping off of the crystal layer from the cooled surface.

Short retention times, low product losses and no phase-transition energies are the advantages of sweating compared to a further crystallization step.

The closed liquid inclusions cannot be affected by sweating. Due to capillary forces, a certain amount of liquid content originally staying in the open pores will also remain in the pores of the layer.

Washing

The post-crystallization treatment washing of the type of a rinsing replaces the adhering residue melt by the washing liquid (Poschmann et al., 1993). In many cases the washing liquid is a pure product, comparable to the reflux in distillation. It could, however, also be feed material. A rinsing process consumes no energy and takes an extremely short time (less than a minute) (Wangnick et al., 1994) with a yield of more than 90 to 95 %. Again no phase-transition energy is required.

A washing of the type of a diffusion washing cannot be executed purely by itself. It is automatically accompanied by a sweating and a washing of the type of a rinsing (Wangnick, 1994; Neumann, 1995). The sweating is unavoidable since the temperature of the melt must be superheated above equilibrium temperature, otherwise a crystallization of the washing liquid on top of the layer would take place immediately. During the diffusion washing - mainly a liquid-liquid diffusion - the highly impure melt moves from the pores into the washing liquid. Due to the remaining temperature gradient over the crystal layer the pores start to close. This could be explained referring to the migration of the liquid inclusions: Besides the temperature gradient the driving force is the concentration gradient between the liquid in the pores and the washing liquid (Neumann, 1995). Apart from the purification also a slight increase in the product mass can be registered.

The process step of a diffusion washing needs time - approx. 20 to 30 minutes -, contaminates the washing liquid with about 10 % of a crystalline layer and requires some energy to maintain the process conditions. Yet compared to a further crystallization step there are advantages concerning energy, time and yield.

Migration of Liquid Inclusions

Due to temperature gradients over the crystal layer thickness, liquid inclusions migrate (see the good literature surveys of e.g. Ulrich et al., 1992; Henning et al., 1995). This migration is mainly caused by diffusion processes within the inclusions and the fact that at the cold side pure material is crystallizing while at the warm side a melting is taking place to insure equilibrium conditions in the inclusions.

During the growth process of a layer the concentration profile over the crystal layer thickness changes: The impurities migrate towards the warmer, the crystal melt interface side. The stronger the thermal insulation properties of a substance are, the steeper the temperature gradient has to be to provide the uniform crystal growth rate, and the purer the created layer will be thereafter. In extreme cases the migration process is as fast as the crystal growth rate or even faster. This is one reason why certain organic acids, e.g. methacrylic acid, can be purified very well in a one-stage layer crystallizing process (Delannoy et al., 1993).

In the case of a diffusion washing the migration process would continue, so that further liquid inclusions could meet pores or end as pores and would finally completely disappear (Neumann, 1995).

Cracks of Crystalline Layers

Whenever there are temperature gradients there will always be changes in the volume of materials. If due to a reduction of temperature the crystalline coat on the outside of a tube is shrinking more than the cooled surface (steel), then strain on the crystal coat is the consequence. If such a crystalline layer cannot take the strain, cracks are unavoidable. The cracks can be observed (Stepanski et al., 1992) and are filling up with melt. The latter has its origin in the boundary layer and is therefore highly contaminated.

Exact parameters for the occurrence of cracks are not known yet. Also lacking is a criterion predicting the appearance of cracks or providing limits to be obeyed in order to avoid cracks in crystalline layers.

Since the structure, length, width and depth of cracks are unknown, the limitation in the greatest achievable purity concerning layer melt crystallization techniques is likewise unknown. (A current field of work of the author's research group).

A crack regarded as a pore is of course subject to the further purification steps of sweating and washing. Everything discussed so far about pores is principally true for cracks as well. However, since the dimensions of the cracks are not explored yet, it is not clear how effective post-crystrallization treatments are with regard to an improvement of the overall purity of a layer.

Summary

In layer melt crystallization there are three main places of impurity inclusions. These are the nucleation zone, the layer itself due to too-fast-growth conditions and the adhering residue melt. It is discussed how post treatments like sweating and the two parts of washing (rinsing and diffusion washing) could reduce the incorporated impurities. Furthermore, the two unavoidable processes taking place during the crystallization are introduced. One (the migration of inclusions) is working for an increase in purity and the other (the occurrence of cracks in the layers) against it. It has not been found out yet how much they affect the overall achievable purification effect of the process. So far they are not completely describable either, so that they cannot be modelled. In future, the effects of the cracks and the migration of the inclusions should be introduced based on a complete description of an overall model for the prediction of the limits of the achievable purification efficiency of a layer melt crystallization process.

Acknowledgement

The author wants to acknowledge the support of the Deutsche Forschungsgemeinschaft for a project which led to a number of results used in this paper.

References

Delannoy, C.; Ulrich, J.; Fauconet M. *Proc.12th Symposium on Industrial Crystallization*, Warsaw **1993**, Vol. 1, 1-049 - 1-054.

Henning, S.; Ulrich, J.; Niehörster, S. *Proc. 3rd International Workshop on Crystal Growth of Organic Materials - CGOM 3*, Washington **1995**.

Kuszlik, A-K. Ph.D. Thesis, University of Bremen, Bremen, **1990**.

Matsuoka, M. In *Science and Technology of Crystal Growth*; Eerden, J.P., von der; Bruinsma, O.S.L., Eds.; Kluwer academic pub., Dordrecht/ NL, **1995**.

Matsuoka, M.; Ohishi, M.; Kasama, Sh. *J. Chem. Eng. Japan* **1986**, *19*, 181-185

Neumann, M. Ph.D. Thesis, University of Bremen, Bremen, **1995**.

Özoguz, Y. Ph.D. Thesis, University of Bremen, Bremen, **1991**.

Poschmann, M.; Ulrich, J. *Proc.12th Symposium on Industrial Crystallization*, Warsaw **1993**, Vol. 1, 1-067 -1-072.

Saxer, K.; Stadler, R.; Ignjatovic, M. *Proc. 12th Symposium on Industrial Crystallization*, Warsaw **1993**, Vol. 1, 1-013 - 1-018.

Scholz, R.; Ulrich, J.; Genthner, K. In *BIWIC 1991* (Bremen Int. Workshop for Ind. Cryst.); Ulrich, J., Ed.; Verlag Mainz: Aachen, **1991**, pp 67-73.

Scholz, R. Ph.D. Thesis, University of Bremen, Bremen, **1993**.

Stepanski, M.; Fischer, O.; Stadler; R. *Symposium GVC Working Party Crystallization* **1992**.

Ulrich, J.; Scholz, R.; Wangnick; K. *J. Phys. D.: Applied Physics* **1993**, *26*, B156-B161.

Ulrich, J. In *Handbook of Industrial Crystallization*; Myerson, A.S., Ed.; Butterworth-Heinemann Series in Chemical Engineering, Boston/USA, **1993**, pp 151-164.

Ulrich, J.; Bierwirth, J. In *Science and Technology of Crystal Growth*; Eerden, J.P., von der; Bruinsma, O.S.L., Eds.; Kluwer academic pub., Dordrecht/ NL **1995**.

Wangnick, K. Ph.D. Thesis, University of Bremen, Bremen, **1994**.

Wangnick, K.; Ulrich, *J. Crystal Research and Technology, 29,* **1994**, *3*, 349-356.

Wintermantel, K. Ph.D. Thesis, TH Darmstadt, Darmstadt, **1972**.

Wintermantel, K. *Chem. Ing. Tech., 58,* **1986**, *6*, 498.

Crystal Growth Kinetics of Complex Organic Compounds

Donald J. Kirwan, Department of Chemical Engineering, U. Virginia, Charlottesville, VA 22903
Ilene B. Feins, Cutting Edge Computer Solutions, Alexandria, VA 22307
Amarjit J. Mahajan, Merck & Co., Inc., Rahway, N.J. 07065

The precipitation of complex organic compounds, such as pharmaceuticals, by non-solvent addition often occurs under high supersaturation conditions. We employed both batch precipitation and a rapid mixing device to measure the growth kinetics of glycine over a wide range of supersaturation. These observations combined with our previous studies on lovastatin and asparagine monohydrate provide a stringent test of theoretical crystal growth expressions. The growth rates in all of these systems were correlated by power law kinetics with orders greater than two which is not in agreement with existing theories.

The recovery, purification, and final product formulation of complex organic compounds in the pharmaceutical, food and speciality chemical industries are often carried out by precipitation using miscible non-solvent addition. These precipitations often occur under high supersaturation conditions. Little information on the nucleation and growth kinetics for the crystallization of such complex compounds in mixed solvent systems exists. Further, only very few experimental studies of crystallization at high supersaturation have been conducted and those have primarily focused on nucleation in inorganic salt systems (Sohnel and Mullin, 1982; Mersmann and Kind, 1988). We have initiated a program of investigation into the fundamentals of crystallization and precipitation of complex organic compounds, particularly pharmaceuticals.

Mahajan and Kirwan (1994) recently summarized available literature information for the nucleation and growth kinetics of organic species as well as reporting their kinetic measurements, over a wide range of supersaturation, for the precipitation of lovastatin and asparagine monohydrate. They employed a grid mixing device (Mahajan and Kirwan, 1993, 1994) which allowed measurement of nucleation and growth kinetics at much higher supersaturation than normally achieved. Extension of such measurements to high supersaturation ranges permits a more rigorous test of available theories for crystal growth as well as providing information relevant to precipitation processes at high supersaturation. Mahajan and Kirwan (1994) found that the kinetic order for crystal growth appeared to be greater than two for the two organic systems studied. Their review of the literature also suggested that the kinetic order of crystallization may be higher when a more extensive supersaturation range is examined and that complex organic compounds exhibit higher orders when crystallizing from solution. However, the limited data available prompted us to explore additional systems to further the understanding of nucleation and growth kinetics at high supersaturation as well as the crystallization behavior of complex organic solutes. Here, we discuss our recent results on the growth kinetics of glycine when precipitated from aqueous solution at 24°C with 2-propanol (final solvent composition 45.2 mass percent 2-propanol).

Materials and Methods

Low supersaturation experiments were carried out as batch precipitations in a water-jacketed, magnetically stirred beaker to

1054–7487/96/0116$12.00/0

which 2-propanol was added. Supersaturation ratios (C/C^*) in these experiments ranged from 1.03 to 1.20. The solubility of glycine (Sigma Chemical, free base, 99% purity) under the precipitation conditions was 27.9 g/kg solvent (Orella and Kirwan, 1991). The upper limit of supersaturation in these experiments was set so that no visible nucleation had occurred at the conclusion of the alcohol addition and that the development of the particle size distribution over time could be readily followed on-line with a particle sensor. The experiment ended when the particle density exceeded that allowable with the device to avoid coincidence. The slurry was continuously recirculated through the sensing tube of a HIAC Royco particle sizing device that operates on the principle of light blockage. The resulting signals corresponding to the number of particles found in each of 16 channels was directly sent to a computer where the particle size distribution was calculated for future analysis. Further details may be found in Feins (1995).

For measurements at high supersaturation ratios (1.60 - 3.1) a rapid, grid mixing device (Update Instruments, Madison, WI) was employed as described by Mahajan and Kirwan (1993, 1994). In this device, the solute-containing stream and the precipitant stream are contained in syringes that are then driven so that solutions flow across a series of fine grids and are mixed in times less than 3 milliseconds (Penefsky, 1988). The mixed solution then enters a holding tube having a precise but adjustable residence time determined by its length and the solution velocity. The slurry leaving the tube is "quenched" with a large volume (35:1) of saturated solution of the same solvent composition. The quenched sample is further diluted and the particle size distribution measured off-line. The upper limit to the supersaturation ratio studied was determined by the solubility of glycine in water and by the available ratio of syringe volumes for precipitant and solute solution. In both the low and high supersaturation experiments, the duration of the experiments was short enough so that the supersaturation ratio had not significantly changed due to loss of solute from solution by precipitation.

From a knowledge of the particle population density distribution function, n, at a series of residence times (**Fig 1a,b**), the variation with time of the crystal size, L, at a particular population density can be found (**Fig. 2a,b**). When the plots of L versus t at various n values are linear and parallel, then the slope may be taken as the particle linear growth rate, G. In using this approach care must be taken to only use values of n that are significant based on the reliability of the particle size counts and that are uninfluenced by agglomeration. In the studies with glycine agglomeration was found to exert an influence on the distributions at longer times. The illustrative results shown in **Fig. 1 and 2** also show the strong influence of supersaturation on glycine kinetics as the residence times where useful measurements could be made varied from seconds to minutes when the supersaturation ratio varied from 1.6 to 1.2.

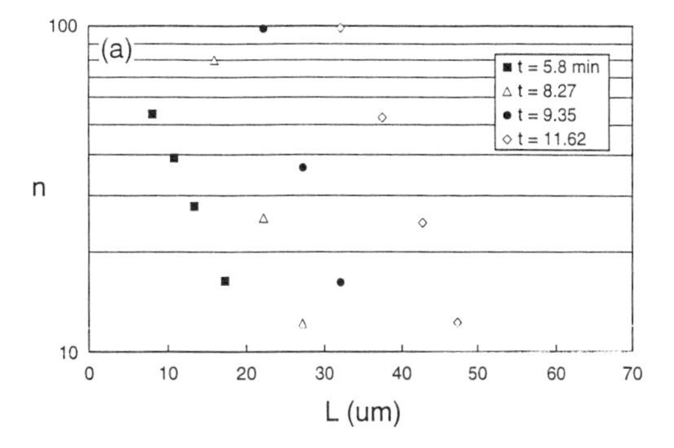

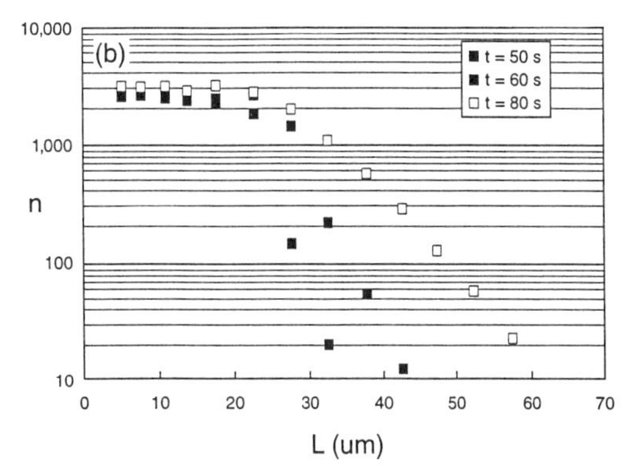

Fig. 1. Particle population density. **a.** $C/C^*=1.2$, **b.** $C/C^*=1.6$.

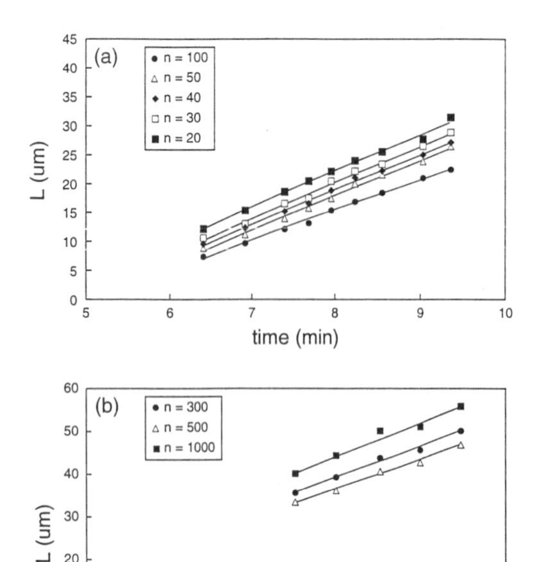

Fig. 2. Crystal size as a function of time. **a.** $C/C^*=1.2$, **b.** $C/C^*=1.6$.

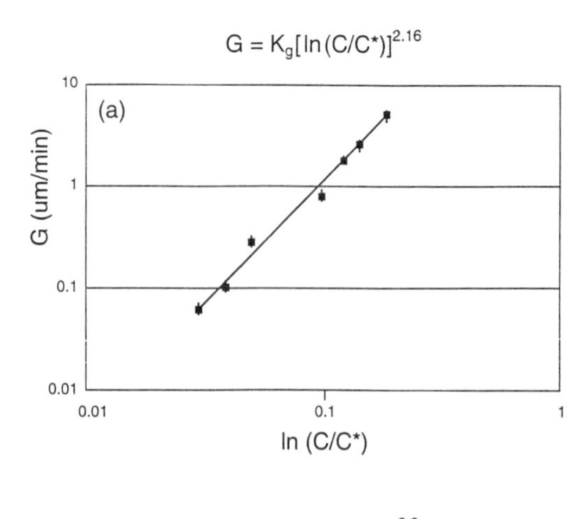

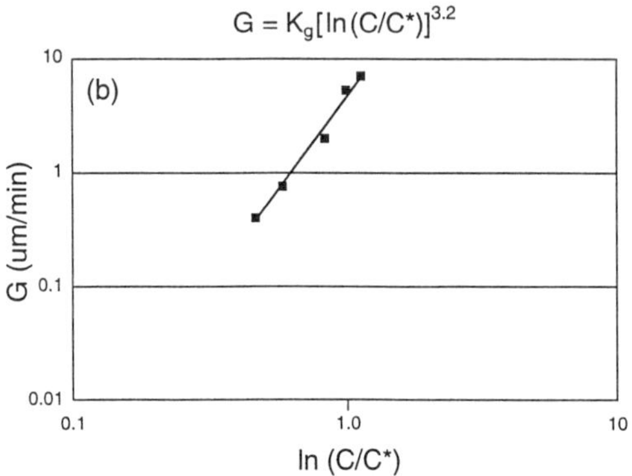

Fig. 3. Linear growth rate of glycine as a function of chemical potential driving force. **a.** batch experiments, low supersaturation range, **b.** rapid mixer, high supersaturation range.

Growth Kinetics of Glycine

Orella and Kirwan (1993) and Mahajan and Kirwan (1994) both employed $\ln(C/C^*)$ as the driving force for correlating crystal growth rate. For ideal solutions $\ln(C/C^*)$ is proportional to the chemical potential difference existing across the crystallizing interface. No data are available for the activity coefficients of glycine in 2-propanol-water systems to permit correction for non-ideality. Orella and Kirwan's use of the chemical potential driving force was based on its ability to bring together growth rate data in mixed solvent systems of different composition while Mahajan and Kirwan noted that the more commonly used relative supersaturation, $\sigma = C/C^* - 1$, is an approximation to $\ln(C/C^*)$ which is not valid at large values of C/C^*. The linear growth rate of glycine as a function of the chemical potential driving force is shown in **Fig. 3a** for the batch experiments at low supersaturation while the growth rate at high supersaturations is shown in **Fig. 3b.** Each experimental point represents the average of a minimum of three experiments at that condition. The standard deviation among runs is approximately the size of the data point shown. The kinetic order appears significantly different when the low and high supersaturation growth rates are correlated separately. However, **Fig. 4** shows the resulting correlation when all growth rate results are considered together. All of the observed growth rates can be correlated with an expression of the form, $[\ln(C/C^*)]^{2.5}$, to a high degree of confidence. This result reinforces the necessity to make growth rate measurements over as wide a range as possible. Unfortunately, as discussed later, most existing kinetic expressions have been developed from data obtained over quite small supersaturation ranges.

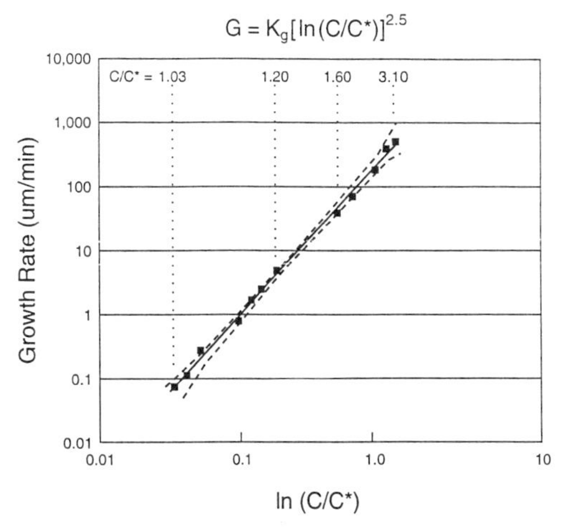

Fig. 4. Correlation of glycine linear growth rate over entire supersaturation range using chemical potential driving force. Dotted lines represent 99% confidence limits.

The growth rates of glycine single crystals growing from an aqueous solution were previously measured by Li and Rodriquez-Hornedo (1992) over a low supersaturation range. As can be seen in **Fig. 5** their growth rate measurements, although on single crystals and in an aqueous rather than a mixed alcohol-water solvent, are in quite good agreement with the measurements reported here. Further, although they reported power law kinetics with an order of 1.3-1.5, it can be seen that the data would, in fact, be compatible with our 2.5 kinetic order. Again, this observation illustrates the necessity of measurement over a wide supersaturation range to establish reliable kinetic expressions.

The glycine growth kinetics can also be correlated to a power law in σ with a kinetic order of 2.15 (**Fig. 6**). Because of the uncertainty of the data and the mathematical relationship between $\ln(C/C^*)$ and $C/C^* - 1$, it would be necessary to have experimentally measured growth rates for supersaturation ratios as high as 5 to clearly distinguish between these two models (Feins, 1995). Experimental limitations prevented us from going beyond about 3.1. Mahajan and Kirwan (1994) in their studies of lovastatin and asparagine monohydrate were able to

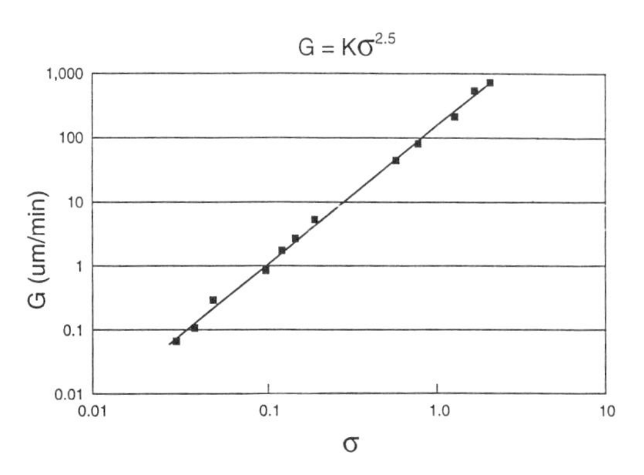

Fig. 6. Correlation of growth kinetics of glycine using a relative supersaturation driving force.

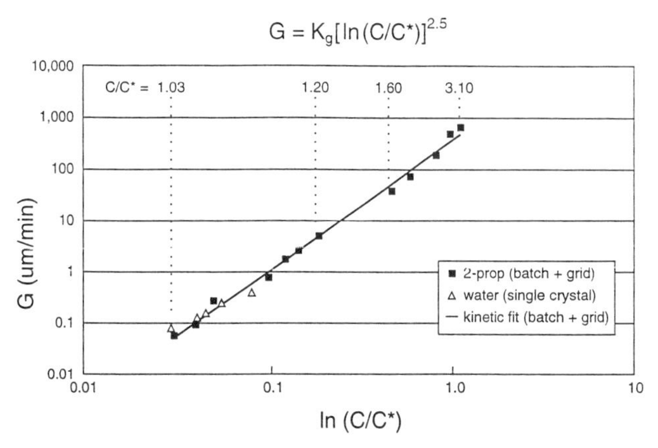

Fig. 5. Comparison of measured glycine kinetics in an aqueous-2-propanol solvent with the results of Li and Rodriquez-Hornedo (1992) in an aqueous solvent.

attain higher supersaturations and showed that the use of the chemical potential driving force was able to correlate their results over their entire superstaturation range while the use of σ was not.

Conventional theories for crystal growth such as the BCF, continuous, and mono and polynuclear growth (O'Hara and Reid, 1973; Mullin, 1993) are not compatible with the observed kinetic order being greater than 2. Further, the BCF and mono and polynuclear models all suggest change in kinetic order to a smaller value as the supersaturation increases. Mahajan and Kirwan (1994) were also able to fit their observed power law kinetics for lovastatin and asparagine monohydrate, which also had growth orders than 2, to Nielsen's surface integration model (Nielsen, 1984) with some degree of reliability.

$$G = K[\sigma+1]^{7/6}\sigma^{2/3}[\ln(\sigma+1)]^{1/6}$$

$$\cdot\exp[-A/\ln(\sigma+1)]$$

The glycine kinetics did not obey this model, however. At present, there does not appear to be a model that can adequately describe growth kinetics over a large supersaturation range.

Comparison to the Growth Kinetics of Other Organic Solutes.

Mahajan and Kirwan (1994) compiled data on the growth kinetics of a number of biochemical solutes and suggested that when a larger supersaturation range was explored the growth kinetics appear to be represented by a kinetic order greater than 2.

Table 1. Growth Kinetics[1] of Organic Solutes

Solute	C/C^* Range	m	Ref.	Conditions
Asparagine Monohydrate	1.05-1.5	2.8	Orella & Kirwan 1993	Cooling & Alcohol Ppt.
Asparagine Monohydrate	1.2-4.1	2.9	Mahajan & Kirwan 1994	Alcohol ppt.
Fumaric Acid	1.06-1.8	3.8	Yamamoto et al. 1989	Cooling
Glutamic Acid	1.5-2.5	7.2	Yamamoto 1989	Cooling
Glycine	1.03-3.1	2.5	This work	Alcohol ppt.
Lovastatin	1.25-8.7	6.7	Mahajan & Kirwan 1994	Ppt. by water
Monsodium Urate	2.9-3.6	2.0-2.8	Burt & Dutt 1989	Cooling
Phenytoin	1.0-6.0	3.3-3.7	Zipp & Rodriquez-Hornedo	pH ppt.

1. Literature growth rates were correlated according to the expression, $G=K\ [\ln(C/C^*)]$

Our results for glycine growth rates are in agreement with these earlier observations. Table 1 summarizes the growth conditions for the those organic solutes whose kinetic orders were observed to be greater than 2. Notice that these observations include crystallization or precipitation induced by cooling, by pH change, and by the addition of a miscible, non-solvent; and they include studies on both single crystals and multiparticle systems. Mahajan and Kirwan (1994) also recorded a number of systems with kinetic orders between 1 and 2, but none of these systems had been studied at larger supersaturations. In the light of our earlier discussion on the difficulty of assessing the true kinetic order when only a small supersaturation is observed, it is interesting to speculate as to what might be the proper growth expressions when larger supersaturation ranges are considered.

The growth kinetics of many salts from aqueous solution have been measured both because of their technological importance but also because they are convenient models systems to test crystal growth theories. It would be of interest, therefore, to examine the crystallization kinetics of such inorganic solutes under higher supersaturation conditions to compare to the results for organic systems that are being developed. The grid mixing device should be quite appropriate for such studies.

Conclusions

1. The growth kinetics of glycine precipitated by 2-propanol over a wide supersaturation range were correlated with a power law model having a kinetic order of 2.5 when using a chemical potential driving force.

2. The unusually high kinetic orders for glycine and other organic solutes previously measured calls into question the applicability of conventional crystal growth theories for these systems.

Symbols

C	Concentration of solute, g / k g solvent
C^*	Solubility of solute, g/kg solvent
G	Crystal linear growth rate, μm/min
K	Growth coefficient, μm/min
L	Characteristic crystal size, μm
m	Kinetic order
n	Population density function, # crystals/ml-μm
t	Time, min.
σ	Relative supersaturation

References

Burt, H.M., Dutt, Y.C., *J. Cryst. Growth*, 1989, *94*, 15-20.

Durbin, S.D., Feher, *J. Cryst. Growth*, 1987, *76*, 583-586.

Feins, I.B., "Crystallization Kinetics of Glycine," M.S. Thesis, University of Virginia, Virginia, 1995.

Li, L., Rodriquez-Hornedo, N., *J. Cryst. Growth*, 1992, *121*, 33-37.

Mahajan, A.J., Kirwan, D.J., *J. Phys. D (Appl. Phys.)*, 1993, *26*, B176-B180.

Mahajan, A.J., Kirwan, D.J., *J.* Cryst. *Growth*, 1994, *144*, 281-290.

Mersmann, A., Kind, M., *Chem. Engr. and Tech.*, 1988, *11*, 264-276.

Mullin, J. *Crystallization*, 3rd Ed., Butterworth-Heinemann, Oxford, U.K. 1993, pp. 202-219.

Nielsen, A.E., *J. Cryst. Growth*, 1984, *67*, 289-293.

O'Hara, M., Reid, R.C., *Modeling Crystal Growth Rates from Solution*, Prentice-Hall, Englewood Cliffs, N.J. 1973.

Orella, C.J., Kirwan, D.J., *Ind. Engr. Chem. Res.* 1991, *29*, 1040-1045.

Orella, C.J., Kirwan, D.J. in *Handbook of Industrial Crystallization;* Myerson, A.S., Ed.; Butterworth-Heinemann, Stoneham, MA, 1993, p. 225.

Penefsky, H.S., *J. Biol. Chem.*, 1988, *263*, 6020-6022.

Sohnel, O., Mullin, J., *J. Cryst. Growth*, 1982, *60*, 239-242.

Yamamoto, H., Sudo, S., Yano, M., Haranao Y., in *Industrial Crystallization '87*, Nyvlt, J., Zacek, S., Eds., Elsevier, Amsterdam, 1989, p. 403.

Zipp, G.L., Rodriquez-Hornedo, N., *J. Cryst. Growth*, 1992, *123*, 247-251.

Precipitation of EDTA from Solution: Effect of Process Conditions on Product Properties

R. Guardani, Department of Chemical Engineering, Polytechnic School, University of São Paulo, Caixa Postal 61548, CEP 05424-970 - São Paulo - SP, Brazil, E-mail: guardani@usp.br

E. P. Fariello, Department of Chemical Engineering, Polytechnic School, University of São Paulo, IPT - Institute of Technological Research, São Paulo - SP, Brazil

A significant amount of ethylenediaminetetraacetic acid (EDTA) is used as a chellating agent in processes for plating of surfaces by metals. In industrial metal plating units, EDTA is recovered by precipitation from spent process solutions. This paper presents the results of laboratory experiments on the effects of operating variables on some of the properties of precipitated EDTA from plating solutions. The results can contribute to the improvement in efficiency of the precipitation process and to the minimization of EDTA in industrial waste solutions.

The range of industrial applications of ethylenediaminetetraacetic acid (EDTA) is considerably wide, from the food and pharmaceutical industries to some processes of the metallurgical industry, such as the plating of surfaces with metals. In this case, EDTA is used as the chellating agent in order to prevent the precipitation of metallic hydroxides under high pH conditions. This study examines EDTA precipitation from solutions used in electroless copper deposition in the production of printed circuit boards.

The global reaction consists of the reduction of copper from a solution containing copper sulfate by formaldehyde, as shown in Eqs. 1 and 2:

$$Cu^{2+} + EDTA^{4-} \leftrightarrow CuEDTA^{2-} \qquad (1)$$

$$CuEDTA^{2-} + 2HCHO + 4OH^{-} \leftrightarrow Cu + 2H_2O + 2HCOO^{-} + EDTA^{4-} \qquad (2)$$

The reaction is carried out at pH values of about 11 (by the use of NaOH) and temperatures of about 80°C. In order to keep the concentration of the solution constant, in industrial processes the reagents (copper sulfate, sodium hydroxide and formaldehyde) are continuously fed to the reactor, while a fraction is withdrawn for treatment. This consists first of removing most of the dissolved copper by adding excess formaldehyde followed by the recovery of EDTA.

A review on processes for the elimination of EDTA-containing effluent solutions from metal plating processes can be found in Erlmann *et al.* (1990). Most of these processes are based on the precipitation route, which is described by a number of published papers, including different equipment configurations (for instance, Yoshida and Sato, 1987; Yoshida and Sato, 1988). These processes consist of lowering the pH with the addition of strong acids (such as sulfuric or hydrochloric acid) until EDTA crystals are formed. This occurs at pH values of less than about 3.

Many industrial processes are carried out batchwise. The spent plating solution from the copper recovery step is fed to a stirred tank to which a strong acid solution is added. The pH of the precipitating medium is kept constant during the precipitation time.

In some units the process is carried out at ambient temperatures, although there is an increase in temperature due to acid dissolution. In other cases, the spent solution is fed at about 40 to 50°C, as a consequence of previous operations, carried out at higher temperatures.

In order to obtain information on the process variables which can affect the efficiency of the EDTA recovery process by precipitation, a study was made on the effects of temperature and precipitation time on some characteristics of the crystals formed. The study was carried out in laboratory scale, using a spent solution from an industrial electroless copper plating plant.

Experimental

The precipitation tests were carried out with the use of a spent solution from electroless copper plating, after recovery of copper. The solution contained 24.8% EDTA, 0.4% HCHO (mass percentages) and less than 1 mg per liter of each of the following elements: Cu, Cr, Fe and Ni. The pH value was 11.7 (due to the presence of NaOH). Concentrated (98%) sulfuric acid was the acidulation agent used.

The tests were carried out in a cylindrical stirred tank made of glass. The tank had a volume of 1 liter and was equipped with a jacket, which was connected to a thermostatic bath. Sulfuric acid was continuously added from a buret, with manual control of flow rate.

The procedure in each precipitation test consisted of feeding the tank with 0.9 liter of the solution and, after stabilization of the temperature, addition of sulfuric acid at an adequate flow rate, in order to achieve a rate of decrease of the pH of about 0.25/minute. Both temperature and pH were continuously monitored during the experiment. As the first crystals were formed (detected by visual observation of turbidity), counting of the precipitation time was started and the addition of acid stopped. Small amounts of acid were intermittently fed along the experiment in order to keep the pH constant. The time under constant pH, defined according to informations from industrial practice, ranged from 1 to 2 hours.

After completion of the precipitation the suspension was filtered and the crystals were left to dry at 25°C until constant mass was reached. After measuring the mass of crystals obtained in each test, samples were collected for measurement of the particle size distribution (by laser diffraction) and for examinations by optical microscope and X-ray diffraction.

Results and discussion

The conditions adopted in the experiments, as well as the characteristics of the precipitated EDTA are presented in Table 1.

The temperature rise due to the addition of sulfuric acid was less than about 3°C in all the experiments.

Table 1. Listing of the EDTA precipitation tests.

Test conditions				Products		
Test No.	Temperature (°C)	Precipitation pH	Time (minutes)	mass (g)	vol. average diameter(μm)	phases present [1]
1	16	1.7	64	17.2	100	α, β
2	16	1.7	75	17.5	86	α, β
3	16	1.5	82	20.3	132	α, β
4	20	1.3	73	17.9	97	α, β
5	21	1.4	80	17.5	81	α, β
6	25	1.4	120	15.6	159	α, β
7	34	1.6	64	16.5	112	α, β
8	41	1.6	1.5	6.8	100	β
9	41	1.6	3	12.6	122	β
10	41	1.4	6	14.2	168	β
11	41	1.7	10	14.0	166	β
12	41	1.4	60	16.5	117	β
13	40	1.6	60	15.9	158	β
14	41	1.6	73	16.4	162	β

[1] - crystalline phases identified by X-ray diffraction: α-EDTA and β-EDTA

The beginning of crystal formation in the tank was observed at pH values between 1.3 and 1.6. This variation was due to the experimental procedure and did not seem to depend on other conditions. The pH during precipitation, however, affects the rate of crystal growth, as can be seen by comparing the mass of crystals collected in tests 2 and 3, tests 4 and 5, or tests 12 and 13.

The precipitation time is apparently long enough to reach completion of the precipitation process. Additional amounts of EDTA can be precipitated by further lowering the pH of the system.

The temperature affects the particle size distribution of the precipitated crystals. Samples collected from experiments made at about 40°C contained much larger crystals than at lower temperatures. This behavior could be also observed by comparing the results of experiments at about 40°C and shorter times.

Differences were also observed in terms of the crystal structure of samples produced at different temperatures. Thus, samples precipitated at about 40°C consisted mainly of β-EDTA, while samples obtained at 34°C or less resulted in X-ray diffraction spectra that correspond to a mixture of α- and β-EDTA.

The existence of two different crystal forms of EDTA was first described by LeBlanc and Spell (1960), who described some of the properties (including X-ray spectra), as well as the conditions for obtaining each form by acid addition to EDTA solutions. According to these authors, β-EDTA is obtained at temperatures near the boiling point of water, while α-EDTA precipitates at lower temperatures. However, no study was made on the transition temperature. In the present work, samples of precipitated EDTA whose X-ray diffraction spectra indicate the presence of both forms were obtained in the tests made at temperatures up to 34°C.

The shape of particles composed by α- and β-EDTA can be compared with the ones containing the β form by the images of samples precipitated at different temperatures. Photographs from EDTA obtained at 16°C (test 3) and at 41°C (test 11) are shown respectively in Figs. 1 and 2. The sample obtained at lower temperature (α- and β-EDTA) consists of agglomerated particles with no regular shape, whereas the one obtained at higher temperature consists of larger individual prisms, together with groups of prisms with the shape of stars. The individual prisms can result from particle breakage due to shocks with the stirrer's

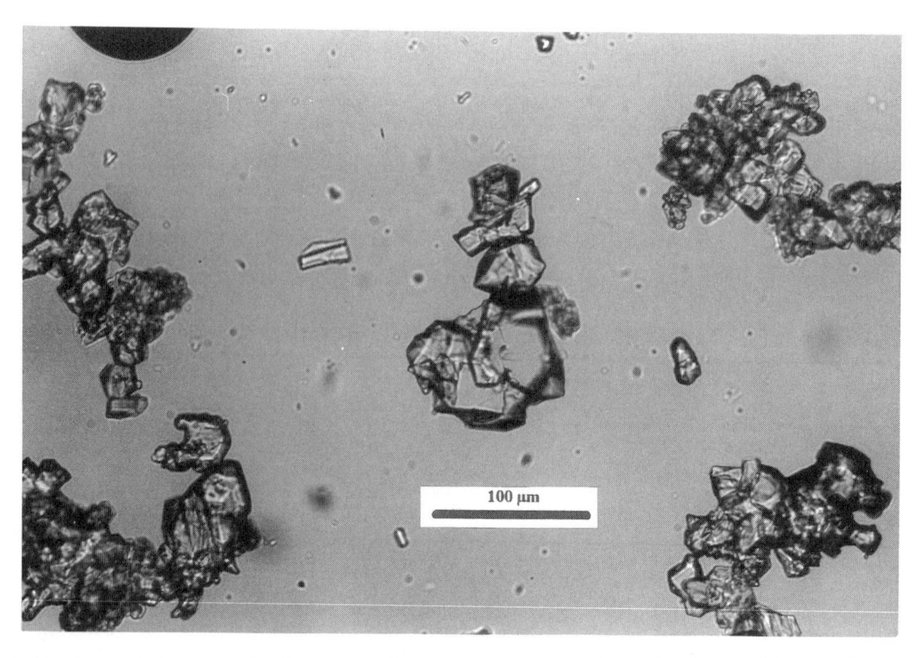

Fig. 1. Photograph by optical microscope from EDTA crystals formed at 16°C (test 3), showing aglomerated particles consisting of α- and β-EDTA.

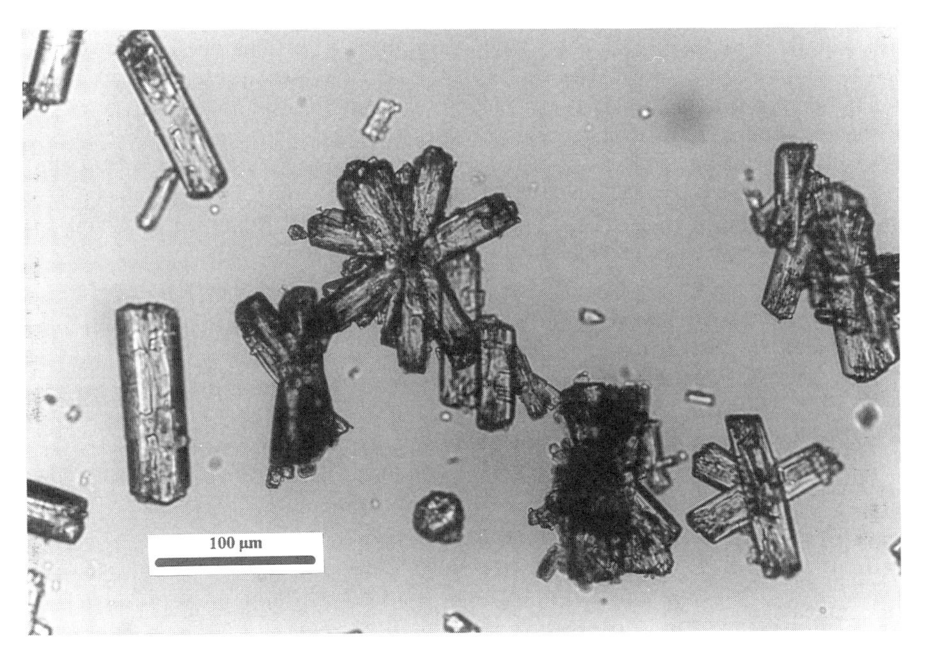

Fig. 2. Photograph by optical microscope from EDTA crystals formed at 41°C (test 11), showing particles consisting of β-EDTA.

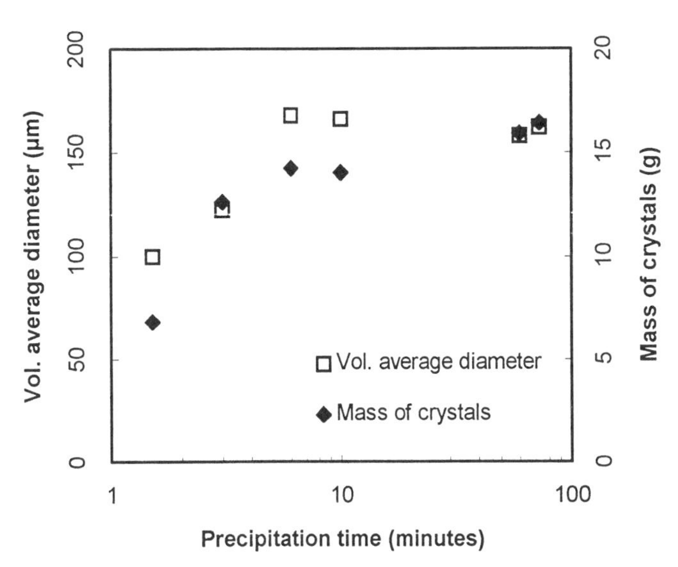

Fig. 3. Evolution of the volume average particle diameter and of the collected mass of crystals along the precipitation time for tests carried out at 40°C.

impeller. This is indicated in Fig. 3, which shows the evolution of the volume average particle diameter and the collected mass of crystals along the precipitation time for tests carried out at about 40°C. For times longer than about 6 minutes the average diameter remained practically constant, while the mass of crystals increased until about 60 minutes.

Under the experimental conditions adopted, further precipitation of EDTA took place only when the pH of the system was lowered, through the addition of sulfuric acid (Table 1, test 12).

Conclusions

Product properties such as crystal form and particle shape and size of precipitated EDTA from metal plating solutions depend on process variables such as temperature and intensity of agitation.

The temperature of the precipitation process should be kept above 40°C, so that larger crystals composed by β-EDTA are produced.

Since larger particles are more easily separated from mother solutions than smaller ones, care should be taken in order to control particle breakage.

The rate of crystal growth is relatively high under the conditions adopted in this study. Thus, long residence times are not necessary in industrial processes.

Acknowledgments

The authors wish to thank the support from E. B. Quitete and C. A. Silva in optical microscope examinations and particle size analyses.

The experimental work was carried out at IPT, Institute of Technological Research, São Paulo, SP, Brazil.

Special thanks are directed to FAPESP, São Paulo Foundation for the Support of Research, for the financial support.

References

Erlmann, W.; Foery, M.; Jantschke, H. *Galvanotechnik* **1990**, *81*(4), 1249-58.

LeBlanc, R.B.; Spell, H.L. *J. Phys. Chem.* **1960**, *64*, 949.

Yoshida, T.; Sato, H. Japanese Patent 63,190,694, 1987.

Yoshida, T.; Sato, H. Japanese Patent 63,297,348, 1988.

Emulsion Solidification: Purification and Crystallisation

R. J. Davey, J. Garside and A. M. Hilton, Dept. Chemical Engineering, UMIST, P.O. Box 88, Manchester, M60 1QD, UK.
J. W. Morrison, ZENECA Fine Chemicals Manufacturing Organisation, Leeds Road, Huddersfield, West Yorkshire, HD2 1SF, UK.

In a model system, based on emulsification, meta chloronitrobenzene, (m CNB), has been crystallised from two component mixtures with its para isomer. As a method of purification this has been found to offer two significant attributes; firstly, it is possible to by-pass the eutectic and so to achieve purification on crystallisation below the eutectic temperature and secondly high degrees of purification can be achieved in a single stage. A mechanism for crystal growth is suggested based on transfer of organic material across the aqueous phase. Further investigation into the use of amphiphillic molecules as emulsifying agents indicates the choice of emulsifier can influence the nucleation of m CNB from a pure m CNB o/w emulsion and may enable control of final crystal size and morphology.

The process of emulsion crystallisation is being investigated in order to optimise crystal purity, yield and physical form. The use of melt crystallisation as a means of purifying organic materials is well known (Sloan, 1988). It is increasingly used by the chemical industry as a result of the availability of new process technology (Mullin, 1993) and commercial and legislative pressures demanding the supply of materials with precisely defined specifications. Currently, progressive freezing techniques are employed in which solidification proceeds on a cooled surface and purity is achieved by a combination of multiple stages and partial melting (Fischer, 1984). As an alternative to this technology we have been examining the process of emulsion crystallisation (Cordiez, 1982) which, since it requires no special equipment or solvent, offers the potential of a low capital and operating cost alternative to melt purification.

The use of emulsified systems as a technique for studying homogeneous nucleation is well known since the pioneering work of Turnbull, (1950) on metals and paraffins. Dispersion of a melt into discrete small volumes allows heteronuclei to be isolated in certain drops and homogeneous nucleation to occur in the remainder. Generally it has been found that, as expected, the smaller the drop size the more difficult nucleation becomes (Skoda, 1963).

By using tailored emulsifying agents, having similar structures to CNB, it is hoped that crystal

1054–7487/96/0127$12.00/0

formation, in pure m CNB emulsions, can be promoted and contained preferably inside the emulsion droplet and in this way final crystal size may be controlled.

Experimental

Purification

For this study we have chosen to crystallise m-chloronitrobenzene, (m CNB), from two component mixtures with its para isomer, (p CNB). The system exhibits simple eutectic behaviour. In the experiments reported here, m : p mixtures of 90 : 10 and 75 : 25 have been emulsified above their melting points, at a weight fraction of 2 % in water using the sulphated alkylethoxolate surfactant, Tensagex, (supplied by Hickson-Manro). At an overall concentration of 0.2 wt. % Tensagex was found to have no effect on the phase diagram as measured by differential scanning calorimetry, (Mettler DSC 30) and stable oil - in - water emulsions were prepared with drop sizes in the range 5 - 10 μm. These were then crystallised by cooling at fixed rates. As expected, at these small drop sizes, the emulsions showed significant undercooling, (typically 30 - 40 °C). In order to over come this problem and perform crystallisations at fixed temperatures a simple seeding technique was used in which the emulsion was cooled to the desired crystallisation temperature and 1 mg/ml of pure m CNB crystals added. The subsequent progress of crystallisation in terms of purity, yield and crystal form was monitored by sampling and use of GC, (Perkin-Elmer 8130, sensitive down to 0.02 % m CNB) and optical microscopy, (Polyvar Met). An emulsion of composition 60.9 % m CNB was also crystallised, (with m CNB seeds), at 15 °C where we would expect a mixture of pure p CNB and eutectic to crystallise, yielding no overall purification. Crystallisation rates and overall purities and yields have also been measured at 28 °C for 90 : 10 m : p CNB emulsions with differing droplet diameter, organic weight fraction and emulsifier content.

Crystallisation

Amphiphillic molecules were used as emulsifiers for pure m CNB in water; specifically chloro and nitro substituted cinnamic acids, (Aldrich Chemical Co.) and alkylphenyl ethoxylates, (Aldrich Chemical Co.). They were dissolved in water at 65 °C at a fixed overall weight % of 0.2; m CNB was present at 2 weight % , (it was first necessary to form the sodium salt for the cinnamic acids). The two phase system was homogenised at 10,000 r.p.m. for 10 mins., resulting in droplet diameters which were independent of emulsifier type and in the range 5 - 10 μm. The emulsions were cooled at a rate of 1°C per min. until crystallisation occured, (recorded by an insitu Brinkman PC 800 Colorimeter). Crystals were viewed under the optical microscope. The degree of activity of the amphiphillic molecules and predicted molecular packings at the m CNB/aqueous interface were obtained by interfacial tension measurements, (Kruss processor tensiometer K12), and the use of the molecular modelling software package

CERIUS2, (Molecular Simulations), respectively.

Results

Purification

Table 1 shows the product purity after 90 mins. for two emulsion compositions crystallised at various temperatures compared to equivalent material crystallised in a single stage, (with no partial melting), on a cold surface. In terms of purity, a single step emulsion crystallisation offers considerable improvement over its progressive freezing counterpart. Previous studies on the melt and solution crystallisation of m CNB have shown that its para isomer is a powerful crystallisation inhibitor, (Chen, 1994) and that its presence in the crystallising environment induces a profound polar morphology in m CNB crystals. Observations of

Table 1. The Effect of Crystallisation Temperature on Final Crystal Purity

Original Droplet Composition m : p CNB	Liquidus Temperature °C	Crystallisation Temperature °C	Final Crystal Composition (After 90 mins.) % m CNB
90 : 10	41	30	99.7
		28	99.3
		28#	90.7
		24	98.2
		22	93.6
		22#	90.4
75 : 25	34	25	99.7
		23	98.4

#Equivalent Data Measured in Single Stage Progressive Freezing onto a Cooled Surface.

the physical form of the crystals here proved very surprising for, as Fig. 1 illustrates, these were always single crystals, at least an order of magnitude larger than the drop size and having polar morphologies typical of crystals grown from melts or solutions in the presence of p CNB.

As crystallisation proceeded the emulsion droplet composition was measured as a function of time, (total crystallisation times of 90 mins.). Liquid phase compositions of emulsions having initial m : p compositions of 90 : 10 crystallised at temperatures of 30, 28, 24 and 22 °C are shown in Fig. 2. The eutectic composition and

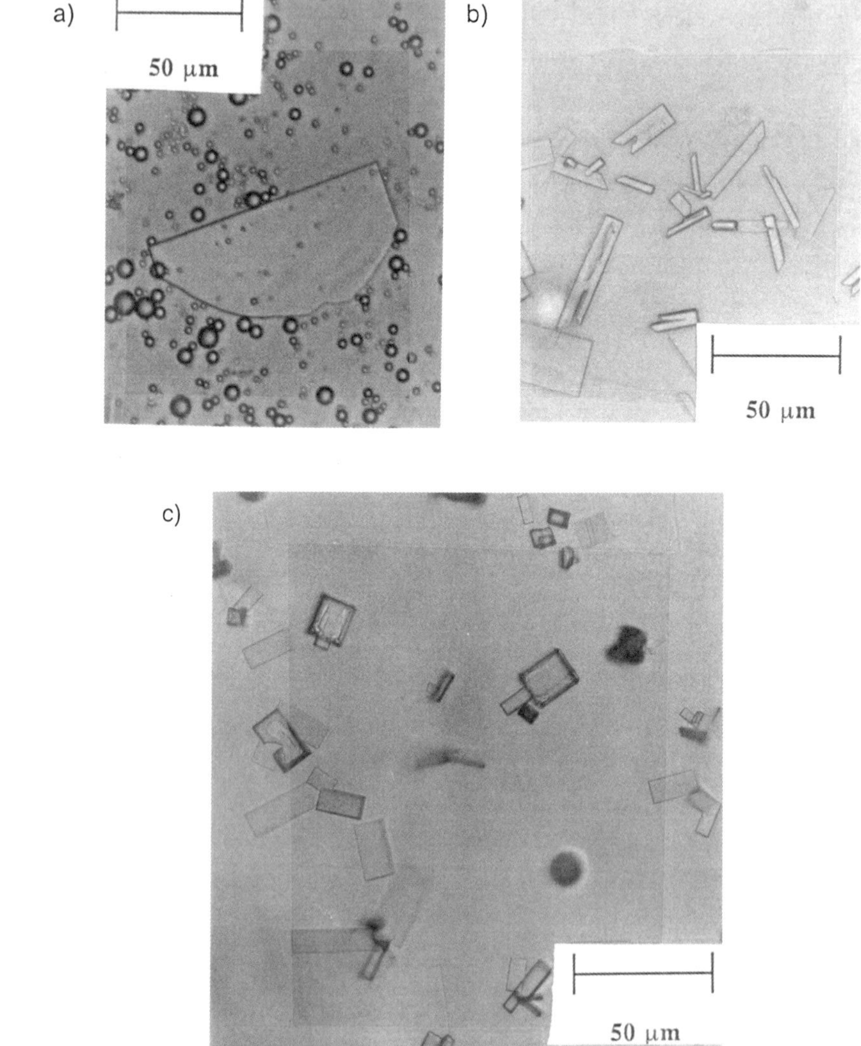

Fig. 1. Crystals of m CNB growing from a) 90 : 10 m : p CNB emulsion at 22 °C, (Tensagex emulsifier). Reproduced with permission from Davey et al 1995. b) m CNB emulsion at 26 °C, (p-nitro cinnamic acid emulsifier) and c) m CNB emulsion at 24 °C, (p-chloro cinnamic acid emulsifier).

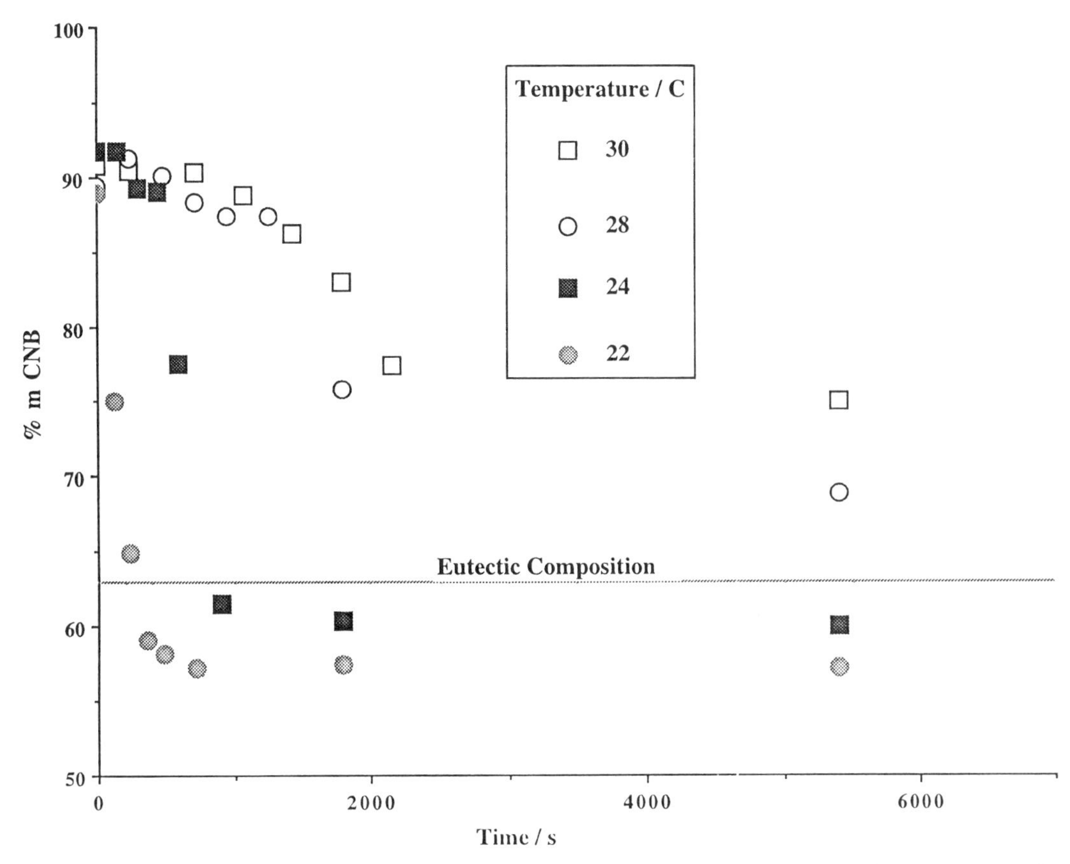

Fig. 2. Time dependence of emulsion droplet composition. Total crystallisation time 90 mins. Reproduced with permission from Davey et al 1995.

temperature, derived from measurements taken during this work, are 61.6 % m CNB and 27.1 oC. At 30 and 28 oC the final liquid phase compositions are somewhat higher than expected from the phase diagram possibly because equilibrium had not been reached. At 24 and 22 oC, however, while the liquid phase might be expected to have the eutectic composition it in fact has a composition richer in p CNB. The final crystal purities were also measured, (Table 1). The solid phase recovered from experiments performed at 24 and 22 oC, (below the eutectic temperature), have purities of 98.2 and 93.6 % m CNB respectively. Crystallisation of the 60.9 % m CNB emulsion at 15 oC, contrary to expectations, yielded a final crystal purity of 65.5 % m CNB and a final liquid phase composition of 55.3 % m CNB.

An increase in organic weight fraction and/or droplet diameter has been found to increase crystallisation rates; this also being true when surfactant weight fractions are reduced. Overall crystal purity has been found to decrease slightly with increasing droplet size and increasing organic weight fraction while there appears to be an overall increase in purity with increasing surfactant concentration, (Table 2); the reverse being true for crystal yield.

Crystallisation

Table 3 shows the crystallisation temperature of m CNB emulsions made up with varying emulsifiers. From this data it is obvious that the choice of emulsifier will have an effect on the crystallisation of the m CNB. In all cases considerable undercooling of the emulsion droplets was still required before nucleation occured, (mpt. m CNB 44.4 °C). Some emulsifiers are however behaving as prefferential nucleators. All the emulsifiers with nitro groups yielded good quality m CNB crystals whose magnitude was of the order of the emulsion droplets, (Fig. 1b). Similarly both isomers of the chlorine substituted cinnamic acid also gave good quality crystals with sizes of the order of the emulsion droplet diameter, however these had undergone some habit modification, (Fig. 1c). As the alkyl chain length increased from 1 to 11 in the alkylphenyl ethoxylates the quality of crystals produced deteriorated; the longest chain producing a polycrystalline agglomeration. Interfacial tension measurements suggested that all the substituted cinnamic acids would be active at the m CNB/aqueous interface, although to a far lesser degree than the alykylphenyl ethoxylates, whose activity increased with chain length.

Table 2 The Effect of Droplet Size, Organic and Surfactant Weight Fraction on Final Crystal Purity, Yield and Emulsion Composition (Original Emulsion Composition 90 : 10 m : p CNB)

Drop Size	Organic Weight Fraction	Surfactant Weight Fraction	Purity	Yield	Yield	Final Emulsion Composition
μm	%	%	% m CNB	Yt	Ye	% m CNB
20 - 50	2	0.20	98.43	71.81	80.96	66.79
5 - 10	2	0.20	99.29	68.05	76.68	68.84
3 - 5	2	0.20	99.23	67.55	76.13	66.69
5 - 10	10	0.80	97.88	86.81	97.84	66.08
5 - 10	4	0.34	99.33	70.11	79.05	67.08
5 - 10	2	2.00	99.49	68.20	76.88	68.94
5 - 10	2	0.10	99.29	68.51	73.85	68.84
5 - 10	2	0.01	95.04	89.00	100.00	64.75

Table 3. Crystallisation Temperature of m CNB Emulsions

Emulsifier	Crystallisation Temperature of m CNB droplets °C
4 - NO_2 $C_6H_4CH=CHCOONa$	26.4
3 - NO_2 $C_6H_4CH=CHCOONa$	23.7
4 - $(C_8H_{17})C_6H_4O(CH_2CH_2O)CH_2CH_2OH$	19.6
4 - $(C_8H_{17})C_6H_4O(CH_2CH_2O)_4CH_2CH_2OH$	15.7
3 - Cl $C_6H_4CH=CHCOONa$	11.3
4 - NO_2 $C_6H_4CONH(CH_2)_5CO_2H$	10.2
4 - Cl $C_6H_4CH=CHCOONa$	9.4
4 - $(C_8H_{17})C_6H_4O(CH_2CH_2O)_{11}CH_2CH_2OH$	4.5

Discussion

Purification

The purification data, taken together with the fact that crystals grow to at least an order of magnitude greater than the drops, indicates, contrary to expectations, that crystallisation does not occur primarily within the drops nor does particle growth occur as a result of drop coalescence. Were these the case the product would comprise aggregates of spheres with sizes similar to the emulsion drop size and purities would be equivalent to those obtained on a cooled surface. It would seem reasonable to assume that nucleation of m CNB results from impacts between seeds and drops and that the resulting secondary nuclei, being dispersed by Tensagex, move out into the aqueous phase. At constant crystallisation temperature the chemical potential of m CNB will be higher in the liquid phase than in the crystalline phase. This means that crystals in contact with an aqueous phase, which is itself in equilibrium with melt, will experience a supersaturation and hence will grow. m CNB will thus be transferred from the melt droplets to the crystals via the aqueous phase. This allows the growth of large single crystals and means that the growing crystals are not in contact with impure melt. The development of the polar morphology results, presumably, from the fact that the p CNB also has some solubility in water.

The consequences of this mechanism are profound since, as the m CNB is transferred to growing crystals the drops will shrink and become richer in p CNB. At any given temperature the final drop compositions should be given by the measured liquidus line in the phase diagram. From Fig. 2 this can be seen not to be the case. Hence, providing the p CNB does not crystallise within the drops, there is no apparent reason why m CNB crystals should not continue to grow as a pure phase and the drops become further enriched with p CNB below the eutectic temperature, taking the liquid phase through the eutectic into the p CNB side of the phase diagram. The data for the liquid and solid phases are consistent with the mechanistic prediction that in this emulsion system the eutectic can be broken as a result of kinetically controlled discrimination.

At larger drop diameters, high organic weight fractions and lower surfactant levels it seems most likely that loss in final crystal purity is largely associated with the state of aggregation of the crystals. For example the significant increase in purification above the critical micelle concentration of Tensagex, (measured in this work to be around 0.02 wt.%), suggests that the surfactant plays a key role not only in stabilising the emulsion but also in subsequent dispersion of the resulting crystals. Thus when there is too little surfactant, due either to operation below the cmc or to increases in the final mass fraction of crystals, the crystals are not dispersed but flocculate due to the hydrophobicity of their surfaces and trap uncrystallised drops thereby lowering the overall purity. The disadvantage of using high surfactant levels is the lower yields achieved due to the stabilisation of smaller drops which are reluctant to crystallise, (Table 2). There is no evidence to suggest surfactant micelles aid the transport of m CNB across the aqueous phase. This is consistent with the work of Kabalnov, (1994), and our own results, (capillary electrophoresis), showing m CNB to have negligible solubility in Tensagex micelles in comparison to its solubility in water.

Larger emulsion drops tend to yield groups of crystals which again trap liquor and are impure; a similar picture is observed at high organic weight fractions.

A mechanism for crystallisation when either the organic weight fraction is high, or the surfactant weight fraction low is suggested. That is, uncrystallised drops collide with already formed crystals and deposit their contents to yield irregular morphologies in an uncontrolled growth process. At high organic weight fractions these collisions become statistically more probable leading to the observed increase in crystallisation rates. This can lead to polycrystalline aggregation and associated purity problems as discussed earlier. Control of final crystal size and morphology in this situation is clearly impossible.

Crystallisation

The hydrophobic portion of an amphiphillic molecule obviously plays an important role in determining its suitability as a template for m CNB nucleation, (Table 3). From molecular modelling it would appear that the nitro substituted amphiphillic molecules are most likely to act as templates since hydrogen bonding is possible to adjacent m CNB molecules. Chloro substituted amphiphillic molecules cannot hydrogen bond in this way. As can be seen with the para nitro benzoyl caproic acid, (4 - NO_2 $C_6H_4CONH(CH_2)_5CO_2H$), however, a nitro group does not automatically provide a good nucleating surface; this molecule is able to form intramolecular hydrogen bonds, which will control the orientation and packing of the molecule at the interface. Molecules such as the cinnamic acids which do not have intramolecular hydrogen bonds are more likely to have their molecular orientation and packing at the interface controlled by the π-π interactions of adjacent phenyl rings; possibly making them superior m CNB templates.

The para substituted nitro group of the cinnamic acid lies planar to the phenyl ring, whereas in the meta isomer the nitro group is twisted out of the plane; this may explain the latters reduced effectiveness as a template for m CNB nucleation.

The hydrophilic portion of the amphiphillic molecule also determines whether or not it can act as a suitable template. This has been discussed earlier in relation to the intramolecular hydrogen bonds but with the alkylphenyl ethoxylates another factor has to be considered. As the alkyl chain length increases the molecule becomes increasingly disordered at the interface, leading to a less close packed monolayer and hence a diminished ability to act as a template for m CNB.

Conclusions

We have shown that the crystallisation of emulsified impure melts can form the basis of a purification process. For the case of m chloronitrobenzene crystallising from mixtures with its para isomer, optimum purity and physical form of product crystals are obtained at low organic phase weight fractions and small, (5 μm), drop diameters. In these systems the surfactant plays a dual role of both stabilising the emulsion and dispersing product crystals into the continuous (aqueous) phase.

Compared to traditional progressive freezing techniques the emulsion process appears, uniquely, to offer excellent purification in a single stage. In addition, purification can be achieved below the thermodynamic eutectic temperature due to kinetic effects in which the crystallisation of impure drops is inhibited by their small size and the increasing levels of p CNB.

Work on amphiphillic molecules as templates for the crystallisation of m CNB from o/w emulsions continues. Results so far indicate the choice of emulsifier will be important in the control of final crystal size, crystal morphology and undercooling required to induce droplet nucleation.

Acknowledgement. This work was supported by ZENECA p.l.c. from their Strategic Research Fund.

References

Chen, B. D., Garside, J., Davey, R. J., Maginn S. J. and Matsuoka, M., *J. Phys. Chem.* **1994**, *98*, 3215.

Cordiez, J. P., Grange, G. and Mutaftschiev, B., *J. Col. Int. Sci.* **1982,** *85*, 431.

Fischer, O., Jancic, S. J. and Saxer, K., *Industrial Crystallisation 84* ; Jancic, S. J. and de Jong, E. J., Eds; Elsevier, Amsterdam, **1984**; p 153 - 157.

Garside, J. and Davey R. J., *J. Chem. Eng. Comm.* **1980**, *4*, 393.

Kabalnov, A. S., *Langmuir.* **1994**, *10*, 680.

Mullin, J. W., *Crystallisation*, Butterworth Heinemann: London, **1993**; p 309 - 323.

Skoda, W. and Van den Tempel, M., *J. Col. Sci.* **1963**, *18*, 568 - 584.

Sloan, G. J. and McGhie, A. R., *Techniques of Chemistry, 18*; Weissberger, A. and Saunders, W., Eds.; Wiley: New York, **1988.**

Turnbull, D. and Cech, R. E., *J. Appl. Phys.* **1950**, *21*, 804.

Davey, R. J., Garside, J., Hilton, A. M., McEwan, D. and Morrison, J. W., *Nature* 1995, *375*, 664.

Solid Layer Melt Crystallization - a Fractionation Process for Milk Fat

Michaela Tiedtke, Joachim Ulrich, Universität Bremen, Verfahrenstechnik / FB 4, Postfach 330440, D- 28334 Bremen, Germany
Richard W. Hartel, University of Wisconsin-Madison, Department of Food Science, 1605 Linden Drive, Babcock Hall, Madison, Wisconsin, 53706, USA

The composition of milk fat varies according to the seasonal differences in available animal food. The separation of milk fat into fractions with different properties is an option to provide the food industry with possibilities to design their products and guarantee a better uniformity and quality. The advantage of solid layer melt crystallization is that no problems due to the handling of solid material occur. An additional filtration step as needed with a suspension crystallization process is not necessary. Experimental results concerning quality and quantity of a high melting fraction are presented. An outlook on the potential of the technology in comparison to the suspension technique is given.

Fractionation of milk fat by suspension crystallization is a commercially available process. It consists of a crystallization step and a filtration step. In this work the potential of the solid layer melt crystallization as a fractionation process for milk fat is investigated.

The advantage of the solid layer crystallization process is the possibility of controlling the crystallization conditions easily because the crystals grow on a cooled surface. The temperature of the liquid is kept slightly above its melting point. No solvents are needed. The low temperature level allows a careful treatment of temperature sensitive materials. At the end of the process the residual melt is drained and the solid layer is molten to obtain the product in liquid state. Handling of solids and encrustion problems do not occur. No additional filtration step is necessary.

Milk fat is a very complex system containing not only more than 200 different triacylglycerols (Gresti 1993, Small 1986), but also lipoids, sterins, free fatty acids, vitamins and flavors.

The driving force for the phase change from liquid to solid is the difference between the melting points of the triacylglycerols which form the milk fat and the temperature at the solid-liquid phase boundary. Crystallization kinetic proceeds in two steps - nucleation and growth. Nucleation depends on the supercooling, the supersaturation, the presence of small particles (nuclei) of either the same or different materials and the roughness of the cooled surface. Growth involves the moving of triacylglycerols from the bulk to the existing crystals or nuclei and then incorporation into their crystal lattice.

During growth, the triacylglycerols, with melting points above the temperature at the solid-liquid phase boundary, form the crystal layer. While the triacylglycerols, with melting points below the temperature at the phase boundary, stay in the liquid phase. To keep the temperature at the phase boundary constant and to compensate the insulation effect of the growing crystal layer, the temperature of the cooling element has to be reduced with increasing layer thickness. Very fast or very slow cooling changes the temperature at the solid-liquid interface in which the melting properties of the growing crystal layer will also change.

In this paper, the results of the experiments concerning the crystal structure, the influence of the Reynolds-Number on the growth, the yield and the growth rate are presented for a one-step fractionation at 30°C. Some data for a fractionation carried out at 20°C will also be given. A model has been developed to describe the growth of the crystal layer.

Materials and Methods

A falling film crystallizer was used in the experiments (Fig. 1). The plant consists of jacketed glass tubes heated by thermostated baths. The cylindrical crystallizer (i.e. heat exchanger, diameter 20mm, length 0.5m, surface area 0.031m^2) was made of stainless steel. Temperature profiles were programmed by a thermostated bath equiped with a control gear. The melt was circulated by a centrifugal pump (0.18kW).

The anhydrous milk fat[1] used has a clear point of 35°C. Before filling into the plant the fat was molten and tempered at 60°C for 30 minutes in order to destroy all the present crystal structures.

Samples were taken from the feed, the residual melt and the crystal fraction. Fatty acid methyl ester profiles (Amer 1985), clear points and solid fat content profiles were determined for the samples.

The structure of milk fat crystals was investigated by observing a thin milk fat crystal layer growing on a cooled glass surface. The crystals have been photographed under the microscope.

The influence of the Reynolds-Number on the growth of the layer was investigated for three

[1] The authors want to acknowledge gratefully the supply of the anhydrous milk fat by the Molkereizentrale Oldenburg-Osnabrück-Ostfriesland eG, Oldenburg, Germany.

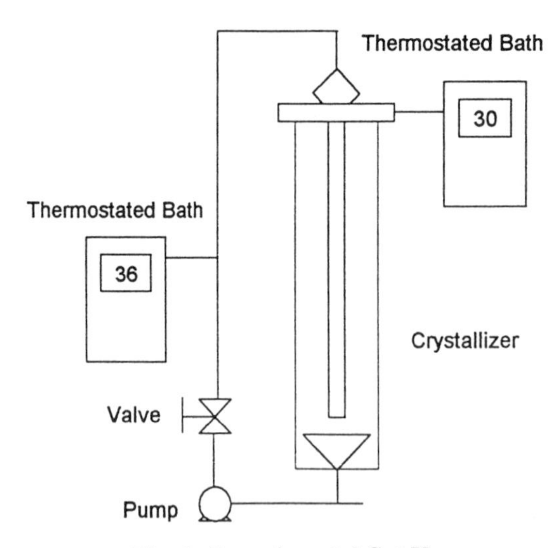

Fig. 1. Experimental Set Up

melt flow rates (2.1kg/min, 0.8kg/min and 0.2kg/min). The crystallization time was 6h for each experiment.

To determine the maximal yield, a certain quantity (ca. 650g) of anhydrous milk fat was placed into the crystallizer and the crystallization process was started. After every 24 hours the process was stopped and the crystalline layer was removed. The crystallization was afterwards restarted. This was repeated untill the liquid hold up of the crystallizer reached its minimum. The duration of these experiments was between 72h and 96h (149h for the fractionation at 20°C). The melt flow rate was 4.8kg/min.

Fractionation at 30°C

Morphology. The crystals formed at 30°C consist of needle like structures which tend to agglomerate (Tiedtke 1994) (Fig. 2). This structure has been classified as β' (Deffense 1991) and has been found in suspension fractionation, too (Grall 1992).

Influence of the Reynolds-Number. An increase of the Reynolds-Number leads to an increase of the mass of product (Table 1).

Increasing the Reynolds-Number by a factor 10 (from 1.2 to 12) leads to a 70% increase in the mass of the product (from 0.79kg/m^2 to 1.34 kg/m^2). There is no influence of the Reynolds-Number evident on the product quality. The clear point does not change.

Melting Properties. The high melting fractions have melting points of 43°C (Table 2). The solid fat content profiles (Fig. 3) are in accordance with the fractions produced at 30°C with suspension technique (Tiedtke 1994) and listed in the data sheets for milk fat fractions (Kaylegian 1993).

Fatty Acid Profiles. The fatty acid profiles can be seen in Fig. 4. For a comparison with data from suspension crystallization (Grall 1992) they were grouped into four categories: Short-chain acids (C4:0-C10:0), medium-chain acids (C12:0-C14:0), long-chain acids (C16:0-C18:0) and unsaturated acid C18:0. No general changes in the composition occured. There was a trend with the short-chained and unsaturated fatty acids migrating to the lower melting fraction and the longer-chained fatty acids to the high melting fraction.

Fig. 2. Crystal Structure of Milk Fat

Yield. The yield is between 0.23 and 0.25 (Table 2). Such a yield was obtained from suspension crystallization, too (Grall 1992). The yield for a 2ℓ-lab-scale plant is between 0.15 and 0.25 while for a 20ℓ-pilot-scale plant the yield is 0.22.

Growth Rate. The growth rate decreases when crystallization time increases because of

Table 1. Influence of the Reynolds-Number

Re [-]	Mass of Product [kg/m^2]
12.0	1.34
4.8	0.95
1.2	0.79

Table 2. Solid Fraction (30°C)

Clear Point [°C]	Yield [-]
43.3	0.23
43.0	0.25
43.1	0.24

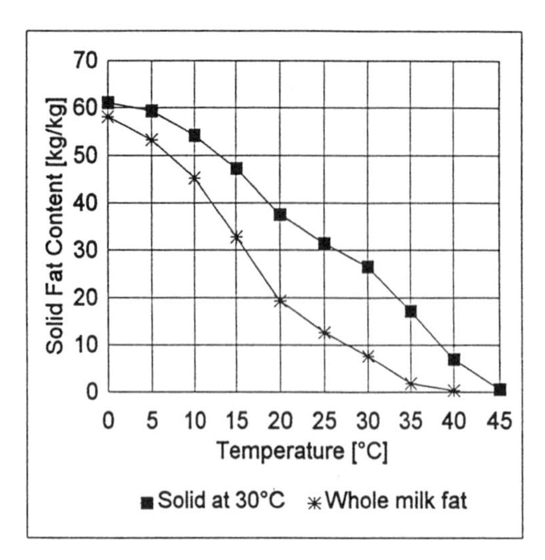

Fig. 3. Solid Fat Content Profiles (30°C)

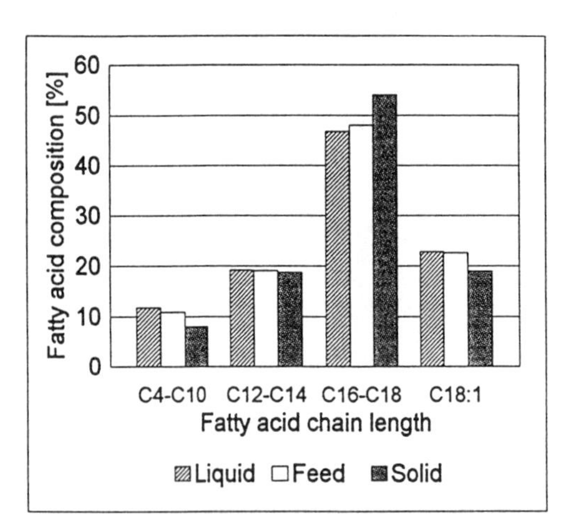

Fig. 4. Fatty Acid Profiles (30°C)

decreasing the concentration of high melting material in the liquid phase (Fig. 5). The growth rate is slightly higher than the growth rates obtained from the suspension crystallization. In the suspension crystallization the growth rates are between $0.4{\cdot}10^{-8}$ and $1.5{\cdot}10^{-8}$ m/s for a 2ℓ-lab-scale plant and $1.2{\cdot}10^{-8}$ m/s for a 20ℓ-pilot-scale plant (Grall 1992).

Description of the Growth of the Layer

The crystallization process is assumed as a first order reaction. Therefore the increase of the

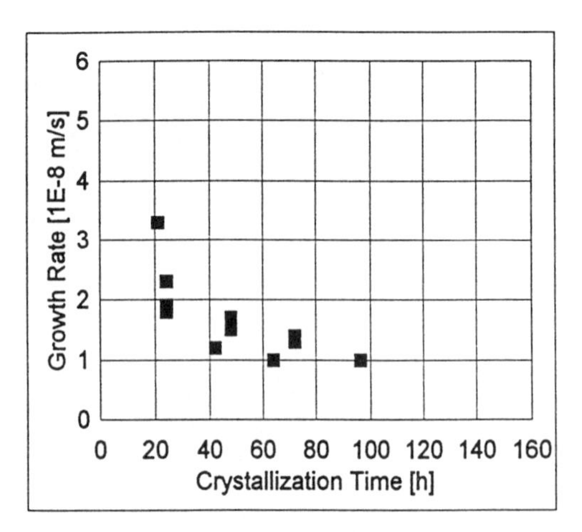

Fig. 5. Growth Rate (30°C)

mass of the solid can be expressed by equation (1).

$$m_{Solid}(t) = m_{Feed} Y_{max} \frac{1-e^{-kt}}{1-Y_{max} e^{-kt}} \qquad (1)$$

The reaction constant k was determined by own experiments. It was found that k is $2.9{\cdot}10^{-6}$ s^{-1} for a fractionation temperature of 30°C and $1.5{\cdot}10^{-6}$ s^{-1} for a fractionation temperature of 20°C. The maximum yield, Y_{max}, can be obtained from literature (Gresti 1993). In Table 3, Y_{max} is listed for fractions received by a multi-step fractionation, at 15, 20 and 30°C, respectively.

Table 3. Maximum Yields (Gresti 1993)

Fraction	Y_{max} [-]	Melting Point [°C]
Solid at 15°C	0.15	15 < MP < 20
Solid at 20°C	0.33	20 < MP < 30
Solid at 30°C	0.27	MP > 30

The literature values are valid for an infinite crystallization time and an ideal separation. In practice the yields are assumed to be smaller. For the solid fraction at 30°C a yield of 0.25 has been found and for a fractionation at 20°C the amount of solid material was 0.47 (maximum: 0.60).

In Fig. 6 equation (1) is plotted together with experimental data. Transformation of the mass of the solid into a layer thickness (equation (2))

$$s(t) = \left[\frac{m_{Feed} Y_{max}}{\rho_{Solid} \pi l} \frac{1 - e^{-kt}}{1 - Y_{max} e^{-kt}} + r^2 \right]^{\frac{1}{2}} - r \qquad (2)$$

and differentiation with respect to time leads to the growth rate as a function of time (equation (3)).

$$v(t) = \frac{m_{Feed} Y_{max}}{\rho_{Solid} \pi l} \frac{k e^{-kt} (1 - Y_{max})}{2 (1 - Y_{max} e^{-kt})^2} \cdot \frac{1}{2} \left[\frac{m_{Feed} Y_{max}}{\rho_{Solid} \pi l} \frac{1 - e^{-kt}}{1 - Y_{max} e^{-kt}} + r^2 \right]^{-\frac{1}{2}} \qquad (3)$$

Inserting the concentration of high melting compounds in the liquid leads to an expression for the growth rate as a function of the concentration (equation (4)).

$$v(c) = \frac{m_{Feed}}{\rho_{Solid} \pi l} \frac{c\, k (1 - Y_{max})}{2 (1 - c)^2} \cdot \frac{1}{2} \left[\frac{m_{Feed}}{\rho_{Solid} \pi l} \frac{Y_{max} - c}{1 - c} + r^2 \right]^{-\frac{1}{2}} \qquad (4)$$

The information of equation (4) is useful for the process design. The optimal end point of a crystallization step can be derived from it.

With this model the growth rate of the layer can be predicted for any fractionation temperature and for any feed stock. By using this model and a heat balance at the phase boundary parameter studies can be performed. E.g. the optimal cooling profile can be found and the influence of the melt flow rate on the heat transfer can be investigated.

Fractionation at 20°C

Fatty Acid Profiles. In Fig. 7, the fatty acid profiles of feed, low melting fraction and high melting fraction are presented. There is a trend that short-chained and unsaturated fatty acids are concentrated in the lower melting fraction while longer-chained fatty acids are migrating to the higher melting fraction. The same shifts in the fatty acid composition are found in corresponding fractions produced by suspension crystallization (Grall 1992).

Yield. At 0.41 - 0.47, the yield for the higher melting fraction (Table 4) is bigger than the yield obtained with suspension crystallization (0.34 - 0.38 for a 2ℓ-lab-scale-plant and 0.33 for a 20ℓ-pilot-scale plant, Grall 1992). The melting point of the low melting fraction is 24.8°C. This means there is still higher melting material in the liquid phase. Therefore it is possible to improve the yield of the higher melting fraction to, at least, a value of 0.5.

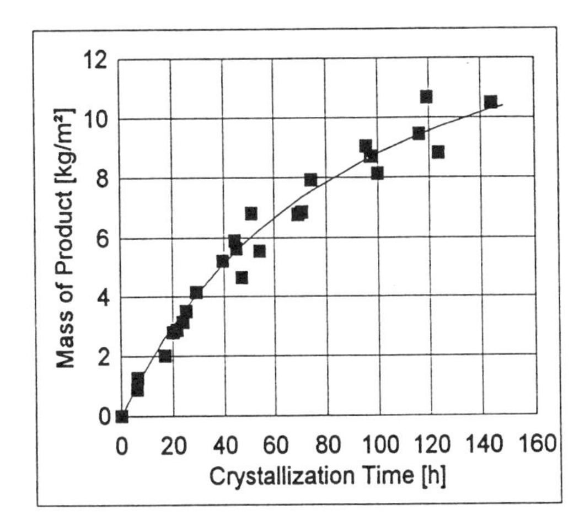

Fig. 6. Mass of Product

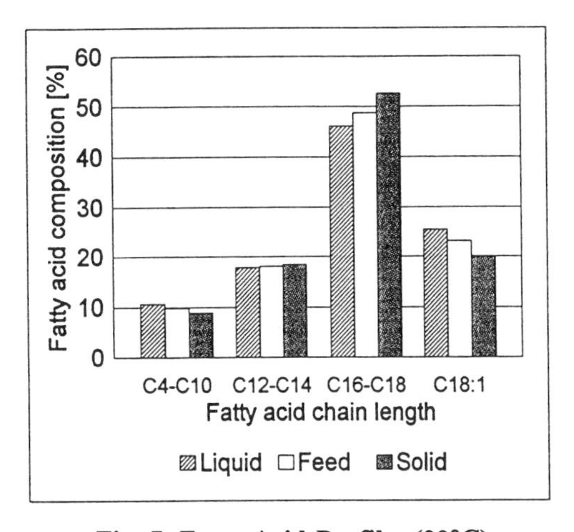

Fig. 7. Fatty Acid Profiles (20°C)

Table 4. Solid Fraction (20°C)

Clear Point [°C]	Yield [-]
39.8	0.41
39.2	0.42
38.4	0.47

Growth Rate. The growth rate decreases when the crystallization time increases (Fig. 8). The growth rate of the crystal layer (fractionation temperature: 20°C) is approximately 70% higher than the growth rate for the fractionation at 30°C. With the suspension technique, the growth rate for a crystallization temperature of 20°C is between $0.8 \cdot 10^{-8}$ m/s and $2.3 \cdot 10^{-8}$ m/s for a 2ℓ-lab-scale plant and $2.5 \cdot 10^{-8}$ m/s for a 20ℓ-pilot-scale plant (Grall 1992). This is the same order of magnitude as it has been found for the mean growth rate of the solid layer crystallization.

With the higher melting fraction a second fractionation at 30°C can be performed. At this temperature, a high melting fraction (HMF, 35°C < CP < 45°C) and a middle melting (MMF, 25°C < CP < 35°C) fraction can be gained.

Outlook

For a multi-step fractionation the solid layer crystallization becomes competitive with the suspension technique. The growth and nucleation rates are functions of the concentration of high melting material in the liquid. Starting the multi-step fractionation with the lowest temperature the fraction with the lowest melting point is separated and removed from the system. Therefore the feed for the next stage contains more high melting material compared to the feed for the corresponding stage of the suspension crystallization process. For this reason nucleation and growth will be faster compared to suspension crystallization technology.

Therefore, an important task for the future is the investigation of a 3-step fractionation (15°C, 20°C, 30°C). In Fig. 9 the flow sheet of a 3-step fractionation for suspension crystallization with data from a 20ℓ-pilot-plant is shown (Grall 1992). With each suspension crystallization step the melting point of the crystal fraction decreases. The crystals become softer and the filtration process becomes more difficult. The yield for the first fraction (solid at 30°C) is 0.22, in which it is very close to the value predicted by theory (0.27). The yield of the second fraction (solid at 20°C) is 0.11. This is only the third part of the theoretical maximum yield (0.33) for this fraction. The nucleation rate became smaller (Grall 1992) for this fractionation. The yield for the third fraction (solid at 15°C) is 0.21 (theoretical maximum yield: 0.15). Here parts of the material which belongs to the fraction solid at 20°C crystallized. This leads to a yield of 0.21. The

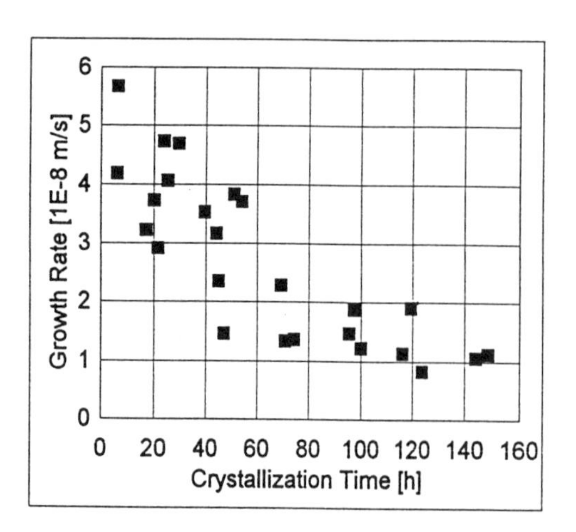

Fig. 8. Growth Rate (20°C)

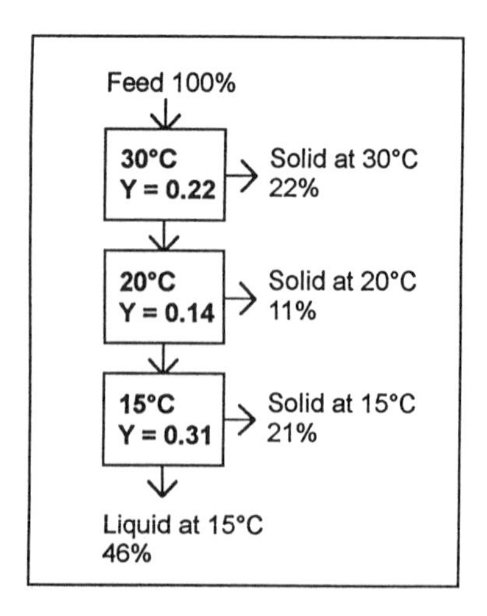

Fig. 9. 3-Step Fractionation (Grall 1992)

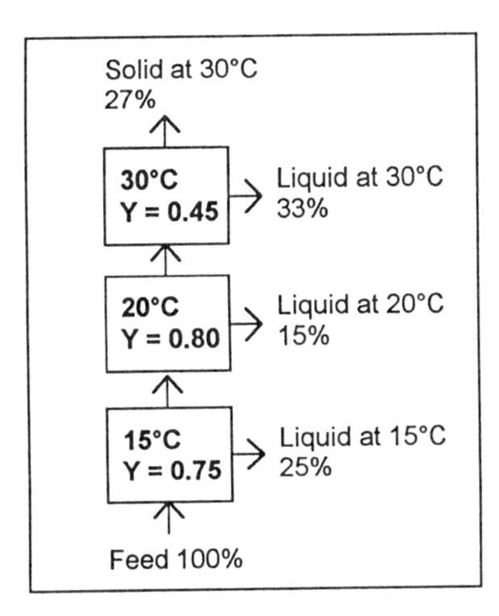

Fig. 10. 3-Step Fractionation (Gresti 1993)

total portion of solid fractions gained from the three crystallization stages is 0.54, in which it is less than the possible maximum yield of 0.75.

Fig. 10 contains a flow sheet for solid layer melt crystallization, which is developed with theoretical data (Gresti 1993), valid only for infinite crystallization times and an ideal separation. Growth rates and real yields for this process have to be quantified in order to evaluate the economical potential of the process.

Conclusions

The quality and quantity of a high melting fraction produced by a one-step solid layer fractionation at 30°C are the same as with suspension crystallization.

First results of a solid layer fractionation at 20°C showed that the yield of solid material obtained is greater compared to a suspension crystallization process (solid layer crystallization: 0.47, suspension crystallization: 0.33). For this reason, a solid layer fractionation at 15°C is expected to have a higher yield in comparison to the yield of the suspension crystallization process of 0.54 (solid at 30°C: 0.22, solid at 20°C: 0.11, solid at 15°C: 0.21).

The advantages of the solid layer technology are based on the multi-step fractionation of milk fat. The multi-step solid layer crystallization starts with the separation of the lowest melting fraction and ends with the highest melting fraction. After each crystallization stage the lower melting fraction is removed from the system. The portion of higher melting material in the feed for the following stage is, therefore, higher compared to the feed for a corresponding stage in a multi-step suspension crystallization process, which allows higher growth rates. Also, it has been found that the separation efficiency of the solid layer crystallization process is higher than the separation efficiency of the suspension crystallization.

The solid layer technology has clear advantages in the multi-step fractionation of milk fat compared to the suspension crystallization. The solid layer technique should, therefore, be a competitive alternative to the suspension technology.

Symbols

c	Concentration	[kg kg^{-1}]
l	Length	[m]
k	Reaction constant	[s^{-1}]
m	Mass	[kg]
ρ	Density of the solid fat	[kg m^{-3}]
Re	Reynolds-Number	[-]
r	Radius	[m]
s	Layer thickness	[m]
t	Time	[s]
v	Growth rate	[m s^{-1}]
Y	Yield	[kg kg^{-1}]

Acknowledgements

The authors want greatfully to acknowledge the support through the Wisconsin Center for Dairy Research at the University of Wisconsin through funding from the National Dairy Promotion and Research Board of the United States and the support of the Senator für Bildung und Wissenschaft of the Freie und Hansestadt Bremen which helped to achieve parts of the presented results.

References

Amer, M.A.; Kupranycz, D.B.; Baker, B.E., JAOCS 62, **1985**, 11, 1551-1557.

Deffense, E.; Fat Sci Technol, **1987**, 13.

Grall, D.S.; Hartel, R.W., JAOCS 70, **1992**, 8, 741-747.

Gresti, J.; Bugaut, M.; Maniongui, C.; Bezard, J., J Dairy Sci 76, **1993**, 7, 1850-1869.

Kaylegian, K.; Hartel, R.W.; Burrington, D., Center for Dairy Research, University of Wisconsin-Madison, August 1993.

Small, D.M., *Handbook of Lipid Research 4, Physical Chemistry of Lipids*; Plenum Press: New York, NY, 1986.

Tiedtke, M.; Niehörster, S.; Ulrich, J.; Hartel, R.W., *4. Bremen International Workshop for Industrial Crystallization BIWIC 1994*, Verlag Mainz: Aachen, 1994.

Tiedtke, M.; Ulrich, J.; Hartel, R.W., *Kurzfassungen der Vortragsgruppen Verfahrenstechnik - Thermische Verfahrenstechnik*, ACHEMA 1994, Frankfurt, 1994.

On the Behavior of Adipic Acid Aqueous Solution Batch Cooling Crystallization

S. Derenzo, IPT - Instituto de Pesquisas Tecnológicas, Chemistry Division, Cidade Universitária 05508-901 - São Paulo - S.P. Brazil
P.A. Shimizu, Universidade de São Paulo, Escola Politécnica, Chemical Engineering Department, 05508-000 -São Paulo - S.P. Brazil.
M. Giulietti, IPT and Universidade Federal São Carlos, Chemical Engineering Department, 13565-905 - São Carlos - SP Brazil.

The batchwise crystallization of adipic acid aqueous solution was performed in a one - liter W - shaped crystallizer, equipped with a draft tube, in the temperature range of 65-20°C, in the time range of 3500-7500s. Four forms of operation were performed: linear and exponential cooling, with and without seeding. Agglomeration was an important mechanism in all modes of operation. Other effects like size and mass of seeds were studied in exponential cooling experiments. The effect of mass of seeds on dominant size was small (~5%), due agglomeration. Larger seeds resulted in larger dominant size of product, due mainly to their sizes. Scanning electron microscopy revealed the occurrence of growth instability, specially in the case of unseeded exponential cooling crystallization.

Adipic acid is an important material in nylon production. It is produced by oxidation of cyclohexanol and is purified by a crystallization step, washing, and a recrystallization step before drying. This recrystallization from an aqueous solution is carried out, in some cases, by batch cooling. The purpose of this work is to examine the behavior of adipic acid crystallization with different cooling and seeding strategies in terms of dominant crystal size as well as crystal morphology. In another paper to be published in a near future a kinetic model to the system will be proposed.

Experimental

Batchwise cooling crystallization of adipic acid aqueous solutions was conducted in a jacketed cylindrical vessel with a volume of one liter, W - bottom. provided with a draft tube, designed according to Nývlt et al., (1985). The suspension was agitated by a glass mechanical stirrer with four flat inclined blades. Solution saturated at 64.5°C were prepared with demineralized water and adipic acid (99.8% purity) in the concentration of 20.11 kg of adipic acid per kg of water. The system was heated to ± 66°C and a cooling curve was imposed, being monitored each three minutes. Batch times were in the range of 3500-7500s.

In the seeded experiments 20 or 34.5 g of crystals in the size of 137 and 460 microns were added to the reactor when the solution was supersaturated. The solution temperature followed a prescribed cooling curve (linear or exponential) up to final value through manual adjustment of the cooling water temperature.

At the end of the experiments crystals were filtered on a buchner funnel and dried at ambient temperature for 2 to 3 days. After drying the material was sieved on a standard series of 14 sieves, in the range of 0.149mm to 1,41mm. From the size distribution the crystal size corresponding to 64.7% of total mass distribution, i.e., the dominant size in the gamma distribution model was determined. For each experiment electron microscopy was performed for the material retained in the fraction 500-590 microns.

The exponential cooling curves were shaped in order to avoid uncontrolled nucleation following the method proposed by Nývlt & Mayrhofer (1988), according to the equation:

$$\frac{T - T_i}{T_f - T_i} = \left(\frac{\tau}{\tau_f}\right)^{\vartheta} \tag{1}$$

where the exponent ϑ is shown on table 1.

Table 1. Relation between the exponent ϑ and the form of cooling.

ϑ	seeding	form of cooling	supersaturation
1	no/yes	linear	varying
3	no	exponential	nearly constant
4	yes	exponential	nearly constant

In eq.(1) the initial temperature (at time zero) was adopted as the nucleation/seeding temperature.

Results

The experiments were classified in four groups, according of the form of operation:

- Linear cooling, unseeded
- Linear cooling, seeded
- Exponential cooling, unseeded
- Exponential cooling, seeded

The results and experimental conditions are presented in table 2, where the apparent growth rate G was calculated according to Nývlt et al (1985) by the equation:

$$G = \frac{L_m - L_s}{\tau} \quad (2)$$

DISCUSSION OF THE RESULTS

The results will be discussed in two ways: crystal quality and dominant size.

Crystal Quality

Figures 1 to 11 are, respectively, electron scanning microscopy views of crystals from experiments 2, 6, 8, 10, 11,13, 14, 16, 17,18 and 19. In general it is possible to recognize, for example on pictures 6 and 7, the presence of growth instability and agglomeration. Both features contribute to entrapment of mother liquor, affecting crystal quality.

Agglomeration

Agglomeration has been observed for crystallization of adipic acid aqueous solution

Table 2. Experimental conditions and results.

GROUP	Number	Experimental			Result		
		Time (s)	L_s (μm)	W_s(*)	L_m (μm	W_f(*)	G (μm/s)
linear cooling unseeded	1	3540	----	----	307	23.256	0.0867
	2	4320	----	----	465	18.716	0.1076
	3	4560	----	----	488	18.447	0.1070
	4	4560	----	----	518	18.665	0.1136
	5	4800	----	----	468	18.563	0.0975
	6	5400	----	----	491	18.752	0.0909
	7	5520	----	----	468	17.881	0.0848
linear cooling seeded	8	5280	460	4.13	578	22.950	0.0223
	9	7080	460	4.13	611	23.038	0.0213
	10	7140	460	4.13	581	23.256	0.0169
exponential cooling unseeded	11	5400	----	----	597	18.994	0.1106
	12	5520	----	----	470	18.855	0.0851
	13	7080	----	----	651	19.014	0.0919
exponential cooling seeded	14	5400	137	2.40	431	21.301	0.0544
	15	5460	137	4.13	416	22.881	0.0511
	16	7230	137	2.40	470	21.042	0.0461
	17	7200	137	4.13	444	22.962	0.0426
	18	5880	460	4.13	565	23.055	0.0179
	19	7500	460	4.13	587	22.546	0.0169

(*) kg of solids/100 kg of solvent

(Klein; Ratsimba, 1991, David et all, 1991). In our seeded exponential cooling experiments, agglomeration was confirmed from the ratio between the number of crystals at the beginning and at the end of the run, N_s/N_f. For this form of operation, the following relation holds (Nývlt, 1991):

$$L_m = L_s\left[\frac{W_f}{W_s}\frac{N_s}{N_f}\right]^{1/3} \qquad (2)$$

From this equation it is possible to calculate the ratio N_s/N_f, that is presented in table 3. When only nucleation and growth takes place, the

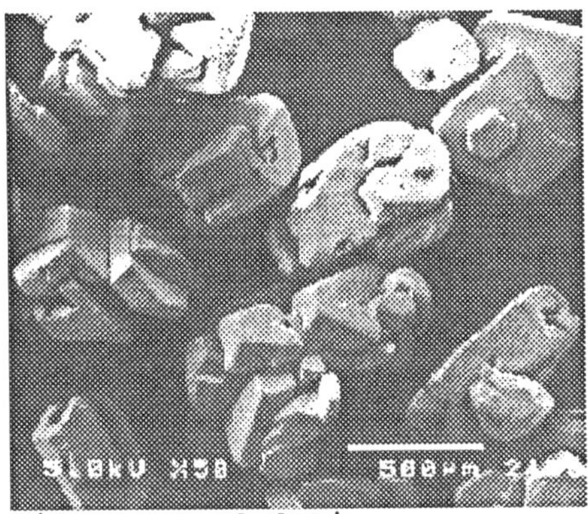
Fig 1-Unseeded Linear cooling, τ=4320s

Fig 2-Unseeded Linear cooling, τ=5400s

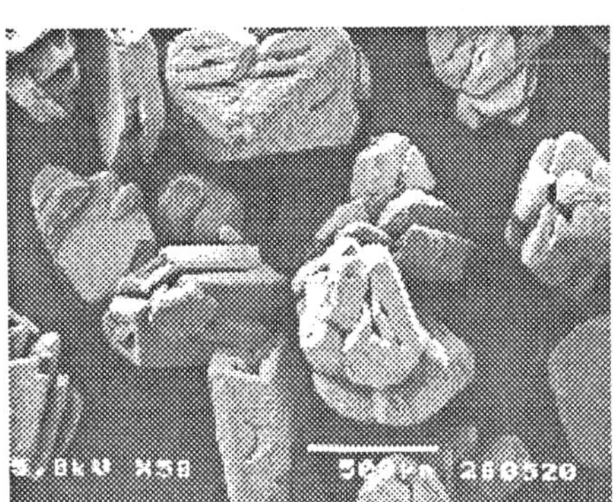

Fig 3-Seeded Linear cooling, τ=5280s

Fig 4-Seeded Linear cooling, τ=7140s

Fig 5-Unseeded exponential cooling, τ=5400s

Fig 6-Unseeded exponential cooling, τ=7080s

Fig 7-Small seeds exponential cooling, τ=5400s

Fig.9-Small seeds exponential cooling, τ=7200s

Fig 11-Large seeds exponential cooling, τ=7500s

number of crystals at the end of the experiment is greater than the initial value ($0 \leq N_s/N_f \leq 1$).

The total number of crystals at the end of the experiments was found to be lower than the initial value for all experiments where small sized seeds were used, indicating that agglomeration was an important mechanism in the process.

Fig 8-Small Seeds exponential cooling, τ=7230s

Fig. 10-Large seeds exponential cooling, τ=5880s

In the experiments 18 and 19, seeded with large crystals, this ratio was lower than 1. In this case the production of new crystals by secondary nucleation was greater than the disappearance by agglomeration. Another indication of this enhanced nucleation is that the apparent growth rate was less than half of those obtained in experiments where small sized seeds were used (experiments 14 to 17, see table 2).

Table 3- Ratio of the number of crystals at the begging and at the end of experiment.

Experiment	L_m	L_s	N_s/N_f
14	431	137	3.508
15	416	137	5.054
16	470	137	4.605
17	444	137	6.123
18	565	460	0.332
19	587	460	0.381

Growth Instability

Growth instability can be a result of the form of operation, that is, the control of cooling velocity of the system, that provides sudden step changes on supersaturation (Myerson, 1993).

As a general rule, crystal quality was better in experiments seeded with small crystals and large batch times, due to the low supersaturation. Enhanced growth instability was found for unseeded exponential cooling experiments as well as in exponential cooling seeded with larger crystals.

In the unseeded exponential experiments the supersaturation reaches the boundary of the metastable zone at the initial stage of the process, and then decreases to a nearly constant value. This fluctuation is likely to be the determining mechanism for the high degree of instability observed. Exponential cooling experiments seeded with larger crystals behaved similarly, as can be deduced from the enhanced nucleation observed (see previous item).

Crystal size

As expected, for short batch times unseeded linear cooling experiments produced small crystals. As batch time increases, their influence is less pronounced. Seeding was an efficient way to produce large crystals.

Unseeded exponential cooling experiments produced large crystals, as for this form of operation, where supersaturation is nearly constant (see table 1). The dominant size were a little higher than for the unseeded linear cooling mode.

Seeding in exponential cooling with small crystals led to crystal sizes smaller than in the absence of seeding. This indicates that the mass of seeds was too large. Doubling the mass of seeds, the dominant size decreased only ~5%, mainly due to the increase in agglomeration (compare N_s/N_f in table 3, experiments 14 with 15 and 16 with 17).

With larger seeds, larger dominant size were obtained, mainly due to the size of seeds themselves, since in these experiments (experiments 18 and 19), nucleation was pronounced (see discussion on agglomeration).

These results show that an optimum mass and size of seed has to be chosen to give an optimum crystal dominant size.

Conclusions

Four forms of operation of batch cooling crystallization were studied. Agglomeration is an important mechanism in adipic acid crystallization. Growth instability creates defects that will trap some mother liquor into the crystals and will result in the presence of water and some impurities that will not be easily removed. Growth instability was reduced by seeding and increasing batch. Large mass of seeds resulted in larger agglomeration. An optimum mass and size of seeds has to be chosen to improve final crystal size and quality.

List of Symbols

G apparent growth rate velocity (μm/s)
L_m dominant size of crystals (μm)
L_s average size of crystals (μm)
N_s number of crystals seeds per cubic meter of solvent (#/m^3)
N_f number of final crystals per cubic meter of solvent (#/m^3)
T Temperature (°C)
T_i Initial temperature (°C)
T_f Final temperature (°C)
W_s Concentration of seeds (kg of adipic acid/ 100 kg water)
W_f Concentration of crystals at the end of experiment (kg of adipic acid/ 100 kg water)
τ Time (s)
τ_c Batch time (s)

Acknowledgments: The authors want to acknowledge Prof. dr. G. M. van Rosmalen from the Laboratory for Process Equipment of Delft University of Technology for the scanning electron microscopy.

References

David, R; Vilermaux, J.; Marchal, P.; Klein, *Chem. Eng. Science*, **1991**, *46*, 1129-1136.

Klein, J.P.; Ratsimba, B. Dificultes et limites de la modelisation en cristllisation organique industrielle, in *Recents Progress en Genie des Procedes: Cristallisation Industrielle et Precipitation*, vol 5. n.18, **1991**, pp.VII-1-VII-11.

Myerson, A .S.; *Handbook of Industrial Crystallization*, Butterworth-Heinemann, **1993**, Stoneham, pp. 73.

Nývlt, J; Söhnel, O; Matuchová M.; Broul, M. *The kinetics of industrial crystallization* Academia Prague and Elsevier, Amsterdam, **1985**, .pp.275.

Nývlt, J.; Mayrhofer, B. *Chem. Eng. Process*, **1988**, 24, pp.217-220.

Nývlt, J. Batch crystallizer design, *in Advances in Industrial Crystallization* Ed. J. Garside; R.J. Davey; A.G. Jones, Butterworth-Heinemann Ltd., Oxford, **1991**, pp.197-212.

A Fullerene Route to Fuel Additives

George W. Schriver, Abhimanyu O. Patil Exxon Research & Engineering Co., P.O. Box 998, Annandale, NJ 08801 USA
Kenneth Lewtas, David J. Martella Exxon Chemical Co., Paramins Technology Division, P.O. Box 536, Linden, NJ 07036 USA,

Wax crystal modifiers are important in the petroleum industry, as dewaxing aids in refining, and in fuels and lubricating oils as pour-point depressants (flow improvers). The variety of fuel compositions and the range of climates in which they are used, along with increasing performance requirements, requires continuing development of flow improver technology. Fullerenes can be viewed as structural building blocks for many types of functional molecules, including fuel and lubricating oil additives. Alkylamino- and alkylfullerenes are effective pour point depressants for both diesel fuel and lubricating oils. Pour point depressions of up to 12°C were observed. Pour-point depression depends on the length of the alkyl chains, with C_{16} the optimum both in oils and in diesel fuel.

The storage, transport and use of petroleum products at low temperatures can cause recurring problems to the various industries and user alike (Zielinski, 1984). These problems occur in crude oils, lubricating oils and distillate fuels (e.g. diesel fuel and heating oil). Difficulties arise because of the unwanted crystallization of the waxy components of the oil or fuel, the higher normal alkanes. In some cases the waxes are removed (e.g. the dewaxing of lubricating oils) however it is desirable to keep them in diesel fuels because of their high cetane index values.

The modern solution to these problems is the use of additives known as flow improvers which, when blended with the oil or fuel at low concentrations (typically at levels of 0.01%) affect the sizes and habits (shapes) of the precipitating wax crystals. Such additives can change the relative stabilities of the crystal faces by various mechanisms to change their habits, thus allowing the temperature operability range of the fuel to be substantially increased (Steere, 1981; Coley, 1966; Brown, 1988; Brown, 1989).

Waxes in Fuels and Lubricating Oils

The normal alkanes in diesel fuel are the most abundant, most anisotropic, least soluble and most crystalline species present. The temperature at which the normal alkanes start to precipitate from the fuel is called the Cloud Point (CP). Below this temperature, platelet crystals form which comprise over 95% n-alkanes. These crystals can readily mat together to gel the fuel or block transfer lines or filters leading to engine failure (Palmer, 1969). Pour Point is often used in both lubricating oils and distillate fuels to indicate the temperature at which the oil or fuel ceases to flow (under normal gravity).

Lubricating oils also contain normal alkanes. Due to the higher boiling ranges of these oils, the waxes present are usually larger and begin to precipitate at even higher temperatures. The wax crystals precipitating from lubricating oils can also gel the oil, leading to oil starvation at critical surfaces and failure to lubricate the metal-metal contact areas. This leads to increased wear or even engine seizure (Rossi, 1994).

Lubricating oil basestocks are usually dewaxed, treated to remove excess wax by precipitation. Dewaxing can be accomplished by cooling or by adding polar solvents. Additives are often added to change crystallization behavior during this process in order to improve filterability and minimize occlusion of non-wax molecules.

In wax crystals, the molecules typically exist in all-extended conformations, lined up in parallel, and all tilted at an angle. The crystal is made up of layers of these sheet-like arrays. Under most growth conditions, addition of new wax molecules is fastest at the edge of one of these layers. Growth in such a direction gives rise to the thin rhombic platelet wax crystals, typically 0.5mm across by 0.01mm thick, observed under ideal conditions. Inhibition of growth along this face will change the shape, or habit, of the crystals as shown in Fig. 1. They now grow prismatically, typically as thin columns, needles or blocks which prevent gelling (Denis, 1987).

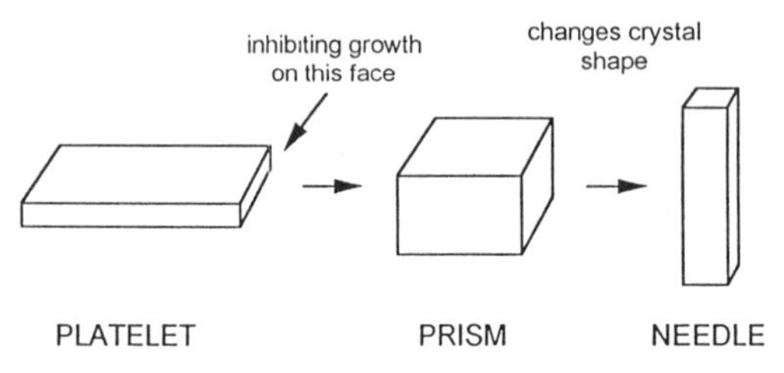

Fig. 1. Habit modification of a wax crystal by a growth modifier.

Commercial Flow Improvers

Crystallization is a kinetic phenomenom, whose thermodynamic driving force is supersaturation, There are many possible ways to slow or speed the crystallization process or to change the habit of the resulting crystals. Additives can be designed to act as hetero-nucleators (Tack, 1986), nucleation suppressers (CP Depressants) (Beiny, 1990; Heraud, 1992), growth inhibitors and habit modifiers (Lewtas, 1991).

It is desirable for an additive to have one or more polymethylene segments which fit into key lattice sites of the growing n-alkane crystal thus co-crystallizing with the native wax and a "foreign" segment which can protrude from the growing crystallizing layer inhibiting the incorporation of more n-alkane molecules by removing the favorable binding energies. This effect slows down the growth in the plane of the dominant (001) face, thus destabilizing the largest face of the crystal. Habit changes caused by a growth inhibitor are shown in Fig. 2.

The polymethylene segment can be in the backbone of random copolymers of relatively low crystallinity such as ethylene-vinyl acetate copolymers (EVA) (Ilnyckyj, 1976). In other polymeric additives, comb polymers (Shibayev, 1971), alkyl side chains are attached to the polymer backbone. Fumarate-vinyl acetate and olefin copolymer additives share this structural principle. It is not sufficient just to attach such a segment. Many other factors need to be taken into consideration, such as how the molecule will interact with the n-alkane crystal structure, the substituent effect upon solubility, etc. These conditions are fulfilled in certain "monomeric" additives (Lewtas, 1988) as well.

Fullerenes

Fullerene-sized carbon clusters were first observed in mass-spectrometric studies in 1984 (Rohlfing, 1984). A year later, the correct

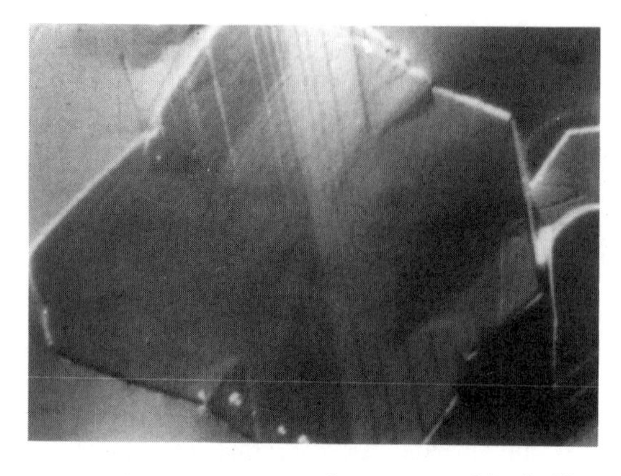

Fig. 2. Wax crystals grown with (left) and without (right) a growth inhibitor.

structure was first assigned to C_{60}, the parent fullerene (Kroto, 1985). For the next 5 years, fullerenes were the subject of a handful of studies by gas-phase chemists and theoreticians.

The situation changed dramatically in 1990, when the preparation of fullerenes in an electric arc was reported (Krätschmer, 1990). The availability of tangible quantities of fullerenes caused explosive growth in fullerene research. A recent search of Chemical Abstracts showed more than 4000 references to fullerenes, including over 200 patents (Krieg, 1994). Some diverse and colorful outgrowths of this research are: superconductivity of doped fullerenes (Hebard, 1991), metal-filled fullerene tubules as molecular wires (Ajayan, 1993), and fullerenes as inhibitors of the HIV protease (Friedman, 1993).

Fullerenes have proven to be highly reactive species, undergoing addition reactions similar to those of electron-poor olefins. The number of available reactions allows great flexibility in construction of fullerene derivatives.

Fullerenes are capable of undergoing multiple reaction. Variations in reaction conditions allows some control over the number of addends and the resulting physical properties of the functionalized fullerene. Further, fullerenes with mixed addends can be obtained. These addends can either be mixtures of closely related moieties or of completely different ones. An example of the former might be addition of amines with alkyl groups of differing length, while the latter might involve adding non-polar alkyl groups and oxidizing to provide polar hydroxyl functionality.

At the current time, fullerenes are still much more research chemicals than commercial materials. The cost of mixed fullerenes (the soluble portion of the soot generated in the arc process) is above $10,000 per pound, and purified C_{60} is about twice as costly. Chemical companies have announced projects aimed at commercialization of fullerenes, and costs should drop significantly as markets develop. In addition to the arc method, several other ways of producing fullerenes are available, each with advantages and disadvantages for commercial production.

Fullerene-Based Wax Crystal Modifiers

In order to influence the growth of wax crystals, an additive must resemble the wax molecule in some aspect so that it can interact with a growing crystal. In addition, it must have properties which differ from the wax so that subsequent deposition of wax molecules can be altered. A fullerene with long-chain alkyl substituents could satisfy these conditions. This is shown schematically in Fig. 3.

The availability of many different functionalization reactions for fullerenes makes it easy to design wax crystal modifiers. Perhaps the simplest addition reaction on fullerenes is the addition of amines (Wudl, 1992). This occurs simply on heating. In this reaction the first addition is more rapid than subsequent ones. This allows the preparation of monoamine derivatives or the sequential addition of two different amines to a fullerene (Patil, 1994; Patil, 1995). For this study, five different alkylamino-fullerenes were prepared by reacting commercially available normal alkylamines with fullerenes (Fig. 4).

It was also of interest to use an all-hydrocarbon pour-point depressant. Fullerenes

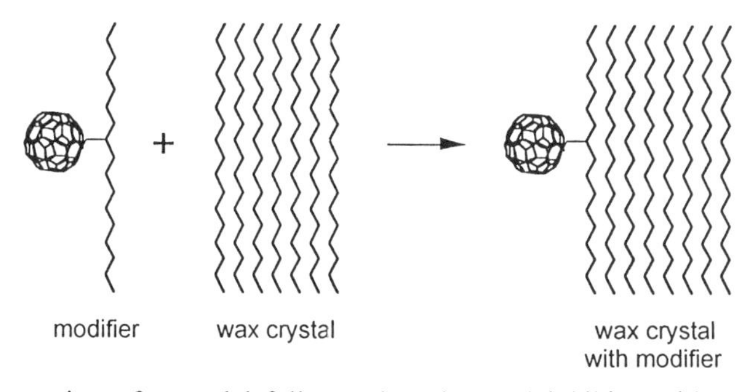

Fig. 3. Interaction of a model fullerene-based growth inhibitor with a wax crystal.

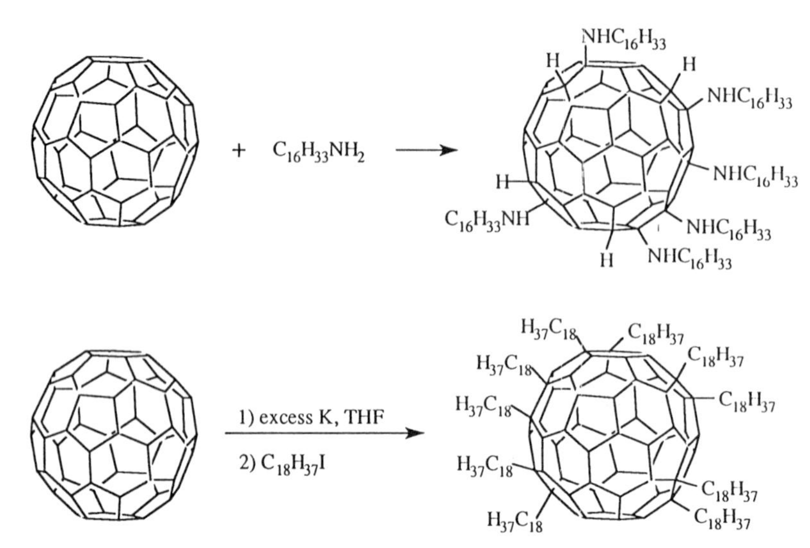

Fig. 4. Syntheses of the model pour-point depressants.

can be alkylated by reaction of their alkali metal salts with alkyl halides (Bausch, 1991. An octadecyl fullerene was prepared by reduction of fullerenes with potassium and alkylation by octadecyl bromide (Fig. 4).

Results

Preparation of Hexadecylaminofullerenes: Mixed fullerenes (ca. 75% C60, 25% C70, <1% higher fullerenes, 0.2g) and hexadecylamine (Armeed 16D, supplied by Akzo, 0.81g) were dissolved in 50 ml toluene. The dark colored solution was stirred for 6 days at 40°C. The solvent was removed under vacuum, and the product was dissolved in chloroform and filtered to remove traces of unreacted fullerenes.

Oil and Fuel Blending
Additives were blended in oil at 65°C for 1 hour and in fuel at room temperature for 1 hour. Additive treat rates were 1000 ppm in oil and 6667 ppm in fuel, both on a weight basis. Low temperature properties of the oils and fuels are: S150N, pour point -12°C; S600N, pour point -9°C; diesel fuel, cloud point -9°C.

Pour Point Measurements
This procedure follows ASTM standard D97-93. The fluid to be tested is placed in a standard test jar with thermometer and cork seal. It is initially heated to 45°C, then transferred to a 24°C bath for cooling. At certain specified temperatures, the test jar is transferred to colder baths. The test jar is checked at 3°C intervals, tilting slightly and looking for movement of the fluid. When no movement of the fluid is detected, the flask is turned to a horizontal position. If movement of the fluid is not detected within 5 seconds, the pour point is judged to have been passed. The last temperature at which fluidity was observed is taken as the pour point.

Discussion

Lubricating oil basestocks were treated with the substituted fullerenes added at 1000 ppm by weight. Diesel fuel was treated at 6667 ppm. Typical commercial additives are used, often in combination, at 50-500 ppm treat rates, depending upon the additive, the nature of the fluid, and the severity of conditions to which it will be exposed. Pour-point depression for the various compounds studied is shown in the Table. The greatest depression observed, 12°C, represents very effective improvement in the flow properties of the fluid. The 9°C pour-point depression observed for diesel fuel also indicates a potent additive.

Within the family of additives, the magnitude of the pour-point depression

Table: Pour-Point Depression By Fullerene Additives (°C)[1]

Carbon Number[2]	Oil Basestock S150N	S600N	Diesel Fuel
C_{12}	0	-6	
C_{14}	-9		
C_{16}	-12	-9	-9
C_{18}	-3	-6	
C_{22}	0	0	
C_{18}[3]	-12		

1 Diesel fuel was treated at 6667 ppm by weight, oils at 1000 ppm.
2 Length of the alkyl chain of the primary alkylamines added to the fullerene.
3 C_{18}-alkylfullerene.

increases and then decreases with increasing chain length. For the S150N basestock, the maximum effect is seen with a 16-carbon chain. For the heavier basestock, the C_{16}-amine again appears to produce the most effective additive. Viewed simply in terms of the waxes present, this is somewhat surprising. Heavier oils typically contain higher molecular weight waxes, and these waxes crystallize at lower temperatures. It might be expected that longer alkyl chains in the additive would better match the larger waxes in the heavier oil, leading to greater pour-point depression. As noted above, many factors and mechanisms control pour point. The exact mechanism of this family of additives has not been elucidated.

One example of a hydrocarbon derivative was tried. This was a fullerene substituted with octadecyl chains. In the S150N basestock, it functioned as well as the C_{16}-amine derivative, and much better than the C_{18}-amine derivative. It may be that the parts of the substituent nearest the fullerene are unable to interact with the wax for steric reasons. If this is so, their chemical nature (polar versus nonpolar, nitrogen versus carbon) may not be very important for the performance of the additive. The present study doesn't cover a wide enough range of compounds to address this unambiguously. Nor are pour points the most precise experimental tools for doing so. By definition, they can only differentiate systems in 3°C units.

In addition to the alkylamino- and alkylfullerenes reported here, many other kinds of functionalized fullerenes are known. Most functionalization methods should be amenable to the addition of long-chain alkyl substituents and should be capable of producing molecules which can cause pour-point depression. Two possibilities are shown in Fig. 5. Esterification or etherification of fullerols (Li, 1993) offers the possibility of incorporating free hydroxyl groups into the fullerene. Transesterification of carbene addition products, (Suzuki, 1991; Isaacs, 1993) allows the introduction of a wide range of different alkyl groups.

Conclusion

Using fullerenes as a building block to attach waxy subtituents to a central core produces effective pour point depressants for lubricating oil basestocks and diesel fuel. The model additives were effective in both light and heavy hydrocarbon fluids. The efficacy of the additive depends upon the chain length of the alkyl substituents attached to the fullerene. In the one case studied, a fullerene with nonpolar, alkyl substituents produced results similar to another with more polar, alkylamino substituents of comparable length.

Acknowledgement

We thank Dr. Glen Miller for providing a sample of the octadecylated fullerene and Dr. Al Rossi for supplying the photographs for Fig. 1.

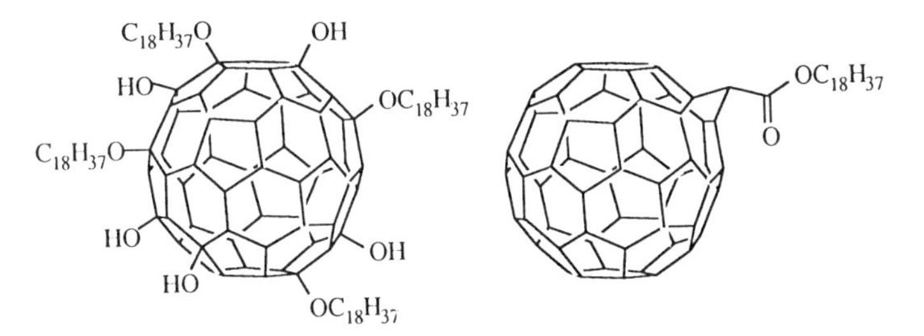

Fig. 5. Examples of other possible structural types for fullerene pour-point depressants.

References

Ajayan, P.M.; Ebbesen, T.W.; Ichihashi, T.; Iijima, S.; Tanigaki, K.; Hiura, H. *Nature* **1993**, *362*, 522.

ASTM *Annual Book of ASTM Standards* vol 05.01

Bausch, J.W.; Surya Prakash, G. K.; Olah, G.A.; *J. Am. Chem. Soc.* **1991**, *113*, 3205.

Beiny, D.H.M.; Mullin, J.W.; Lewtas, K. *J. Cryst. Growth* **1990**, *102*, 801.

Brown, G.I.; Tack, R.D.; Chandler, J.E. *Society of Automotive Engineers, Technical Paper Series* **1988**, No. 881652.

Brown, G.I.; Lehmann, E.W.; Lewtas, K. *Society of Automotive Engineers, Technical Paper Series* **1989**, No. 890031.

Coley, T.; Rutishauser, L.F.; Ashton, H.M,. *J. Inst. Pet.* **1966**, *52*, 173.

Denis, J. *Rev. Inst. Français Pet.* **1987**, *42*, 385.

Friedman, S.H.; DeCamp, D.L.; Sijbesma, R.P.; Srdanov, G.; Wudl, F.; Kenyon, G.L. *J. Am. Chem. Soc.* **1993**, *115*, 6506.

Hebard, A.F.; Rosseinsky, M.J.; Haddon, R.C.; Murphy, D.W.; Glarum, S.H.; Palstra, T.T.M.; Ramirez, A.P.; Kortan, A.R. *Nature* **1991**, *350*, 600.

Heraud, A.; Pouligny, B.; *J. Colloid Interface, Sci.*, **1992** *153*, 378.

Ilnyckyj, S.; Cole, C., U.S. Patent 3,961,916, June 8, 1976.

Isaacs, L.; Diederich, F. *Helv. Chim. Acta* **1993**, *76*, 2454.

Krätschmer, W.; Lamb, L.D.; Fostiropoulos, K.; Huffman, D.R.; *Nature* **1990**, *347*, 354.

Krieg, J.F. personal communication, 1994.

Kroto, H.W.; Heath, J.R.; O'Brien, S.C.; Curl, R.F.; Smalley, R.E. *Nature* **1985**, *318*, 162.

Lewtas, K.; Lehmann, E.W.; Tack, R.D.; Rossi, A., Eur. Patent 0,261,957, March 30, 1988.

Lewtas, K.; Tack, R.D.; Beiny, D.H.M.; Mullin, J.W. in *Advances in Industrial Crystallization;* Garside, J.; Davy, R.J.; Jones, A.G. Eds.; Butterworth-Heinemann: Oxford, U.K., 1991, pg. 166.

Li, J.; Takeuchi, A.; Ozawa, M.; Li, X.; Saigo, K.; Kitazawa, K. *J. Chem. Soc., Chem. Commun.* **1993**, 1784.

Palmer, R.C.; Batchelor, M.A.; Batchelor, J. *Inst. Chem. Eng, Inst. Chem. Ind. Cryst. Symp.* **1969**, *4*, 179.

Patil, A.O.; Schriver, G.W.; Lundberg, R.D., U.S. Patent 5,292,813, March 8, 1994.

Patil, A.O.; Schriver, G.W. *Macromol. Symp.* **1995**, *91*, 73.

Rohlfing, E.A.; Cox, D.M.; Kaldor, A. *J. Chem. Phys.* **1984**, *81*, 3322.

Rossi, A. *Society of Automotive Engineers, Technical Paper Series* **1994**, No. 940098.

Shibayev, V.P.; Petrukhin, B.S.; Plate, N.A.; Kargin, V.A. *Polym. Sci. USSR (Eng)* **1970**, *12*, 160. Platé, N.A.; Shibaev, V.P.; Petrukhin, B.S.; Zubov, Yu.A.; Kargin, V.A. *J. Polym. Sci., A-1* **1971**, *9*, 2291.

Steere, D.E.; Marino, J.P. *Society of Automotive Engineers, Technical Paper Series* **1981**, No. 810024.

Suzuki, T.; Li, Q.; Khemani, K.C.; Wudl, F.; Almarsson, Ö. *Science* **1991**, *254*, 1186.

Tack, R.D., Eur. Patent 0,183,447, June 4, 1986.

Wudl, F.; Hirsch, A.; Khemani, K.C.; Suzuki, T.; Allemand, P.-M.; Koch, A.; Eckert, H.; Srdanov, G.; Webb, H.M. in *Fullerenes: Synthesis, Properties and Chemistry fo Large carbon Clusters,* Hammond, G.S.; Kuck, V.J. eds.; ACS Symposium Series, vol. 481 American Chemical Society: Washington, DC, 1992; pg 161.

Zielinski, J.; Rossi, A.; Stevens, A. *Society of Automotive Engineers, Technical Paper Series* **1984**, No. 841352.

Thermodynamic Properties of Supersaturated Solution and Their Use in Determination of Crystal Growth Kinetic Parameters

Soojin Kim and Allan S. Myerson
Department of Chemical Engineering, Polytechnic University, 6 Metrotech Center, Brooklyn, New York 11201

The crystal growth rates of glycine and β-succinic acid were expressed in terms of the fundamental driving force of crystallization calculated from the activity of supersaturated solutions. The kinetic parameters were compared with those from the commonly used kinetic expression based on the concentration difference. From the viewpoint of thermodynamics, rate expressions based on the chemical potential difference provide accurate kinetic representation over a broad range of supersaturation. The rates estimated using the expression based on the concentration difference coincide with the true rates of crystallization only in the concentration range of low supersaturation and deviate from the true kinetics as the supersaturation increases.

Crystallization is an important separation and purification process for organic materials in many areas such as the chemical, pharmaceutical, petrochemical, and food industries. The kinetics of crystallization is essential information required for the design of any crystallization equipment. Numerous experimental studies of the kinetics of crystal growth and nucleation of organic materials have been published in the literature. However, a standardized approach in correlating experimental results to the theoretical or semiempirical models to describe the kinetic processes has not been established. Consequently, estimating rates of crystallization from different expressions often leads to inconsistent results.

The driving force for crystallization is the degree of supersaturation which has been commonly expressed as the difference in concentration between the supersaturated and saturated solutions. This practice of expressing the crystallization rate as a function of the concentration difference causes confusion and inconsistency: even dimensionless supersaturations calculated from different concentration units result in different numerical values that are not proportional to one another, and thus different sets of kinetic parameters could be evaluated from a given set of experimental data.

It is well known that the fundamental driving force of crystallization is the difference between the chemical potential of the supersaturated solution and that of the solid crystal face, which can be used independently of units (Mullin and Söhnel, 1977). Using the concentration difference in place of the fundamental driving force of crystallization is based on the assumption that the solute activity of a supersaturated solution can be closely approximated by the concentration, which may cause serious errors in evaluating the true kinetics of crystallization.

Unfortunately, the kinetic expression using the fundamental driving force seldom has been used in crystallization practices, because there had been virtually no experimental data of activity in supersaturated solutions. Recently, efforts to study thermodynamic properties of supersaturated solutions have been carried out by a number of researchers utilizing an electrodynamic microparticle levitator (Cohen et al., 1987; Na et al., 1994, 1995). Water activities of many aqueous supersaturated solutions have been measured using a single micron-sized solution droplet that is electrically levitated and continuously weighed as the concentration is increased by a slow evaporation of water. Solute activity of the supersaturated solutions can be computed from the water activity data using the Gibbs-Duhem relation; hence, we can establish a direct relationship between the chemical potential and the concentration of a supersaturated solution.

From this relationship, many kinetic data reported in the literature can now be expressed in terms of the chemical potential difference providing kinetic expressions of more exact and thermodynamically accurate form. The purpose of this paper is to investigate the effects of using the fundamental driving force in the expression of crystallization kinetics and to quantify the errors associated with the conventional use of the concentration-based driving force on the actual rates of crystallization.

Driving Force for Crystallization

The fundamental driving force for crystallization of an organic material can be expressed in dimensionless form:

$$\frac{\mu - \mu^*}{RT} = \ln\left(\frac{a}{a^*}\right) = \ln\left(\frac{\gamma}{\gamma^*}\frac{c}{c^*}\right) \qquad (1)$$

where c is the concentration, a is the activity, and γ is the activity coefficient, and * indicates the properties at saturation.

The use of the dimensionless concentration difference,

$$\sigma = (c-c^*)/c^* \qquad (2)$$

as the driving force in place of $\Delta\mu/RT$ is justified only in the case that meets the following conditions:

i) $\gamma \approx \gamma^*$, thus $\ln(a/a^*) \approx \ln(c/c^*)$,
ii) $\sigma << 1$, thus $\ln(\sigma+1) \approx \sigma$.

Condition i) will be satisfied only for an ideal solution or for a solution whose supersaturation is small enough so that its concentration is almost the same as that of the saturated solution. There are many circumstances in crystallization practices in which a high supersaturation level ($\sigma > 0.1$, for example) is encountered thus condition ii) is violated. Examples are primary nucleation and precipitation processes where relatively insoluble materials are produced as a result of reaction. Therefore, using σ in place of $\Delta\mu/RT$ is an approximation that is inadequate in a number of practical situations.

Solute Activity

Solute activity in an aqueous solution is computed from experimental water activity data as a function of concentration. Through the Gibbs-Duhem equation relating the chemical potentials of solvent and solute, the water activity, a_w, can be expressed in terms of solute concentration, m in molality and the activity coefficient, γ, for example (Robinson and Stokes, 1959):

$$-\frac{1000}{MW_w}\, d(\ln a_w) = m\, d\big(\ln(\gamma m)\big) \qquad (3)$$

upon integration of which we get the relation:

$$\ln(\gamma/\gamma^*) = \phi - \phi^* + \int_{m^*}^{m} \frac{\phi - 1}{m}\, dm \qquad (4)$$

where MW_w is the molecular weight of water, and ϕ, the osmotic coefficient, is defined as

$$\phi = -\frac{1000}{MW_w}\frac{\ln a_w}{m}.$$

Therefore,

$$\ln(a/a^*) = \ln(m/m^*) +$$

$$\phi - \phi^* + \int_{m^*}^{m} \frac{\phi - 1}{m}\, dm. \qquad (5)$$

Crystallization Kinetics Expression

For correlation purposes, a simple semi-empirical power law equation of the form

$$G = k\,\sigma^n \qquad (6)$$

has been used frequently for the expression of surface integration or overall rate of crystal growth. The kinetic expression in terms of the fundamental dimensionless driving force, as shown in eq 1, is:

$$G = k'\left(\frac{\Delta\mu}{RT}\right)^{n'}$$

$$= k'\left(\ln\left(\frac{a}{a^*}\right)\right)^{n'} = k'\left(\ln\left(\frac{\gamma}{\gamma^*}\frac{m}{m^*}\right)\right)^{n'}. \quad (7)$$

The kinetic orders n and n' would be the same only in the concentration range of very low supersaturations, where $\gamma/\gamma^* \approx 1$ and $\ln(m/m^*)$ can be approximated as σ. The difference between experimentally obtained n and n' will be greater for the data of a kinetic experiment conducted at a higher supersaturation range. Thermodynamically, crystallization rate is determined by the constant power of the fundamental driving force; therefore, true kinetics of crystallization can be described by eq 7 only. The kinetic expression (eq 6) based on constant parameters, k and n, is applicable only in the restricted range of σ in which the experimental data were obtained to determine the parameters and cannot be used for the prediction of crystallization rates over a broader range of concentration.

Kinetics of Crystal Growth

Experimental data of the crystal growth rates of glycine at 20°C and β-succinic acid at 27.3°C reported in the literature (Li and Rodríguez-Hornedo, 1992; Mullin and Whiting, 1980) were used for our analyses. First, the supersaturation data expressed in various units were converted to be in a consistent unit of molality (mole of solute/kg of water). The data of growth rates were converted to contain the unit of kg of crystal/(m^2 of surface area • sec).

The water activity data of glycine and succinic acid solutions reported by Na et al. (1995) were used to calculate the solute activity as a function of concentration. Details of the experiments can be found elsewhere (Na, 1993). The water activity expressed as a polynomial function of molality was used in the numerical calculation of eq 4 to acquire the activity coefficient ratio as a function of concentration. The fundamental driving force $\Delta\mu/RT$ was calculated from the activity coefficient function determined at 25°C. Using the activity coefficient function at 25°C for the data at 20°C and 27.3°C can be justified by the fact that the activity coefficients of many binary mixtures have very small dependence of temperature in a moderate temperature range (Prausnitz, 1986).

The calculated ratio, γ/γ^* is plotted as a function of dimensionless supersaturation, σ (molality-based) in Fig. 1. The ratio remains very close to unity up to the supersaturation degree of σ =0.01; as σ increases, the activity coefficient deviates either positively (glycine) or negatively (succinic acid) from that of the saturated solution.

Fig. 2 shows the calculated fundamental driving force, $\Delta\mu/RT$ (= $\ln(a/a^*)$) as a function of σ. Although it is not apparent from the plots, both curves have a linear section with slope 1 in the region of σ below 0.01. The decreasing

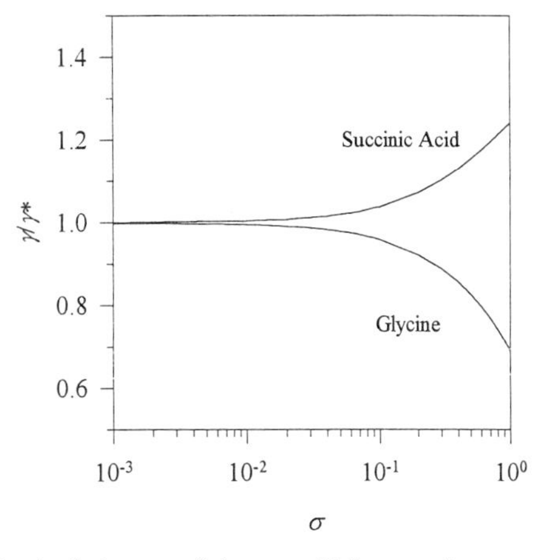

Fig. 1 Solute activity coefficient ratio versus dimensionless supersaturation (molality-based).

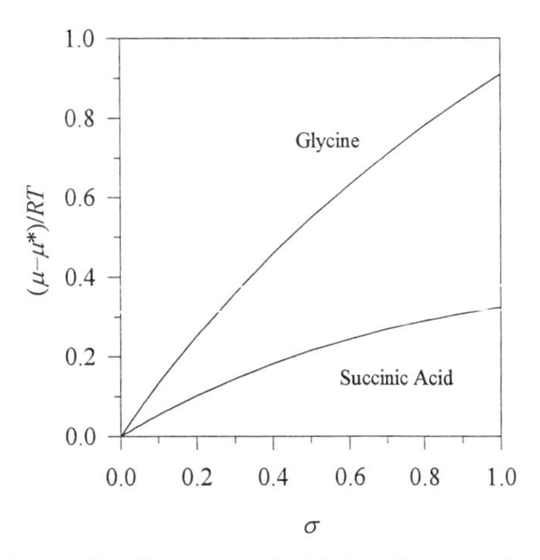

Fig. 2 The fundamental driving forces calculated as functions of dimensionless supersaturation (molality-based).

slope of the curvature at high σ (above 0.5) is due mainly to the fact that $\ln(\sigma+1)$ becomes relatively smaller than σ as σ increases.

Figs. 3 (a) and (b) show the experimental data points of the growth rate versus σ and versus $\Delta\mu/RT$. The kinetic parameters for the rate expressions of the forms eq 6 and eq 7 were determined from linear regressions of the logarithmic plots. The evaluated kinetic parameters are listed in Table 1. It should be noted that the kinetic parameters k_r and n are good for only the experimental range of σ indicated in the table, whereas k_r' and n' are valid for all ranges of supersaturation. The values of n and n' are close for the data measured at the range of very low σ (less than 0.01), in which the fundamental driving force and concentration driving force remain more or less proportional to each other. For example, the difference between of n and n' for succinic acid (measured at $0.007<\sigma<0.05$) is smaller than 0.9%. On the other hand, glycine (measured at $0.08<\sigma<0.01$) shows a greater difference between n and n' (approximately 2%).

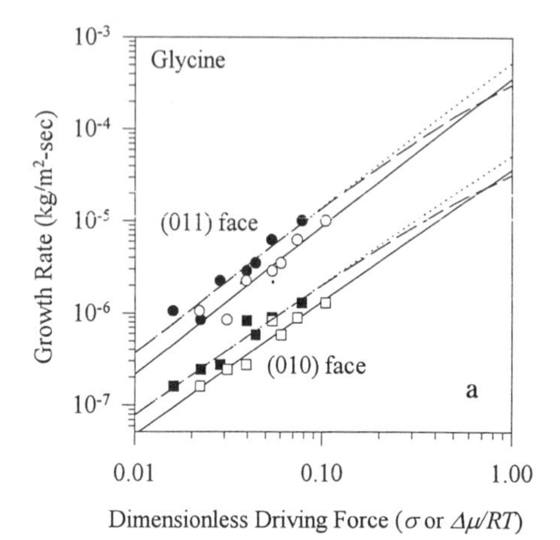

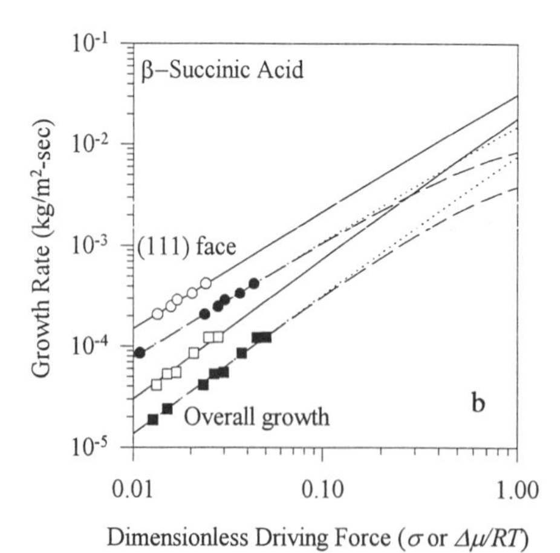

Fig. 3 The growth rates of **(a)** glycine at 20°C and **(b)** β-succinic acid at 27.3°C as functions of dimensionless driving force σ (filled symbols) or $\Delta\mu/RT$ (hollow symbols). The straight solid lines are the calculated rates from the expression $G=k_r'(\Delta\mu/RT)^{n'}$. The broken solid lines are the same rates plotted against σ. The dotted lines represent the rates based on $G=k_r\,\sigma^n$. (Experimental data adapted from Li and Rodríguez, 1992 and Mullin and Whiting, 1980.)

The solid lines (extrapolated from the growth rate data with respect to $\Delta\mu/RT$) shown in Figs. 3 (a) and (b) represent the true kinetics of crystal growth over all ranges of the driving force. The true rates of growth determined from the fundamental driving force were plotted against σ (broken solid lines). These are the growth rates that would be measured in kinetic experiments conducted over a wide range of σ. The plots clearly show that different values of the kinetic parameters, k_r and n would be obtained from experiments conducted at different ranges of supersaturation. The dotted lines, which were extrapolated from the growth rate data with respect to σ, are included only to illustrate the inadequacy of the rate expression based on σ (eq 6).

In general, the order n obtained from the experimental data of low σ (<0.01) can be considered a close approximation of the true kinetic order n'. We are interested in how n (determined from experiment) deviates from n' as σ increases. The true growth rate G can be expressed as:

$$G = k'(\Delta\mu/RT)^{n'} = k(\sigma)\,\sigma^{n(\sigma)} \qquad (8)$$

where k' and n' are constant, but $k(\sigma)$ and $n(\sigma)$ are changing functions of σ. Taking the logarithm of both sides produces:

$$\log k' + n'\log(\Delta\mu/RT) = \log k(\sigma) + n(\sigma)\log\sigma. \qquad (9)$$

Eq 9 indicates that $d(\log(\Delta\mu/RT))/d(\log\sigma)$ is equal to the ratio of the kinetic orders, $n(\sigma)/n'$.

Table 1 Kinetic parameters of crystal growth evaluated from experimental data.

Compound		Glycine		β-Succinic Acid	
References		Li and Rodríguez, 1992		Mullin and Whiting, 1980	
Face		(011) face	(010) face	(111) face	Overall growth
σ range of experiments		0.01–0.08	0.01–0.08	0.01–0.04	0.007–0.05
$G=k_r\sigma^n$	k_r	$5.4\cdot10^{-4}$	$5.3\cdot10^{-5}$	$1.53\cdot10^{-2}$	$7.73\cdot10^{-3}$
	n	1.58	1.41	1.15	1.38
$G=k_r'(\Delta\mu/RT)^{n'}$	k_r'	$3.6\cdot10^{-4}$	$3.7\cdot10^{-5}$	$3.15\cdot10^{-2}$	$1.84\cdot10^{-2}$
	n'	1.61	1.44	1.16	1.39

Therefore, in the logarithmic plots of $\Delta\mu/RT$ versus σ (shown in Fig. 4), slopes of the tangent lines at varying σ should be equal to $n(\sigma)/n'$. It is easily seen from the figure that the plots have the tangent lines with slopes very close to unity when $\sigma < 0.01$, and the slopes decrease with increasing σ. The ratio $n(\sigma)/n'$ determined from the slopes with varying σ is plotted in Fig. 5. In the crystallization of both glycine and succinic acid considered here, $n(\sigma)$ is within 5% below n' in the range of σ smaller than 0.1. As indicated in the figure, the range of σ (greater than approximately 0.2) at which $n(\sigma)/n' < 0.9$ can be considered as the concentration region where $n(\sigma)$ significantly deviates from n'.

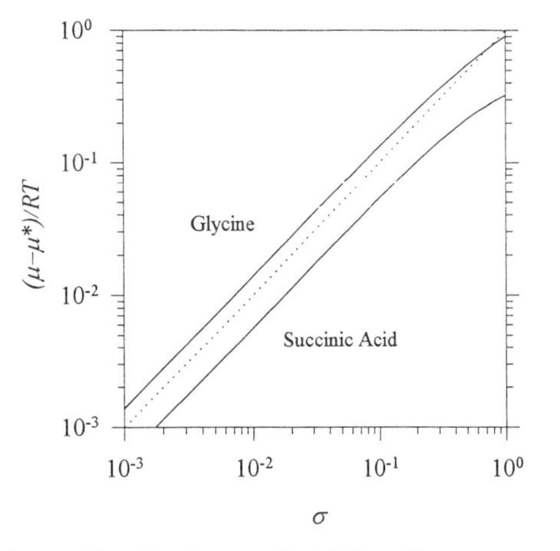

Fig. 4 The fundamental driving force versus dimensionless supersaturation based on molality.

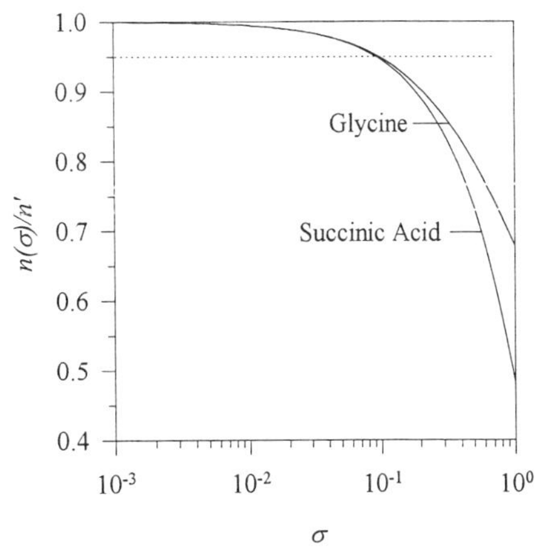

Fig. 5 $n(\sigma)/n'$ determined from the tangent lines of the plots in Fig. 4 versus σ.

Conclusions

Crystal growth kinetics of glycine and β-succinic acid were investigated in terms of the fundamental driving force of crystallization based on the chemical potential difference. The chemical potential of supersaturated glycine and succinic acid solutions were calculated as a function of concentration from the water activity of supersaturated solutions. The kinetic expressions based on the chemical potential difference are thermodynamically exact representations of the true kinetics of crystallization. The commonly used kinetic expressions based on concentration difference are in significant error at high supersaturation range, in which many crystallization processes are commonly operated.

Acknowledgment

Support of this work through NSF Grant CTS-9020233 and NASA Grant NA68-960 is gratefully acknowledged.

References

Cohen, M. D.; Flagen, R. C.; Seinfeld, J. H. *J. Phys. Chem.* **1987**, 91, 4563-4574.

Li, L.; Rodríguez-Hornedo, N. *J. Cryst. Growth* **1992**, 121, 33-38.

Mullin, J. W.; Söhnel, O. *Chem. Eng. Sci.* **1977**, 32, 683-686.

Mullin, J. W.; Whiting, M. J. L. *Ind. Eng. Chem. Fundam.* **1980**, 19, 117-121.

Na, H. S. Ph. D. Thesis; *Polytechnic University*: Brooklyn, New York, **1993**.

Na, H. S.; Arnold, S.; Myerson, A. S. *J. Cryst. Growth* **1994**, 139, 104-112.

Na, H. S.; Arnold, S.; Myerson, A. S. *J. Cryst. Growth* **1995**, 149, 229-235.

Prausnitz, J. M.; Lichenthaler, R. N.; de Azevedo, E. G. *Molecular Thermodynamics of Fluid-Phase Equilibria*; Pren tice

Hall: Englewood Cliffs, NJ, **1986**.

Robinson, R. A.; Stokes, R. H. *Electrolyte Solutions*, 2nd Ed. Butterworths: London **1959**.

The Migration of Liquid Inclusions in Solid Layers

Sabine Henning, Joachim Ulrich, Silke Niehörster, Universität Bremen, Verfahrenstechnik/FB4, Postfach 330 440, D-28334 Bremen, Germany

For economical production rates in melt crystallization processes high crystal growth rates are required. High growth rates, however, are the reason for impure mother liquor inclusions in crystalline layers. A high temperature gradient at the phase interface between the layer surface and the warm melt as well as high temperatures at the phase interface lead to migration rates of the liquid inclusions during the growth of the crystalline layer. Due to the migration of impure droplets a shift of concentration profil over the layer thickness occurrs. Experimental results of the movement of liquid inclusions in a diffusion cell are represented and discussed. The results should be used to model the migration of liquid inclusions in solid layers of melt crystallization processes.

Solid layer melt crystallization processes are mainly used to separate and purify organic substances or to concentrate compounds. The noticeable development of these processes is based on the increasing industrial demand in high pure organic substances. The design of a technical apparatus for a melt crystallization process and the model to calculate the efficiency in advance needs a good knowledge especially of growth kinetics. Additionally also solid-liquid phase transformation and the achievable separation effect is significant. As the physical mechanisms have not been fully described so far, industrial layer crystallization is generally designed and operated more or less by experience.

In order to achieve economical production rates, high growth rates are required. Higher crystal layer growth rates than those given by the 'purity criterions' (Mullins and Sekerka, 1964, Wintermantel, 1972, 1986), however, result in dendritical growth. The dendritical growth leads to undesirable local inclusions of impure melt in the crystal layer. The efficiency of separation by crystallization decreases sharply.

The driving force of solid layer melt crystallization processes is the temperature difference between the cooled phase interface and the equilibrium temperature of the warm feed melt. From literature (Scholz, 1993, Wangnick, 1994) it is known that concentration profiles over the thickness of layer exist. First experimental evidence shows that the change in the concentration observed in the crystal layer attributes to a migration of liquid inclusions. This migration is mainly induced by a high temperature gradient at the phase interface between the cooled layer surface and the warm melt as well as high temperatures at the phase interface.

Existing purification models neglect the migration of liquid inclusions. Experimental results, however, show that the movement of entraped liquid impurities have to be considered as a part of a crystallization - purification - process.

Inclusions in Crystal Materials

Single crystals and solid layers in general contain impurities which are solid, liquid or gaseous. These impurities covered by solid material are called 'inclusions'. There is a differentiation between primary and secondary (mainly liquid) inclusions. In this sence primary inclusions are formed during the growth of the crystalline material. Secondary inclusions are formed after the growth of the material. Often they are a result of crystals cracking due to internal stresses created during growth (Mullin, 1993).

The existence and the movement of inclusions in single crystals or crystal layers was already investigated by a number of scientists. Whiteman (1926) was the first who observed the migration of brine inclusions through arctic sea ice due to seasonal temperature changes. This phenomenon was further studied by many other authors (Harrison, 1965; Hoekstra et al., 1965; Kingery and Goodnow, 1963; Jones, 1973; Yamazaki et al., 1987). Temperature gradient techniques for removing solvent inclusions from solution-grown single crystals are observed (Chase and Wilcox, 1966). Besides various driving forces like thermal gradients and accelerational fields for the migration of brine droplets are studied (Cline and Anthony , 1972).

An interesting review of the topic 'inclusions' according to the latest findings in 1955 was published (Deicha, 1955).

For the purification of metallic products the effect of moving liquid inclusions was used in the 'temperature gradient zone melting techniques' studied by e.g. Schildknecht (1964) or Pfann (1966).

Another focal point of a number of investigations was the consideration of how inclusions are formed (Brooks et al., 1968; Belyustin and Fridman, 1968; Denbigh and White, 1966; Williams, 1981; Myerson and Saska, 1984; Yamamoto, 1939). It was shown that lower supersaturation at the centre of a crystal face than at the edges can be an answer of the question why cavities are often formed at a face centre (Bunn, 1949; Humphreys-Owen, 1949). A theoretical model of the development of a mother liquor inclusion has been given (Murata and Honda, 1977; Sato, 1988; Dzyuba, 1983).

Myerson and Kirwan (1977) found that growth instabilities on crystal faces can also be responsible for the entrapment of impure melt in single crystals or crystalline layers (see e.g. Fig. 1: dendritic growth).

One explanation for the migration of liquid inclusions through solid crystal layers induced by a temperature gradient was given by Wilcox (1968) and is well illustrated in Fig. 2 by Scholz (1993).

Step (a) shows the starting point of a moving droplet on its way through a crystal layer. On the basis of a temperature difference between the cooled surface and the warm melt at the layer surface (solid-liquid interface) there is a temperature gradient over the width of the droplet, too. Figure (b) shows the decreasing size of the solvent inclusion caused by crystallization of a part of the pure component at the colder side x_{cry}. The result is an increase of impurity concentration in the droplet. By adjusting the equilibrium between local temperature and concentration of the entraped melt by melting a part of the crystalline material at the warmer side (x_{melt}) the liquid inclusion is growing again (c).

On condition that the migraton of liquid inclusions through crystal layers (v_{mig}) is diffusion driven a simple approach to describe the migration of liquid inclusions mathematically was gi-

Fig. 1. Dendritic growth in melt crystallization

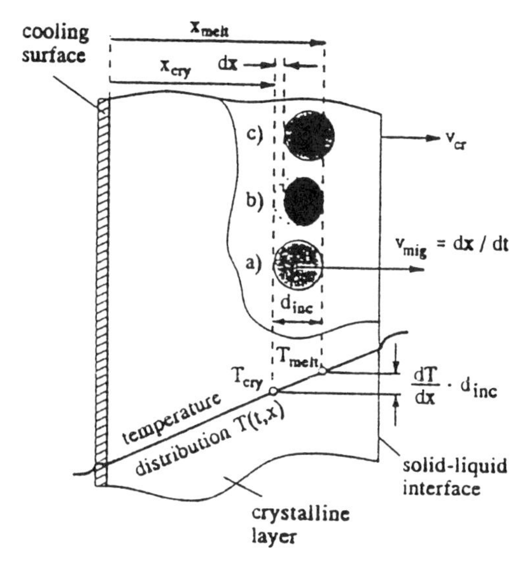

Fig. 2. Migration of a liquid inclusion in a crystalline layer

ven (Wilcox, 1968). Scholz and Wangnick (1993) developed this equation further to improve the existing crystal layer purification models.

Experimental Results

Experimental results are obtained in a 'diffusion cell' (Neumann, 1995) as shown in *Fig. 3*. This equipment consists of five discs positioned on top of each other with two small drillings (like cylindrical pores) through all of them. The bottom and the top of the diffusion cell are temperature controllable. The discs and the chambers are rotatable against each other (2). At normal position (1) the drillings in the discs and in the chambers make up a small channel over the total height of the diffusion cell. Due to different temperatures at the top (high temperature) and the bottom (low temperature) of the cell there is a linear temperature gradient over the length of the channel.

The experimental procedure is as follows: The bottom pores are filled with impure melt and crystallized. The above discs and the top chamber are turned against the bottom chamber, then they are filled with pure melt and turned back to start position after having been crystallized. The driving force for a movement (a diffusion) of the (liquid) impurities through the channel from the bottom to the top (in the direction of the warmer side) is given by the temperature difference between the upper and the lower chamber.

The migration of liquid inclusions through the diffusion cell was examinated by the variation of three parameters:

- impurity concentration (1 / 5 / 15 wt-%),
- temperature difference between the top and the bottom and (15 / 35 / 55 K) and
- duration of the experiment (24 / 48 / 72 / 96 hours).

A total view of experimental results of migration of liquid impurities through the pore channels filled with crystalline material is shown in Fig. 4. In the diagram a summary of all parameters is represented. The values for the temperature differences are given by different temperatures at the top (warm side) and at the bottom (cold side) of the diffusion cell. At starting point the impure melt is located in the bottom chamber. After different process times the impurity migrates to higher positions (from disc 5 towards disc 1) in the direction to the warm upper side. For interpretation it is important to know that the demonstrated results only show the highest position (maximal distance of migration) in the equipment which is reached by a part of liquid inclusions. There is no information given about the impurity concentration in the single discs.

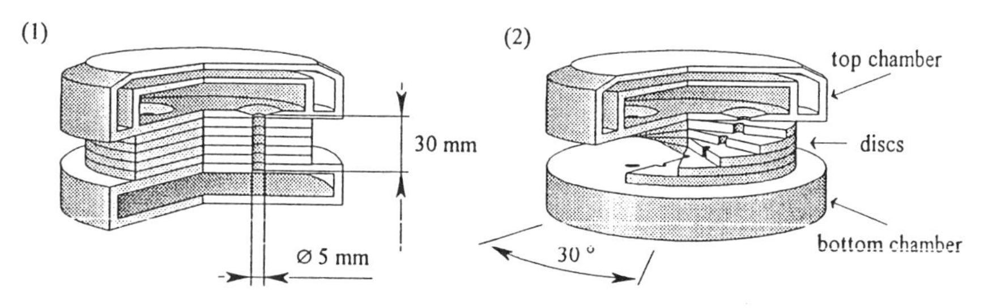

Fig. 3. Schematic description of the diffusion cell

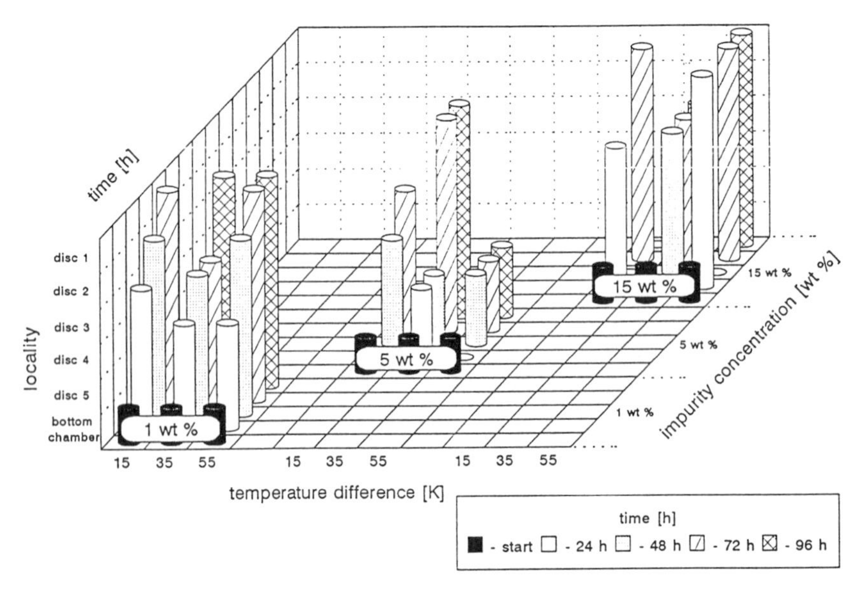

Fig. 4. Total view of experimental results

Achieved migration distances for different starting impurity concentrations are comparable. There is no clear difference in movement established between the lowest (1 wt %) and the highest (15 wt %) impurity concentration. But there is a significant decrease in the migration rate for impurities starting at concentrations of 5 wt %.
A variation of the temperature difference between top and bottom of the diffusion cell leads to highest migration distances for experiments with lowest (15 K) and highest (55 K) temperature gradients.
An increasing process time usually induces expected higher migration distances covered by liquid inclusions. There are a few exceptions in the experimental results (migration stops at a maximal point independent of process time).

The influence of the temperature gradient (dT / dx) in relation of the migration rates (v_{mig}) is shown in detail in Fig. 5 to Fig. 7. In the diagrams the migration rate is calculated by the quotient of the migration distance of the liquid inclusions (x_{mig}) and the duration of the experiment.
For impurity concentrations of 1 and 15 wt % it is shown that lowest and highest temperature gradients produce highest migration rates. So higher start concentrations usually lead to higher migration rates. Results for starting impurity concentrations of 5 wt % show the contrary.

A detailed description of the migration of liquid inclusions through the pore channels is given in Fig. 8. In this diagram it is shown that the concentration profile from the starting point turns into a distribution of the inclusions with time. The evaluation of the concentration is given by the concentration in the single disc (pore) c_{cr} normalized by the starting concentration in the bottom chamber $c_{cr,o}$. The values for the concentrations in the single pores are determined by an analysis of the samples.

With an increasing process time the impurity concentration at start position (bottom chamber $\rightarrow x_{mig} = 0$ mm) decreases. The enrichment of the pure crystalline material in the above discs ($x_{mig} = 0$ to 30 mm) with impurities gives clear evidence for the migration of liquid inclusions through the pore channels. After a process time of 96 hours a big part of impure material migrates from the cold bottom chamber ($x_{mig} = 0$ mm) to the warm top disc 1 ($x_{mig} = 30$ mm).

Discussion

The reason for the highest migration rates at lowest as well as highest temperature gradients (Fig. 4, Fig. 5, Fig. 7) has its origin in the following fact: A high temperature difference between the bottom and the top of the diffusion

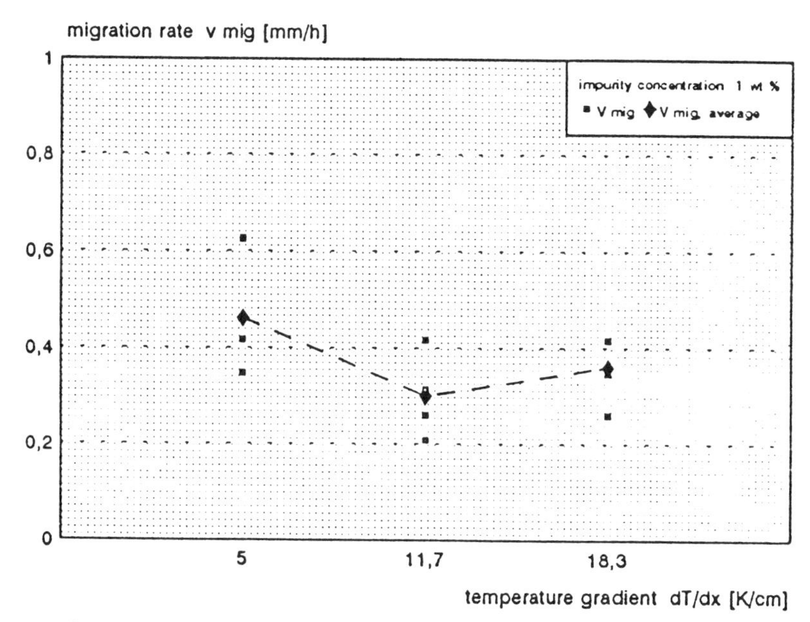

Fig. 5. Migration rates versus temperature gradients (impurity concentration 1 wt %)

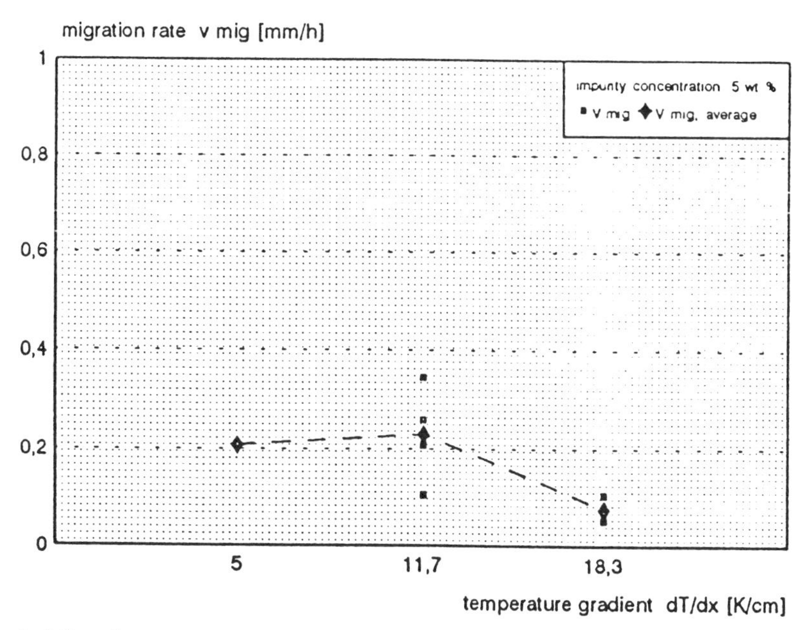

Fig. 6. Migration rates versus temperature gradients (impurity concentration 5 wt %)

cell results in a low temperature level. This has to be the case since the temperature of the melting point is always fixed. The conclusion is therefore: High temperature gradients as well as high local temperatures at the position of inclusions lead to high migration rates in crystalline layers !

Literature data and observations refering to solid layer melt crystallization processes (Scholz, 1993; Wangnick, 1994) are confirmed. For cry-

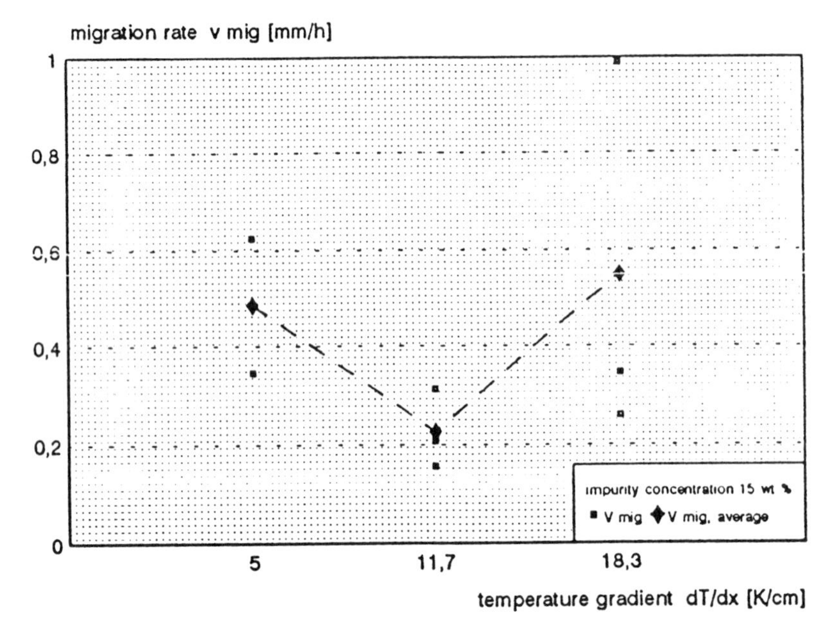

Fig. 7. Migration rates versus temperature gradients (impurity concentration 15 wt %)

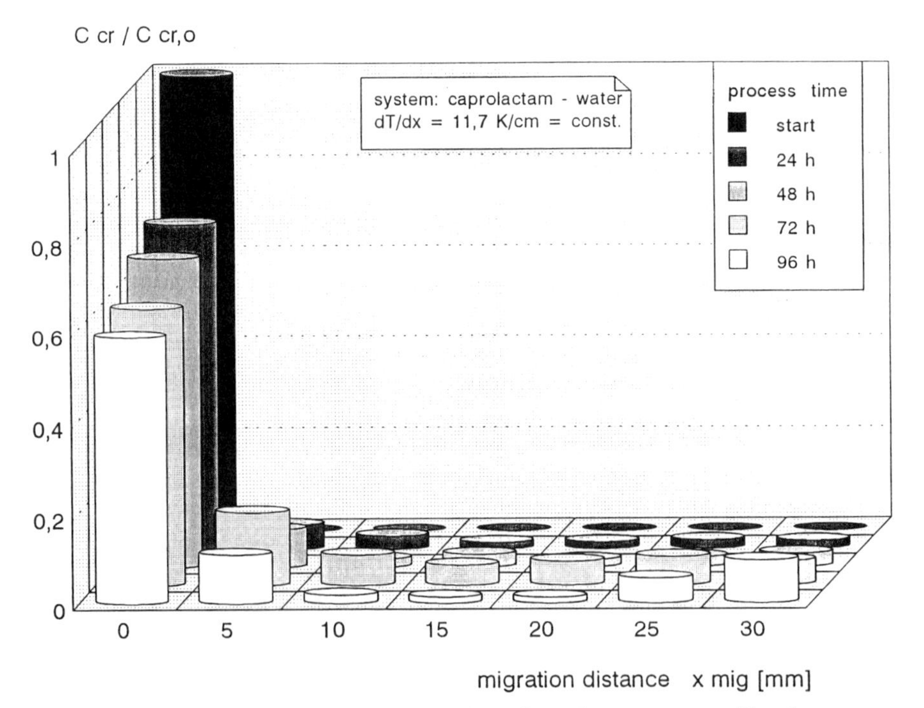

Fig. 8. The migration of liquid inclusions through a pure crystalline layer

stallization processes it has to be considered that the parameters 'high temperature gradient' and 'high temperature level' will not be working against each other like they do in the experiments with impurity start concentration of 5 wt % (Fig. 4, Fig. 6). The migration rate is a function of the impurity concentration over the pore length. Higher concentration gradients usually lead to higher migration rates of liquid inclusions through crystalline materials.

It has been generally shown that a shift of the concentration profil over a crystal layer takes place while the layer is growing. The represented changing in the distribution of the impurity concentration (Fig. 8) is caused by the migration of liquid inclusions. The experimental proof for the migration is to be found in:

1.) The values for the received migration rates are in the area of those literature data which obtained migration rates (0,015 mm/h $\leq v_{mig} \leq$ 4 mm/h) for liquid inclusions through single crystals as well as through crystal layers.

2.) The optical evaluation of the cylindrical crystalline samples out of the pores (discs) shows a clear migration of coloured impurity not in single droplets but as a piston. Therefore investigated migration rates are an average value of a multitude of small single liquid inclusions building a moving impurity front.

Experimental uncertainties or errors result for example from the following facts:

- inaccuracies due to optical evaluations (Fig. 4, Fig. 5 - 7),
- evaluation of maximally distances covered by the impurities without considering the mass distributions,
- different inclusion diameters,
- different structures of the cystalline material,
- an interrupted contact between the impure material in the bottom chamber and the pure crystalline material in the upper discs.

Up to now the migration of liquid inclusions is described by a simplified mathematical approach which neglects parameters like inclusion diameters, layer structures, etc. In the area of high temperature gradients there is a great difference between theoretical calculated and experimental results. An improvement of mathematical models is necessary to calculate the migration step further in detail .

Summary

For economical production rates in solid layer melt crystallization processes high crystal growth rates are required. High crystal growth rates lead to liquid inclusions. The migration of liquid inclusions in crystal layers is a proven fact. High temperature gradients as well as high temperature levels at the position of the inclusion lead to high migration rates of inclusions through the crystalline material. Migration rates are also influenced by the height of the impurity concentration .

Purification models for a description of the separation efficiency of solid layer melt crystallization processes do not consider the migration of liquid inclusions. Existing mathematical approaches to describe the impurity movement are strongly simplified. Therefore a description under consideration of the migration of the liquid inclusions for solid layer melt crystallization processes is desirable.

Nomenclature

c_{cr}	impurity concentration in the crystalline material (end of the process)
$c_{cr,o}$	impurity concentration in the bottom chamber (startposition)
d	diameter
dT/dt	linearized temperature gradient in the crystal layer
t	time
T	temperature
v_{cr}	growth rate of crystal layer
v_{inc}	migration rate of liquid inclusions
v_{mig}	migration rate of liquid inclusions
x	radial coordinate
x_{mig}	distance of migration

Subscripts

cr	crystal layer
cry	crystallizing side of a liquid inclusion
inc	inclusion
melt	melting side of a liquid inclusion
mig	migration
o	start position

Acknowledgement

The authors gratefully acknowledge the support of the 'Deutsche Forschungsgemeinschaft DFG' which made this research possible.

References

Belyustin, A. V.; Fridman, S. S. Sov. *Phys. Crystallography* **1968,** *13,* 298-300.

Brooks, R.; Horton, A. T.; Torgesen, J. L. *J. Crystal Growth* **1968,** *2,* 279-283.

Bunn, C. W. *Discussions of the Faraday Society* **1949,** *5,* 132-144.

Chase, A. B.; Wilcox, W. R. *J. Am. Ceram. Soc.* **1966,** *49,* 460.

Cline, H. E.; Anthony, T. R. *J. Crystal Growth* **1972,** *13/14,* 790-794.

Deicha, G. *Lacunes des Cristeaux et leurs Inclusions Fluides;* Masson, Paris, 1955.

Denbigh, K. G.; White, E. T. *Chem. Ing. Sci.* **1966,** *21,* 739-754.

Dzyuba, A. S. *Sov. Phys. Crystallography* **1983,** *28,* 111-112.

Harison, J. D. *J. Appl. Phys.* **1965,** *36,* 3811.

Hoekstra, P.; Osterkamp, T. E.; Weeks, W. F. *J. Geophys. Res.* **1965,** *70,* 5035-5041.

Humphreys-Owen, S. P. F. *Proc. of the Royal Soc.* **1949,** *A197,* 218-237.

Jones, D. R. H. *J. Crystal Growth* **1973,** *20,* 145-151.

Kingery, W. D.; Goodnow, W. H. *Ice and Snow;* Kingery, W. D., Ed.; Chap. 19; M. I. T. Press; Cambridge, 1963.

Mullin, J. W. *Crystallization;* 3rd ed.; Butterworth-Heinemann Ltd.; Oxford, 1993.

Mullins, W. W.; Sekerka, R. F. *J. of Applied. Physics 35* **1964,** *2,* 444-451.

Murata, Y.; Honda, T. *J. Crystal Growth* **1977,** *39,* 315-327.

Myerson, A. S.; Kirwan, D. *J. Ind. and Eng. Chem. Fund.* **1977,** *16,* 414-424.

Myerson, A. S.; Saska, M. *AIChEJ* **1984,** *30,* 865-867.

Neumann, M. Ph.D. Thesis, University of Bremen, Bremen, 1995.

Pfann, W. G. *Zone Melting;* Metallurgical Research Laboratory; New York, 1966.

Sato, K. Jap. *J. of Appl. Phys.* **1988,** *19,* 1257 -1264 and 1829-1836.

Schildknecht, H. *Zonenschmelzen;* Weinheim Verlag Chemie, 1964.

Scholz, R. Ph.D. Thesis, University of Bremen, Bremen, 1993.

Scholz, R.; Wangnick, K. *J. Phys. D: Appl. Phys.* **1993,** *26,* B156-B161.

Wangnick, K. Ph.D.Thesis, University of Bremen, Bremen, 1994.

Whiteman, W. G. *Am. J. Sci.* **1926,** *11 (62),* 5th series, 126-132.

Wilcox, W. R. *Ind. Eng. Chem.* **1968,** *60,* 13 -23.

Williams, A. J. Explosives Research Establishment, Waltham Abbey (private communication), 1981.

Wintermantel, K. Ph.D. Thesis, TU Darmstadt, Darmstadt, 1972.

Wintermantel, K. *Chem. Ing. Tech. 58* **1986,** *6,* 498-499.

Yamamoto, T. *Sci. Papers of the Inst. of Phys. Chem. Res. Japan* **1939,** *35,* 228-289.

Yamazaki, Y.; Kashima, K.; Toyokura, K. 10th Symp. on Ind. Cryst., Bechyne, CSFR, 1987.

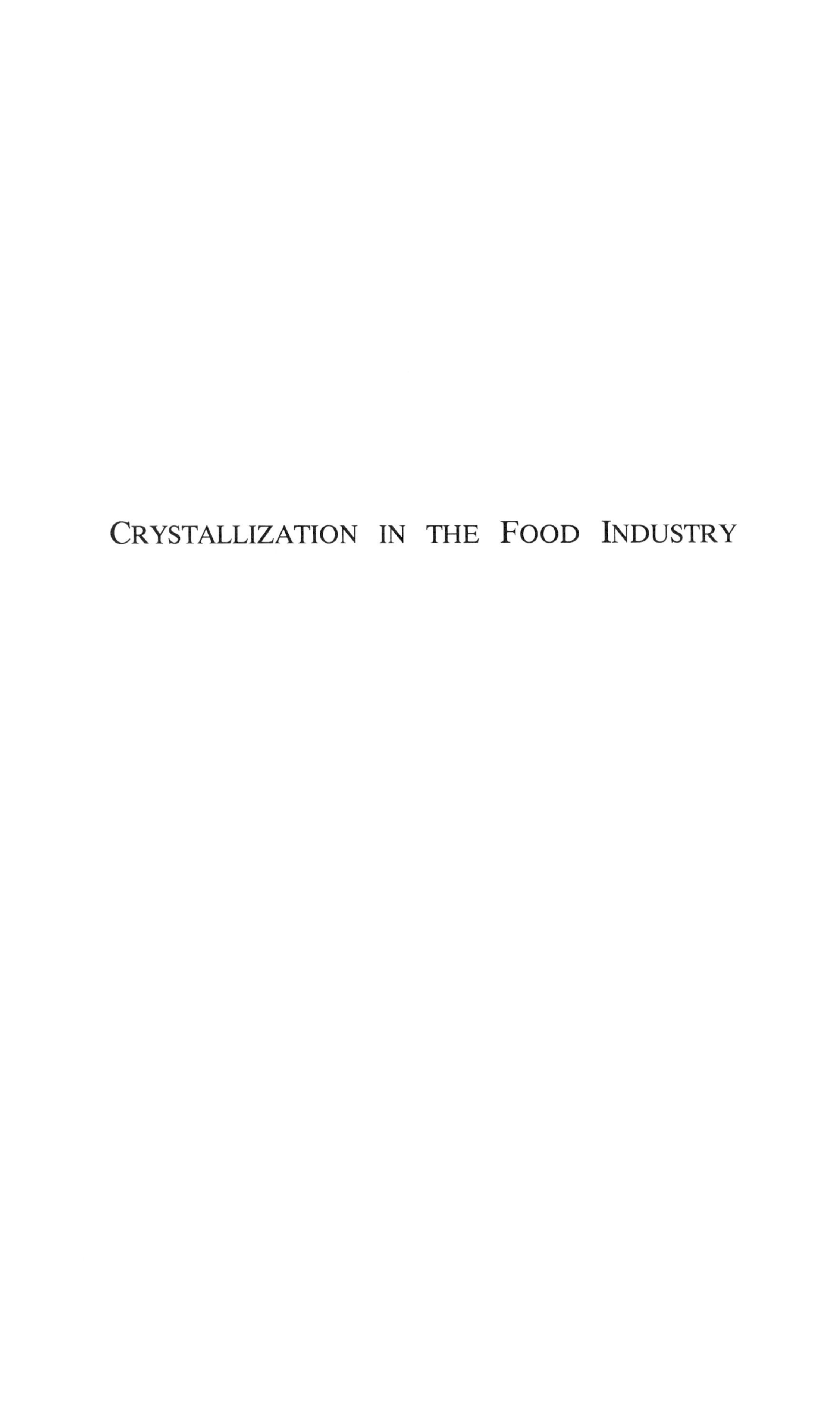

Crystallization in the Food Industry

CONTROLLING CRYSTALLIZATION IN FOODS

Richard W. Hartel, Department of Food Science, University of Wisconsin, Madison, WI 53706

Formation of crystals during processing and storage of foods plays a critical role in product texture, quality and shelf-life. In some foods, crystals are desired to give the appropriate product quality. In other foods, the occurrence of crystals is a defect which detracts from product quality. By understanding the principles of crystallization processes, food manufacturers formulate and process foods to control formation of crystalline structure.

Foods are primarily comprised of water, sugars, proteins, starches, salts and other minor components. Each of these main substances can crystallize, and affect the structure and texture of foods. The food manufacturer must control the composition of the food and the conditions under which they are processed in order to control crystallization processes for these products. In contrast with other industries, crystallization does not lead to a separation process, but rather, crystallization is a phenomenon that occurs within the food to affect its quality.

Examples of foods that contain ice crystals include ice cream and frozen foods. For sugars, many candies (fondant, cream, fudge, etc.) contain sugar crystals, as do sugar coated cereals, and icings. Control of fat (lipid) crystallization is important in chocolates, compound coatings, butter, margarine and peanut butter.

Crystallization often occurs during the food manufacturing process. Controlling crystallization to obtain the proper number, size and shape of crystals impacts product quality (Hartel, 1993). Usually, a large number of very small crystals is desired to provide a smooth texture. Ice cream and fondant or creams are good examples of foods where a large number of small crystals is desired. Controlling crystallization requires an understanding of kinetic rates of nucleation and growth, which are both affected by formulation (or composition) and processing parameters (temperature, drying, etc.).

Sometimes, foods are produced in such a way that a metastable, supersaturated state is formed during processing. For example, hard candies are sugar glasses with extremely high sucrose supersaturation. Any moisture uptake or temperature increase causes these candies to crystallize or grain. This is a negative quality attribute, as dye and flavor molecules, which are evenly dispersed throughout the glassy state, accumulate around grain boundaries and cause unsightly product with reduced flavor (Minifie, 1989).

Another crystallization problem in foods occurs due to changes in crystalline structure during storage. A crystal size distribution that has reached some quasi-equilibrium with the surrounding matrix undergoes slow, but inevitable, change as the crystal surface approaches a more favorable equilibrium (lower surface energy). This ripening, or recrystallization, occurs in such products as ice cream and frozen foods (Fennema, 1973). A smooth ice cream can become extremely coarse (due to ice recrystallization) over a period of a few weeks during storage at normal freezer conditions (Donhowe et al., 1991).

The food manufacturer must understand the basic principles of crystallization and the driving forces for change in order to control quality and shelf-life of the product. Several specific examples will be discussed here, including formation of sugar crystals in thin films, production of a sucrose fondant, and recrystallization of ice in ice cream.

Phase Behavior

Phase equilibria in foods determine the driving force for crystallization, either during manufacture or during storage. However, the phase behavior of a food product is often difficult to determine or predict due to the range of components (from lipids to sugars). In some products, simple mixtures can be used to estimate the phase boundaries. An example of a phase diagram for a simple mixture of two sugars (lactose and sucrose) in water, supplemented by a glass transition curve, is shown in Figure 1 (Hartel, 1993). Such a phase or state diagram approximates the conditions in caramels or the unfrozen portion of ice cream.

In Figure 1, several different phase or state boundaries are depicted. The solubility values for pure lactose and sucrose are shown, along with the freezing curve for the solvent, water. Note that the freezing point of water decreases as the sugar content increases, due to freezing point

1054–7487/96/0172$12.00/0

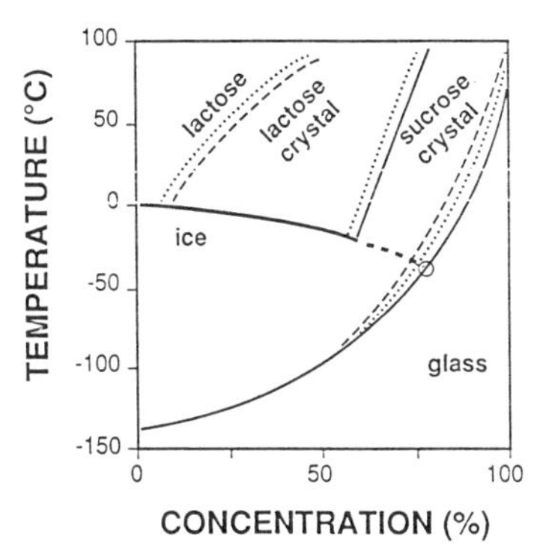

Figure 1. Phase/state diagram for sucrose and lactose mixtures. Dotted lines indicate mixed system conditions. (Reproduced with permission from Hartel. Copyright 1993 Institute of Food Scientists).

depression of the sugars. This is important during processing of frozen foods, where an increase in concentration of the unfrozen phase during freezing causes a shift on the phase diagram.

Not all boundaries on this diagram phase equilibria. The boundary between a concentrated liquid phase and an amorphous glassy state is also shown, as the glass transition curve. This curve extends from the glass transition for pure sugar (no water) down to the glass transition for pure water (at about -135°C). For pure lactose, the glass transition is about 100°C while, for pure sucrose, the glass transition temperature occurs at about 65°C (Roos et al., 1991). The glass transition curve should be viewed as a metastable limit, where below this limit, the system exists in a metastable state. Above the glass transition curve, the system becomes unstable and changes occur.

The phase boundaries shown in Figure 1 are not independent. The solubility of sucrose is affected by the presence of lactose, and vice-versa (Nickerson et al., 1972; Hartel et al., 1991). These effects are indicated by the dashed lines, showing the reduction in solubility for the mixed sugars. In addition, the glass transition curve for the mixture falls somewhere between the pure glass transition curves (Roos and Karel, 1991). Since both sugars have the same molecular weight in this example, the freezing point depression curve is unaffected by the mixture of sugars. For complex food products, the actual phase/state diagram may be quite difficult to obtain or estimate.

Crystallization Kinetics

In order to control crystallization in foods, the kinetics of nucleation and growth must be known for the conditions existing during processing and storage. Induction time and rate of nucleation, as well as rate of crystal growth must be determined. For complex foods, these can be difficult to ascertain over the wide range of conditions important in processing and storage.

Typical nucleation and growth rate kinetic curves are shown in Figure 2. Here, crystallization rates increase initially as driving force increases (either increasing concentration or decreasing temperature), but then decrease again to zero at the glass transition point. At this point, viscosity is sufficiently high that crystal formation or growth can not occur. Thus, there is a maximum in the rate curves for nucleation and growth. For products where formation of a large number of small crystals is desired (as in many confections, frozen foods and ice cream), nucleation should occur at the peak of the rate curve. This results in the large number of crystals needed for maintaining small average size, and smooth consistency. Processing conditions must be chosen so that nucleation does not occur prior to attaining this point in temperature and concentration. For example, fondant, a highly crystallized sugar syrup which contains many small crystals (ideally less than 10-15 μm), is cooled rapidly to the optimal temperature for nucleation on a drum cooler and then agitated to promote spontaneous nucleation. Should nucleation occur at some other point on the rate curve (either above or below the optimal point), fewer crystals would be formed, each of which could grow to larger size. At some point, these crystals become detectable on the palate, and result in a product with a coarse or grainy texture.

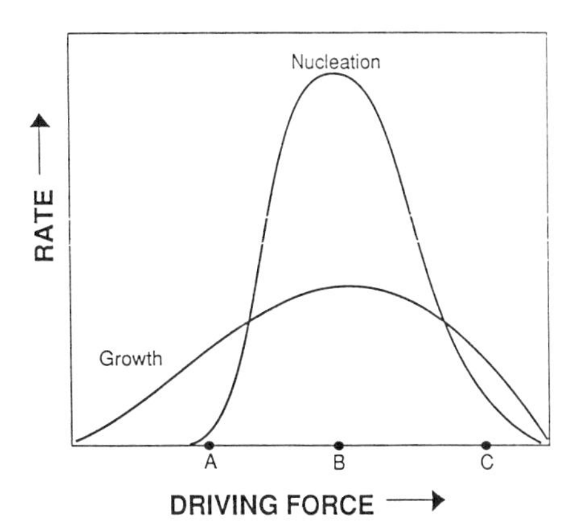

Figure 2. Generalized kinetic curves for nucleation and growth. (Reproduced with permission from Hartel. Copyright 1993 Institute of Food Scientists).

For other products where crystallization must be avoided, the induction time prior to nucleation is the important piece of rate data. Figure 3 shows the induction times necessary for onset of sugar crystallization (in a pure system) at 30°C for different concentrations. Induction time rapidly falls to zero at supersaturations of 1.2 or so indicating that spontaneous nucleation occurs. However, as the glass transition is approached, at about 95% concentration for 30°C, induction times increase again. At concentrations above the glass transition point, nucleation is effectively inhibited by the lack of mobility at the extreme viscosity. Both processing conditions (e.g., agitation rates) and formulation of the product determine the shape of these rate curves.

Both process parameters and formulation factors have significant effect on these kinetic rate curves. Typically, addition of other components or impurities (in a formulation) causes an inhibition of nucleation and growth rates (Mullin, 1972). The only exception to this is addition of particulate material, which may promote nucleation by providing nucleation sites. However, very little quantitative data exists for predicting nucleation and growth kinetics for all but the simplest foods. Process parameters that influence crystallization kinetics include rate of cooling, temperature, agitation, and drying rate. The effect of these parameters on crystallization kinetics must be understood to allow control of crystallization during food manufacture. Storage parameters, such as temperature and relative humidity, also influence changes in crystal size distribution after the food product has been made.

Crystallization in a Sucrose Fondant

Fondant is a highly crystallized sucrose syrup used for production of icings and creams, and when flavored, as filling for chocolate-coated confections (Minifie, 1989). A sucrose solution is evaporated to high concentration, usually in a thin-film evaporator to minimize color changes, before cooling to crystallization temperature. In one fondant processing system, cooling occurs on a drum cooling device, where a thin film of sucrose syrup is cooled from within the drum. Some additional evaporation occurs during cooling as well. The cooled and concentrated mass is then input to the crystallizer, which is a pin working unit. At the crystallization temperature, severe agitation or beating of the sugar mass causes rapid nucleation. Many crystals are formed at this condition, at the optimal temperature for nucleation.

Control of conditions, such that the relative rates of nucleation and growth promote

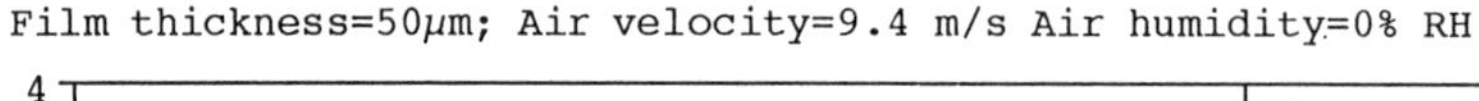

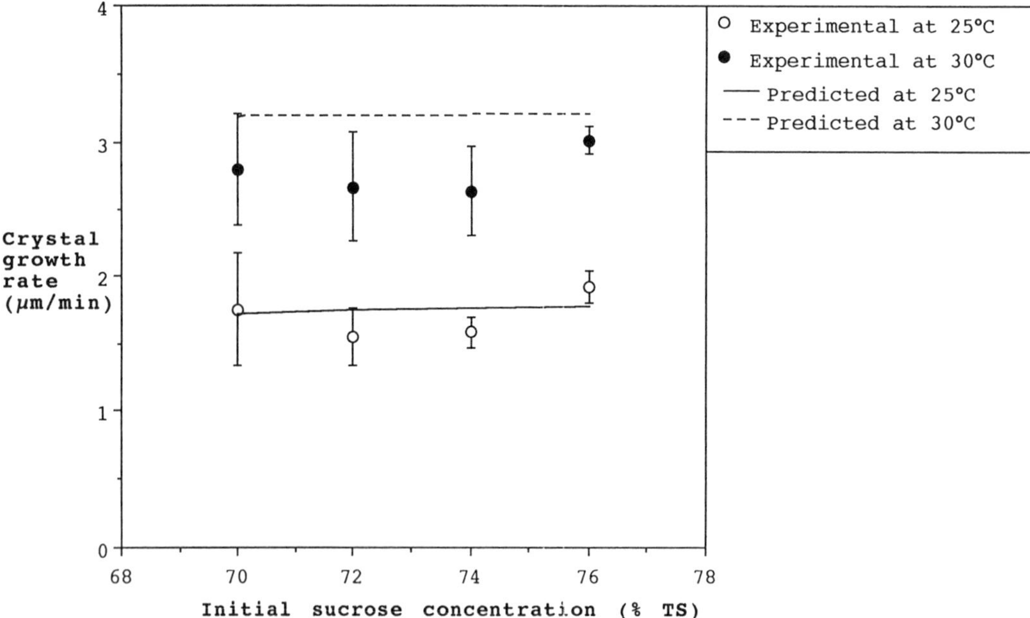

Figure 3. Effect of initial concentration on crystal growth during drying of sucrose solution in thin films. Film thickness = 50μm; Air velocity = 9.4 m/s; Air humidity = 0% RH.

desupersaturation primarily due to nucleation, result in production of a size distribution with many small crystals. These small crystals provide texture to the fondant without being detectable in the palate. Note that sugar crystals should generally be smaller than 15 μm to be below the detection limit (Minifie, 1989).

These fine crystals in fondant are prone to recrystallization during storage, where smaller sugar crystals grow into larger ones as the system approaches a more global equilibrium. Regions of higher curvature in crystals have slightly lower equilibrium with the surrounding solution and eventually will become smoother. The result is that a distribution of many, small crystals becomes coarser over time, and may result in detectable grittiness in the fondant. This process of recrystallization will be discussed in greater detail in the section on ice crystals in frozen desserts.

Crystallization of Sugar During Drying of Thin Films

In some food applications, a relatively thin (anywhere between 10 to 200 μm) film of sugar syrup is deposited on a food surface and crystallization controlled to give the desired product characteristics. Many cereals, for example, are coated with a thin sugar film that provides a frosty appearance when properly crystallized. In addition, the quality of panned candies (sugar coated pieces) is dependent on controlling sugar crystallization in a thin film as the piece is tumbled in a rotating panning device. A thick sugar shell is built up through consecutive processes of spraying with sugar syrup, and allowing each layer some time to dry before applying another syrup layer. Some pharmaceuticals are sugar coated in a similar procedure. Drying of the thin sugar film occurs at the same time as crystallization is taking place, so that each process significantly influences the other. In order to control the panning process, a better understanding of the competitive effects of drying and crystallization is needed.

During crystallization with drying, concentration of the solution phase is driven by both processes. Drying forces the concentration to increase as water molecules are removed from the entire film, whereas crystallization causes the concentration to decrease as sugar molecules are removed from solution. Yet concentration controls both processes through viscosity and diffusivity, and strongly influences crystallization through supersaturation levels. The rate of panning thus depends on how concentration changes during the process, which in turn depends on the rates of drying and crystallization.

Experimental results on crystallization and drying in thin sugar films have been confirmed by simulation of a computer model to solve the simultaneous heat and mass transfer equations coupled with a crystallization kinetic expression (Shastry, 1994). Thin sucrose films (about 50 μm) were produced on a microscope stage apparatus that allowed control of film thickness, initial sucrose concentration, air velocity across the film, and air temperature and humidity. Seed crystals were added to promote crystal growth, and growth rate characterized, as change in crystal size with time, using videomicroscopy. Initial growth rates were linear at all conditions, although growth slowed considerably as crystallization approached completion.

A computer model, solved by simple finite difference technique, was used to simulate the experimental crystallization and drying process. Heat and mass transfer equations, accounting for changing diffusivity with concentration and film shrinkage according to the method of Okazaki et al. (1974), were solved simultaneously with predetermined crystal growth kinetic expressions to predict initial crystal growth rate and concentration profiles throughout the film.

Figure 3 shows that experimental and predicted values for initial growth rate are quite similar for temperatures of 25 and 30°C, and a range of initial sucrose concentrations. Growth rates at 30°C were substantially higher than at 25°C, whereas initial sucrose concentration (from 70 to 76%) did not affect growth rate. This latter result is most likely due to the rapid increase in concentration due to drying. In fact, predicted concentration profiles show that concentration increased rapidly at first due to drying, followed by a slow decrease due to crystallization. Figure 4 schematically shows the shape of these curves and demonstrates that both drying and

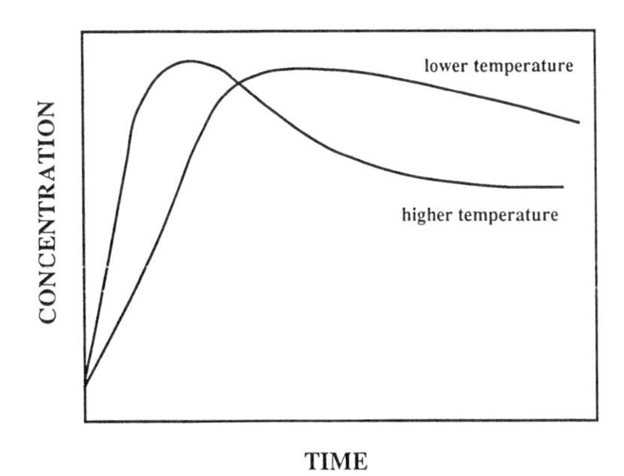

Figure 4. Schematic depiction of change in sucrose concentration during drying and crystallization in thin sugar films.

crystallization occur more rapidly at higher temperature. This would be expected based on the dependency of diffusivity on temperature.

Experimental and predicted results also show that, under the conditions studied, external air conditions have no effect on crystallization and drying. That is, changes in air velocity and relative humidity have no effect on initial growth rates or drying rates. Thus, we conclude that internal mass transfer limits the drying process under these conditions, and that external mass transfer processes are sufficiently rapid as to have negligible effect.

Control of Ice Crystals During Manufacture and Storage of Frozen Desserts

The quality of ice cream, other frozen desserts and frozen foods in general, is highly dependent on the size of ice crystals formed initially and the extent to which they change during distribution storage. Many small ice crystals are desired to provide the smooth consistency of a high quality frozen dessert (Arbuckle, 1989), while small crystals in frozen foods provide improved integrity of the thawed product (Fennema et al., 1973).

During manufacture of ice cream and frozen desserts, nucleation is controlled at low temperatures and agitation to initiate high rates of nucleation. Vaporizing ammonia on the outside of the barrel of a scraped surface freezer supplies refrigeration to the ice cream. Low temperatures (about -30°C) at the inside of the freezer barrel cause rapid nuclei formation. Scraper blades clean these nuclei off the wall on a regular basis and disperse them into the warmer interior of the freezer. There, the ice nuclei ripen into irregular shaped, block crystals. It has been estimated that there are about 4 x 10^9 crystals per liter of ice cream (Berger et al., 1979), with an average size that lies somewhere around 35 μm, depending on the method of manufacture. About half of the available water is frozen when the ice cream exits the scraped surface freezer at a typical draw temperature of -6°C. At this point, ice cream is packaged into appropriate containers for further processing.

Ice cream is hardened for 24 hours in a deep freezer, with air temperatures of -30°C to bring the ice cream temperature down to about -18°C (Arbuckle, 1989). Due to the changing equilibrium (due to freezing point depression), some additional ice crystal growth occurs in the hardening room. Ice crystal size increases to about 45 μm, and it has been estimated that between 75 and 80% of the water is now frozen.

For distribution and storage, ice cream may be held at temperatures anywhere from -20 to -5°C. Temperature variations occur as product is transported and shelved, and temperature cycling occurs at any point in the distribution system due to refrigeration cycles. These temperature swings cause equilibrium shifts in ice cream that have significant effect on ice crystal size (Arbuckle, 1989). The increase in size upon storage of frozen desserts is called recrystallization and arises due re-equilibration of the ice crystal distribution. Significant recrystallization occurs, however, even at constant temperature (Donhowe, 1993).

Some recent work has quantified the kinetics of ice recrystallization in ice cream at different thermal conditions (Donhowe, 1993). The effects of ice cream composition (formulation) on recrystallization kinetics have also been studied (Hagiwara and Hartel, 1995).

Ice cream samples were either stored in half-gallon containers in temperature controlled cabinets or were exposed to controlled temperature variations on a microscope slide (Donhowe, 1993). Mean size of the distribution of ice crystals was followed with time, using image analysis of photomicrographs of ice crystals taken in a refrigerated glove box apparatus, to measure recrystallization rate. A linear relation was found between mean size and time to the one-third power for all conditions studied. Storage temperatures were between -20 and -5°C, with thermal fluctuations between constant temperature (±0.01°C) and ±2°C. Recrystallization rate was then defined as the slope of this line, and given in units of $\mu m/day^{1/3}$.

Figure 5 (4-25a) shows how recrystallization rate depended on thermal and storage conditions (Donhowe, 1993). Accelerated recrystallization conditions (on the microscope slide) gave the most rapid recrystallization rate when temperature fluctuated (at ±1.0°C), but gave the slowest recrystallization rate when temperature was constant (±0.01°C). Storage in half-gallon containers (bulk storage) gave intermediate results under both conditions. Fluctuating temperatures gave significantly faster recrystallization rates than constant temperature, but significant recrystallization still occurred even at constant temperature. The rate of recrystallization increased dramatically as storage temperature increased. In fact, storage at -20°C gave recrystallization rates of only a few μm per $day^{1/3}$.

More recent work has shown the effects of different composition on recrystallization rate in ice cream (Hagiwara et al., 1995). In a study similar to that described above, the effects of different sweeteners on recrystallization rate was investigated. Recrystallization was found to vary with the type of sweetener used, and depended primarily on the amount of water frozen at the

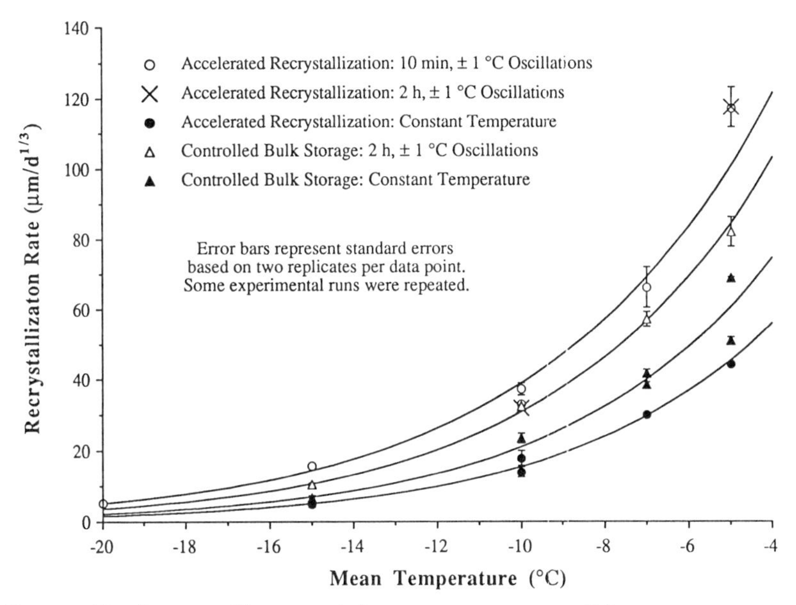

Figure 5. Recrystallization rate of ice crystals in ice cream stored at different temperatures and storage conditions.

storage temperature. Ice creams made with sweeteners that gave a large freezing point depression had less water frozen at a given temperature and were more prone to recrystallization. Conversely, ice creams made with sweeteners that gave higher freezing point had more water frozen at a given temperature, and were less likely to show recrystallization. The addition of a commercial stabilizer made with a blend of locust bean gum and carrageenan inhibited recrystallization, but the extent of inhibition was dependent on the storage temperature and the type of sweetener used.

Summary

Through these three examples, the complexity of crystallization processes in foods has been shown. There is much yet to learn in order to adequately control crystallization (of sugars, ice and lipids) in foods, to provide finished products with high quality.

References

Arbuckle, W.S., *Ice Cream*, 4th Ed., Van Nostrand Reinhold: New York, NY, 1986.

Berger, K.G.; White, G.W., in *Food Microscopy*, (Vaughn, J.G., ed.), Academic Press: New York, NY, 1979.

Donhowe, D.P., PhD Dissertation, University of Wisconsin, Madison, WI, 1993.

Donhowe, D.P; Hartel, R.W.; Bradley, R.L., *J. Dairy Sci.*, **1991**,*74(10)*, 3334.

Fennema, O.; Powrie, W.D.; Marth, E.H., *Low-Temperature Preservation of Foods and Living Matter*, Marcel Dekker: New York, NY, 1973.

Hagiwara, T.; Hartel, R.W., *J. Dairy Sci.*, **1995** (in press).

Hartel, R.W., *Food Technol.* **1993**, *47(11)*, 99-107 .

Hartel, R.W.; Shastry, A.V., *Crit. Rev. Food Sci. Nutr.*, **1991**, *30(1)*, 49.

Minifie, B.W., *Chocolate, Cocoa, and Confectionery*, 2nd Ed., Van Nostrand Reinhold: New York, NY, 1989.

Mullin, J.W., *Crystallisation*, Butterworth: London, 1972.

Nickerson, T.A.; Moore, E.E., *J. Food Sci.*, **1972**, *37*, 60.

Okazaki, M.; Shioda, K.; Masuda, K.; Toei, R. *J. Chem. Eng. Japan.* **1974**, *7(2)*, 99-104.

Roos, Y.; Karel, M., *Food Technol.* **1991**, *45(12)*, 66-71, 107.

Carminic Acid as a Fluorescent Probe of Sugar Solution Composition

Robert J. Richards, Beatrice A. Torgerson, Department of Chemical Engineering Michigan State University, East Lansing, Michigan 48824 USA
Kris A. Berglund, Departments of Chemical Engineering and Chemistry, Michigan State University, East Lansing, MI 48824 USA

Emission spectroscopy was evaluated as means to probe the solute concentration of carminic acid-doped sugar solutions. Fluorescence emission spectra of carminic acid in aqueous solutions of glucose and sucrose were analyzed using excitation at 310 nm. Carminic acid in aqueous glucose solutions showed an exponential dependence of the peak intensity ratio (PIR) of the 594 nm band to that of 450 nm band. Additionally for sucrose solutions excited at 310 nm, there was a strong dependence of the PIR between the species at 594 nm to that of the emitting species at 450 nm.

The molecular interactions that occur in supersaturated solutions are essential to crystal nucleation and growth. Much work on understanding these phenomena has been done in recent years. Previous investigators have studied the nature of molecular association and solute organization in supersaturated solutions using a variety of techniques. (Myerson and Sorrel, 1982, Myerson and Chang, 1984) undertook diffusion studies in a variety of electrolytes (NaCl, KCl) and non-electrolytes (urea, glycine, glycine-valine) indicating the formation of aggregates in solution. (Larson and Garside, 1986, Larson and Garside, 1986) observed concentration gradients in isothermal columns of supersaturated citric acid, potassium sulfate, urea and sodium nitrate, which suggested solute structuring.

Attempts to characterize solute-solvent interactions in aqueous sugar solutions have been carried out using several methods. Allen et al. (1972) performed studies with sucrose similar to Garside and Larson which showed that isothermal columns of supersaturated solutions contained concentration gradients. Sugget and Clark (1976) have used NMR and dielectric relaxation techniques to identify the interaction of water in sugar solutions, namely the presence of associated water and bulk water. Richardson, et al. (1987) used ^{2}H and ^{17}O high field NMR to show water mobility in sucrose solutions as a function of concentration. Similarly, Hills (1991) investigated the hydrogen exchange kinetics and water reorientation dynamics in aqueous glucose solutions.

The use of extrinsic fluorescent probes provides another method to determine solution structure. A recent investigation by Chakraborty and Berglund (1993) using trace amounts of pyranine in aqueous sugar solutions showed that relative amounts of associated water and bulk water could be detected by monitoring the relative amount of protonated and deprotonated forms in solution. By relating the emission intensities of the protonated and deprotonated forms, they showed that relative amounts of associated and bulk water can be determined in a given solution. The use of pyranine has been demonstrated to work well for the studies of water activity in aqueous sugar solutions, but it is only a D and C approved dye, thus it is not allowed for direct application in foods.

In this study we investigated the use of carminic acid, a naturally occurring, food dye (The Merck Index, 1989) as a probe of sugar solutions. Its ground state electronic spectrum is known to be influenced by pH (Schwing-Weill, 1986) and we found similar behavior in the excited state. As can be seen from Fig. 1 (The Merck Index, 1989), carminic acid has a glucose group in its structure, which may allow it to interact more easily with sugar molecules in an aqueous solution. The glucose structure may also allow it to function as a tailor-made impurity (Black et. al., 1986), which will be discussed in more detail below. These features suggest that carminic acid is a reasonable candidate indicator of solution properties.

Previously, carminic acid has been used to determine trace quantities of Boron in solution (Calins, and Stenger, 1956, Anarez et. al., 1985), and it has also been used in the spectrofluorimetric determination of several transition metal ions (Kirkbright et al., 1966, Poluektov et al., 1970, Pilipenko and Schevchenko, 1989). Carminic acid has also been used as a staining agent in histography (Stockert et al., 1990). It has also received considerable attention in the food industry as a red coloring agent (Riboh, 1977). When

1054–7487/96/0178$12.00/0

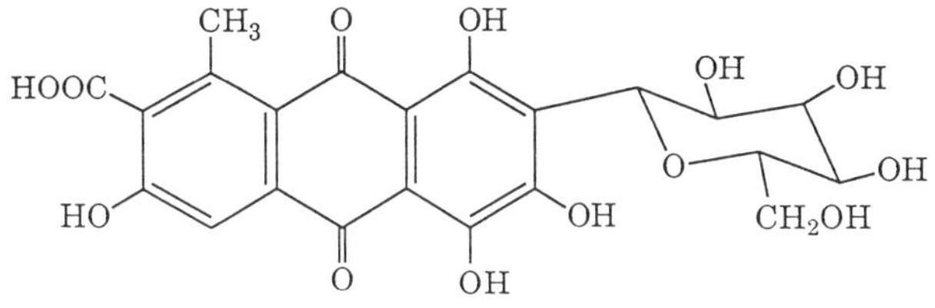

Carminic Acid

Fig. 1. Chemical structure of carminic acid.

complexed with aluminum, carminic acid forms carmine, which can give color in the range from pale strawberry red to a dark blackcurrant red (Lloyd, 1980). Carminic acid undergoes ester hydrolysis in alkali solutions (Rasimus et al., 1995), which accounts for its instability above pH 7.5. Allevi and coworkers completed (Allevi et al., 1991) the first total synthesis of carminic acid.

Carminic acid was also chosen because it is a glycoside, thus it may be able to function as a tailor-made impurity. A tailor-made impurity (Black et al., 1986, Klug, 1993) is a special class of additive designed to interact with selected crystalline faces. Such impurities have structural characteristics similar to the host molecule which allow them to be more easily incorporated into the lattice during crystallization. Carminic acid has a glucose group that makes it a candidate as a tailor-made impurity (Black et al., 1986) in sucrose and glucose solutions. Generally, a tailor-made impurity is added to inhibit crystal growth and/or change the crystal morphology. This is due to dissimilarities between host and impurity that disrupt the normal crystallization process. In our study, we were able to use a very small amount of acid and still obtain intense spectra. With low probe concentration, (2.5 ppm) the change in crystal interface conditions would be kept to a minimum, but the high emission intensity would still allow monitoring of the crystallization process conditions at the crystal interface and the bulk solutions.

Materials And Methods

Absorption spectra were obtained using a Perkin Elmer Lambda 3A UV/Vis Spectrophotometer equipped with a Perkin Elmer R100A recorder. Emission spectra were obtained with a Spex 1681 FLUOROLOG 2 0.22m Spectrometer. Four sided Suprasil quartz cuvettes with a 1 cm pathlength were used to contain the samples in both absorption and emission. To aid in sugar dissolution, a Bransonic 220 sonic bath was used. All solution pH measurements were taken with an Orion Research digital analyzer / 501 and a combination electrode.

Ultra pure grade RNase-free sucrose (Baker Chemicals), reagent grade anhydrous glucose powder (Baker), carminic acid (Aldrich), and HPLC grade water (Fisher) were used. A stock solution of 4 x 10^{-4} M carminic acid was prepared with spectroscopic grade methanol (Malinckrodt). For emission fluorescence studies, solutions of 2.5 ppm carminic acid were prepared. For absorption studies, solutions of 5 ppm were made. Before each solution was prepared, the amount of carminic acid necessary for a given concentration was calculated. The proper amount of stock solution was then calculated and placed into a 20 ml vial with a precision pipette. The stock was allowed to evaporate, thus giving the proper amount of probe. Each vial containing dry carminic acid was set aside and kept away from light.

The sugar solutions were prepared to weigh exactly 10 grams. (+0.01%) The sucrose samples varied in weight percent from 20% to 75%, with a saturation point of 67% at 25 C, whereas the glucose solutions varied from 30% to 75%, with a saturation point of 48% at 25 C. The samples were prepared in 10% increments up to 60%. Above 60%, they were incremented by 5% in order follow changes in the supersaturated regions more closely. In preparation it was absolutely necessary to ensure complete dissolution of both sucrose and glucose solutions by heating and agitation.

After each solution was dissolved, it was quickly transferred into a vial with pre-evaporated carminic acid stock solution, and allowed to cool to room temperature. To insure homogeneous mixing of carminic acid, each cooled solution was reheated in the water bath for one minute, and allowed to cool again to room temperature. Solutions that were not reheated had visible and unwanted concentration gradients. Once the solutions reached room temperature, spectrophotometric readings were taken.

Fluorescence emission of carminic acid in aqueous glucose and sucrose solutions was measured using an excitation wavelength of 310 nm and was recorded from 400 nm to 700 nm.

Excitation and emission slit widths of 2 mm were used for all spectra. For each sample, a background spectrum was taken of the corresponding sugar-water solution. Samples were background corrected by subtracting the fluorescence emission spectrum of each aqueous sugar solution .

Absorption spectra of carminic acid-doped sugar solutions were collected in the 250 nm to 600 nm range. For both sugars, a scanning speed of 120 nm/min. was used. The pH of each solution was measured.

Results And Discussion

The absorption spectra of carminic acid in aqueous sugar solutions were determined first to ascertain excitation wavelengths and also to determine the species that existed in the ground state. Based on the work done by Schwing-Weill (1986), the visible absorption spectra indicate that carminic acid can exist in a fully protonated, a singly deprotonated, or a doubly deprotonated state indicated by LH_3, LH_2^-, and LH^{-2}, respectively. In samples containing less than 60% glucose, carminic acid exists as a mixture of LH^{-2} and LH_2^- states, whereas in solutions with concentrations higher than 60% glucose, carminic acid exists in the LH_3 species. Comparing these spectra to that of buffer solutions with similar pH, it appears that visible absorption spectral changes in the ground state are a function of glucose solution pH. The UV region of the absorption spectrum shows an equilibrium between a species absorbing at 325 nm and a species absorbing at 280 nm that is a function of glucose concentration. The assignment of these bands is as yet undetermined.

There is little or no variance in the visible absorption spectra of carminic acid as a function of sucrose concentration. When these spectra are compared to those of buffer solutions with comparable pH's, they appear to have spectra that is not dependent on pH. The sucrose solution pH changes by two units as a function of sucrose concentration with no corresponding change in visible absorption spectra. In a larger sense, it implies that the carminic acid molecule in the ground state exists in an environment that is independent of sucrose concentration. Similar to carminic acid in aqueous glucose solutions the UV absorption in aqueous sucrose solutions spectra show an equilibrium between the species absorbing at 325 nm and the species at 280 nm that is a function of sucrose concentration.

The emission spectra of carminic acid in both sugars were collected using an excitation wavelength of 310 nm. This excitation wavelength was chosen based on the UV-visible spectra of the sugars which showed isosbestic points at 310 nm in the ground state.

Tentative band assignments of carminic acid in the excited state were made based on the visible emission spectra in buffer solutions. In 1M HCl, the fully protonated state (LH_3*) emits at 594 nm. In 1M NaOH, the fully deprotonated state (L^{-3}*) emits at 645 nm. It also appears that the LH_2^{-*} species emits at 610 nm and 450 nm, and the LH^{-2*} species emits at 580 nm. A full photochemical characterization of carminic acid is presently being undertaken, but for the purposes of this study we used the fore-mentioned band assignments. Similar to pyranine in sugar solutions (Chakraborty and Berglund, 1993), carminic acid indicates an equilibria between different microenvironments which can be used to determine sugar concentration, especially in the supersaturated region.

In Fig. 2, the emission spectra of carminic acid in solutions of varying glucose concentration using an excitation of 310 nm are shown. The PIR between 594 nm and 450 nm show a much stronger exponential dependence than that seen between 594 nm and 610 nm. There also appears to be a clear isosbestic point at 525 nm, indicating an equilibria between two species of carminic acid as a function of glucose content.

According to Hills (1991) there are two types of water that can exist in an aqueous glucose solution, namely, bulk water and solvation water. Bulk water refers to water that is not hydrogen bonded to a glucose molecule. Solvation water refers to water that is bound to a glucose molecule. The bulk water is capable of proton exchange, whereas the solvation water is not. Using a context similar to Chakraborty and Berglund (1993), we can refer to a micro environment where there is either a presence or absence of exchangeable protons in the vicinity of carminic acid. The presence of exchangeable protons would cause carminic acid to exist as the LH^{-2*} species, thus emitting at 610 or 450 nm, whereas the absence of exchangeable protons would cause carminic acid to exist as the LH_3*, thus emitting at 594 nm.

In undersaturated aqueous glucose solutions at room temperature, Hills (1991) calculated that there are 10 water molecules which surround a solvated glucose molecule. This model assumes that each glucose hydroxyl group is hydrogen bonded by two water molecules in solution. Since the solvation water is incapable of proton exchange, carminic acid residing near the glucose molecule would exist in the LH_3* state and thus emit at 594 nm. Carminic acid residing in the bulk water would be in an area suitable for proton exchange, thus it would emit in the LH^{-2*} state of 610 nm or 450 nm.

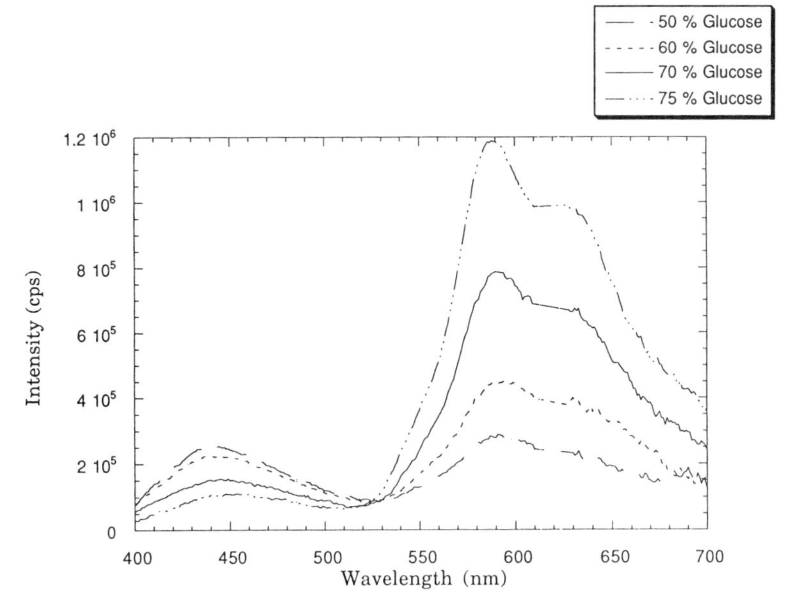

Fig. 2. Emission spectra of carminic acid in aqueous glucose solutions. Carminic acid concentration is 2.5 ppm and the excitation is 310 nm.

Figure 3 shows the PIR of the 594 nm band to the 450 nm band. This ratio shows strong dependence of PIR on glucose concentration. The emission seen at 594 nm represents solvation water and emission seen at 450 nm represents bulk water. At around 50% concentration, there is a change in slope of the PIR. This occurs at about the saturation point of glucose in water at 25°C. At this point, there is no longer enough water in solution for each respective glucose molecule to be hydrogen bonded by 10 water molecules.

Since carminic acid is a known pH indicator (The Merk Index, 1989, Schwing-Weill, 1986), the pH's of each glucose solution were recorded. The pH readings were the same for samples

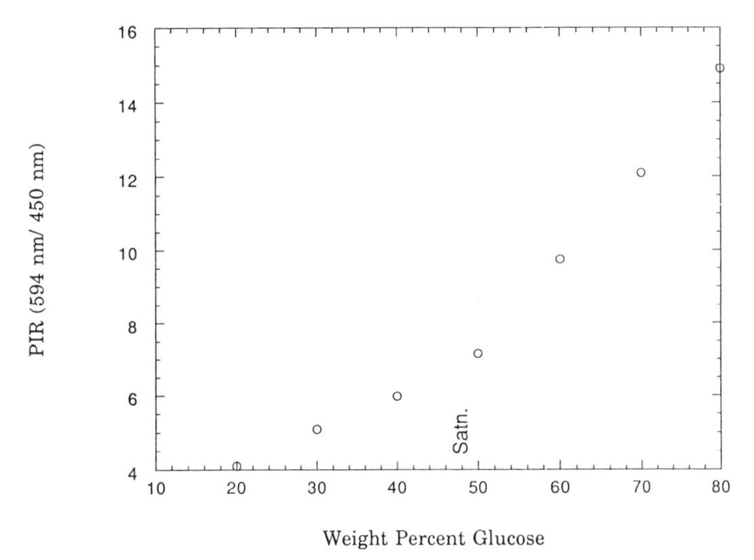

Fig. 3. Emission peak intensity ratio (PIR) between 594 nm and 450 nm in aqueous glucose solutions. Excitation wavelength is 310 nm and carminic acid concentration is 2.5 ppm.

with and with out carminic acid. The results indicate that the pH decreases linearly from 6.4 (30% glucose) to 4.4. (75% glucose) Thus, it could be possible to assume that the change in emission spectra with respect to glucose concentration could be a function of pH and not glucose concentration. By comparing the emission spectra of the glucose solutions to emission spectra of buffer solutions with similar pH's, spectral differences are seen between glucose solutions and buffer solutions. Fig. 4 shows the PIR's (594 nm to 450 nm) of carminic acid-doped glucose solutions along with the PIR's of buffered carminic acid solutions of similar pH. From these results it is clear that in the supersaturated region there is no effect of pH, and that the spectral changes seen are due to glucose content.

Previous work with sucrose has shown that there are two types of exchangeable protons in aqueous sucrose solutions.(Suggett and Clark, 1976) This again would imply that there are two types of micro environments in which the carminic acid can exist, one as a protonated species and the other as a deprotonated species. Richardson et al. (1987) showed through NMR water mobility studies in aqueous sucrose solutions that samples up to 40 weight percent sucrose contained water that was primarily free water, which is capable of proton exchange. At about 40 weight percent, the water mobility drops sharply, thus causing the proton exchange ability to decrease. In the first region, water would have the ability to exchange protons, thus carminic acid would exist in the deprotonated state. In the second region, there would be less likelihood for water to be able to exchange protons, thus, carminic acid would be more likely to exist in the protonated state. Based upon this and the emission spectra of carminic acid in aqueous glucose solutions results discussed earlier, it appeared that carminic acid-doped sucrose solutions would give similar results to that of glucose.

The emission spectra of carminic acid in sucrose solutions excited at 310 nm are given in Fig. 5. These samples showed large changes in spectral characteristics as a function of sucrose content.

Fig. 6 gives the PIR between bands at 594 nm and 450 nm. Assuming that the emission at 594 nm represents the LH_3* form of carminic acid and the emission at 450 nm represents LH_2^{-*} form of carminic acid, then Figure 6 illustrates how the lack of bulk water forces carminic acid to exist in the fully protonated form as the sucrose concentration is increased.

As with glucose, the possibility of a pH effect on the emission spectra of sucrose solutions was investigated. The pH's of each respective sucrose solution were taken. They range from 8.6 (20% sucrose) to 6.5. (75% sucrose). Comparison of the PIR of carminic acid in buffer solutions to those obtained from sucrose solutions at the same pH (Fig. 7) shows that there is no dependence of PIR on pH, thus it appears that the spectral changes are dependent on sucrose concentration.

Summary And Conclusions

Trace amounts of carminic acid (2.5 ppm) can be used as a fluorescent probe to measure

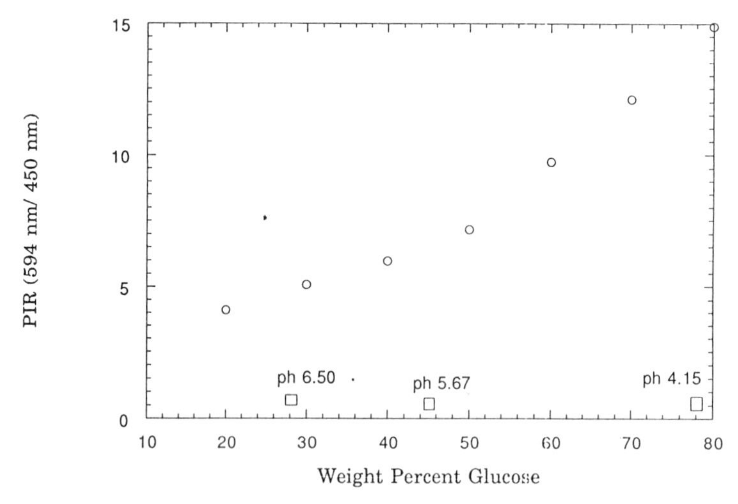

Fig. 4. Emission peak intensity ration (PIR) comparison between carminic acid emission in aqueous glucose solution and buffer solutions. Diamonds and squares represent glucose and buffer solutions, respectively. Carminic acid concentration is 2.5 ppm.

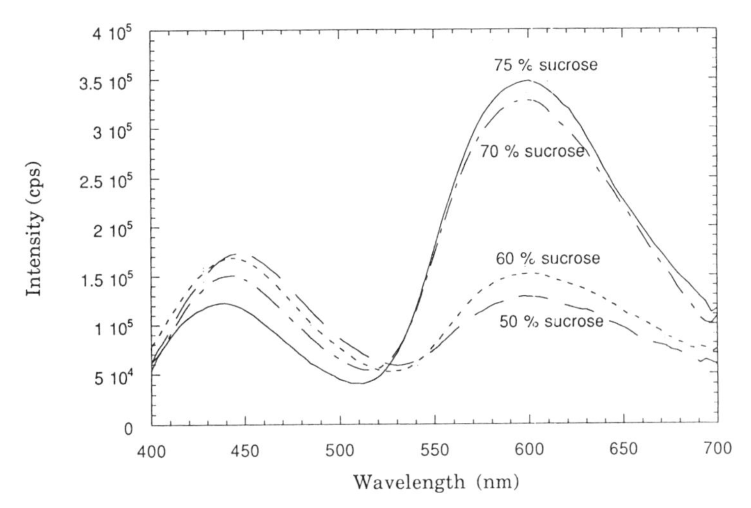

Fig. 5. Emission spectra of carminic acid in aqueous sucrose solutions. Carminic acid concentration is 2.5 ppm and the excitation is310 nm.

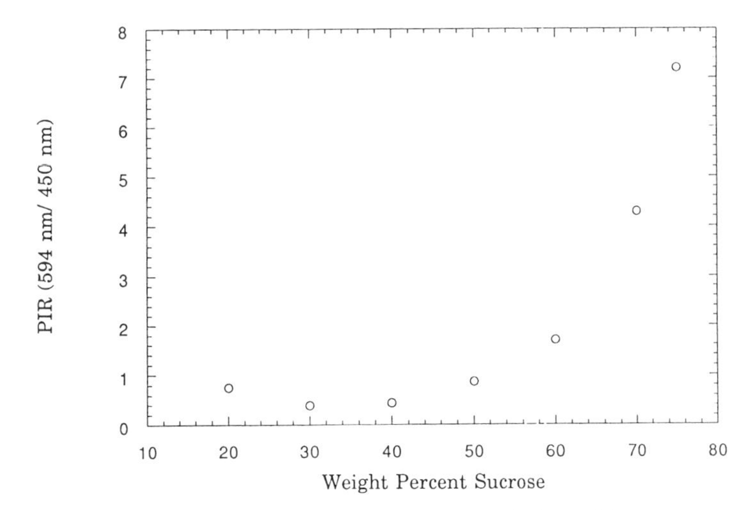

Fig. 6. Emission peak intensity ratio (PIR) between 610 nm and 450 nm for carminic acid in aqueous sucrose solutions. Excitation wavelength is 310 nm and carminic acid concentration is 2.5 ppm.

solute concentration in aqueous glucose and sucrose solutions in a manner similar to that done with pyranine. The possibility of a pH effect was considered, with spectral differences being attributed to solution sugar content.

For an excitation of 310 nm, the PIR between 594 nm (LH_3*) and 450 nm (LH_2^{-*}) shows a very strong exponential dependence upon glucose concentration. In aqueous sucrose solutions an excitation of 310 nm also results in a strong dependence of the PIR between 594 nm (LH_3*) to 450 nm (LH_2^{-*}) on sucrose concentration.

The visible absorption spectra of carminic acid in aqueous glucose solution indicate that carminic acid exists in the LH_2^- and LH^{-2} states in undersaturated solutions and in the LH_3 state in supersaturated solutions. The visible absorption spectra of carminic acid in aqueous sucrose solutions indicate that carminic acid exists in the LH^{-2} and L^{-3} with no change as a function of concentration. Ultraviolet absorption spectra showed an equilibrium between species at 325 nm and 280 nm for aqueous solutions of both sugars, but these species are yet to be identified.

Carminic acid appears to be a good candidate to determine glucose and sucrose concentration as a trace fluorescent probe. It has a high fluorescence intensity at low

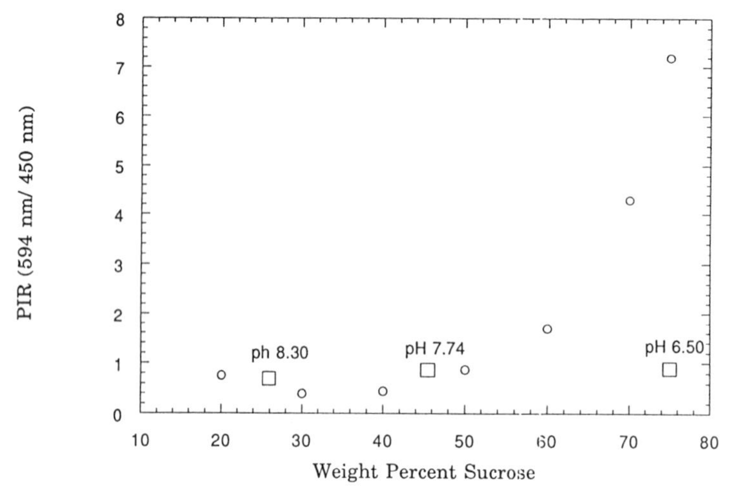

Fig. 7. Emission peak intensity ratio (PIR) comparison between carminic acid emission in aqueous sucrose solutions and buffer solutions of comparable pH. Circles and squares represent sucrose and buffer solutions, respectively. Carminic acid concentration is 2.5 ppm and excitation is 310 nm.

concentrations (2.5 ppm) making it easy to use, and a benign effect upon ingestion, making it acceptable for use in food and drugs. More work, both theoretical and experimental needs to be done to verify exactly how and why carminic acid acts in sugar solutions.

Acknowledgments

Funding was provided by the National Needs Fellowship Program and the Cooperative State Research Service of the United States Department of Agriculture, Grant Number 90-34189-5014. Additional Support was provided by the Crop and Food Bioprocessing Center/Research Excellence Fund at Michigan State University. We are also grateful to the National Science Foundation for support through Grant CTS 94-07563. Salary support for K.A.B. from the Michigan Agricultural Experiment Station is also acknowledged.

References

Allen, A. T. et al. *Nature* **1972**, 235, 236.
Allevi, P et al. Journal of the Chemical Society Communications, 1991, 1319.
Anarez, J. ; Ferrer, A.; Rabadan, J. M.; Marco, L. *Talanta*, **1985**, 32, 1156.
Black, S. N.; Davey, R.J.; Halcrow, M. *Journal of Crystal Growth,* **1986**, 79 , 765.
Calkins, R. C.; Stenger,V. A. *Analytical Chemistry*, **1956**, 28 , 399.
Chakraborty, R.; Berglund, K. A., *Journal of Crystal Growth*, **1993**, 130, 587.
Hills, B. P. *Molecular Physics* ,**1991**, 72, 1099.
Kirkbright, G. F.; West, T. S.; Woodward, C. *Talanta* **1966**, 13, 1637.
Klug, D. L. in: Handbook of Industrial Crystallization, Ed. A. S. Myerson (Butterworth-Heinemann) **1993**, 82.
Larson, M. A.; Garside, J. *Chemical Engineering Science* **1986**, 41, 1285.
Larson, M. A.; Garside, J. *Journal of Crystal Growth* **1986**,76, 88.
Lloyd, A. G. *Food Chemistry,* **1980**, 5, 91.
Myerson, A. S.; Chang Y.C. in: Industrial Crystallization '84, Ed. S. J. Jancic and E. J. Jong (Elsevier, Amsterdam, **1984**) p. 27.
Myerson, A. S.; Sorrell, L.S. *AIChe Journal* **1982**, 28, 778.
Pilipenko, A. T.; Shevchenko, T. L. *Zhurnal Analiticheskoi Khimii*, **1989**, 7, 1262.
Poluektov; Lauer, R. S.; Sandau, M. A. *Zhurnal Analiticheskoi Khimii* **1970**, 25, 2118.
Rasimus, J.; Blanchard, G.; Berglund, K., to be published.
Riboh, M. *Food Engineering*, **1977** 66.
Richardson, S. J.; Baianu, I. C.; Steinberg, M. P. *Journal of Food Science* **1987**, 52, 806.
Schwing-Weill, M. J. *Analusis* **1986**, 14 , 290.
Stockert, J. C.; Llorente, A. R.; Castillo; P. Del; Gomez, G. *StainTechnology*, **1990**, 65, 299.
Suggett, A.; Clark, A. H., *Journal of Solution Chemistry* **1976**, 5, 1.
The Merck Index, Eds. S. Budavari, M. J. O'Neil, A. Smith and P. E. Heckelman (Merck & Co. Inc., **1989**) p 280.

Melt Crystallization Behavior of Long Chain Milk Fat Triglycerides

Frank Z. Yang, Rebeca Baca-Diaz, Danping Shen, Armand Boudreau, and Joseph Arul
Department of Food Science and Technology, Université Laval, Sainte-Foy, Quebec G1K 7P4

Fractionation of butter fat by crystallization into plastic and oil fractions is a promising approach to improve the utilization of butter fat. The objective was to study the crystallization behavior of the long chain triglycerides (LCT). LCT was produced by short path distillation. Crystallization experiments were carried out on bench scale with and without agitation. The melt was cooled by pumping it through a stainless steel coil to induce nucleation, and crystals were grown in a temperature controlled beaker. Solid and liquid phases produced by crystallization were characterized by dropping point, DSC melting profile, solid fat content, and fatty acid and glyceride compositions. Average crystal growth rates at different conditions were obtained by analyzing crystal size distribution. Results showed that agitation had a significant effect on crystallization of the long chain triglycerides. Solid phase yield with agitation increased significantly compared to that without agitation. Supercooling level of 1 °C and flow rate of 30 ml/min through the coil were found satisfactory in terms of solid phase yield and filterability of the crystals though optimal supercooling level seemed to be 3 °C. A comparison of crystallization behavior of the long chain triglycerides to that of whole butter fat is also discussed.

Demand for milk fat, especially in the form of butter, has been steadily declining for more than two decades because of its poor spreadability as table fat and its perception as a saturated fat. The dairy industry is thus facing a serious problem of a large surplus. Butter fat imparts pleasing flavor to dairy and non-dairy foods, but it is a broad-spectrum fat exhibiting a broad range of plasticity and melting behavior which limits its utilization in various food-fat formulations. Fractionation of butter fat into liquid and solid fat fractions which differ markedly from one another in composition and technical properties could catalyze the utilization of butter fat. Various fractionation processes have been investigated (Boudreau and Arul, 1993): melt crystallization, supercritical CO_2 extraction (Arul *et al.*, 1987), short-path distillation (Arul et al., 1988). Among these processes, melt crystallization is desirable from the standpoint of melting property which is the basis for fat selection in food fabrication, since crystallization separates triglycerides based on melting point. Melt crystallization is also attractive because modified milk fat retains its unique flavor (Makhlouf *et al.*, 1987). There is a vast literature describing studies on fractionation of butter fat by melt crystallization. Important factors affecting milk fat crystallization are crystallization temperature (Voss *et al.*, 1971; Mulder and Walstra, 1974; Tirtaux, 1987; Foley and Brady, 1984; Amer *et al.*, 1985; Grönlund *et al.*, 1988; Deffense, 1992), rate of cooling (deMan, 1968; Schaap and Rotten, 1976; Deffense, 1987; Shi *et al.*; 1991), rate of agitation (Black, 1975; Fouad *et al.*, 1990; Grall and Hartel, 1992), and method of separation of melt slurry (Deffense, 1987; Alex *et al.*, 1994; Breeding and Marshall, 1995). Studies on multi-step fractionation of butter fat by melt crystallization also were reported (Badings *et al.*, 1983; Makhlouf *et al.*, 1987; Deffense, 1987).

The **purpose of this study** was to investigate crystallization behavior of long chain triglyceride (LCT) fraction generated by short path distillation from its melt, compared to whole butter fat; and to examine possible advantages in terms of technical properties, if any, over whole butter fat crystallization.

1054–7487/96/0185$12.00/0

Material and Methods

Commercial grade anhydrous winter butter fat without any fine protein particles (Ault Foods Ltd., Mitchell, Ontario, Canada) was used for distillation. LCT fraction was obtained by short-path distillation at 265 °C and a pressure of 60 μm Hg.

Cooling Profiles LCT fraction was melted, stabilized at about 65 °C for about one hour, and then cooled to 45 °C; subsequently was transferred to a 250 ml beaker placed in a programmable Haake R water bath, and cooled down further to 25 °C. Cooling rates of both 0.1 °C/min and 0.5 °C/min were studied respectively. Temperatures in the melt were recorded every five minutes.

Crystallization All crystallization experiments were carried out with LCT fraction on bench scale with and without agitation. The experimental setup is shown in Fig. 1. After melting in a large beaker at 80 °C and maintained at this temperature for at least one hour for destroying crystal memory, the molten LCT was cooled by pumping through a stainless steel coil (6,28 m × 4,58 mm i.d.), which was placed in the LAUDA-RM20 water bath, to induce nucleation. After the temperature was stabilized at the outlet of the coil, LCT was introduced into three temperature controlled crystallizers (250 ml beakers) for crystal growth. In these crystallization experiments, temperature of crystallization (33°C, 34°C, 35°C, 36°C), flow rate through the coil (20mL/min, 30mL/min, 40mL/min, 50mL/min), agitation speed (0, 100rpm, 200rpm) were varied.

Samples were taken at 10-min intervals in the first 90 minutes, then at 30-min intervals until the completion of the experiment, and crystal images were observed and acquired by photomicroscopy with a polarized-light microscope (Carl Zeiss) equipped with a black and white TV camera. Information related to nucleation and crystal growth was later extracted from those stored images using image analysis system VIDAS 21 (Kontron Elektronik GmbH, Germany).

For crystallization with agitation, a G. K. Heller HST 20 stirrer (Glas-Col, Terre Haute, Indiana), mounted with a spiral-type impeller, was used to provide agitation during the course of crystallization.

Crystallization experiments were completed after 4 hours and the slurries were filtered by use of a 11-cm Buchner funnel with Whatman #4 filter paper under approximate 630 mm Hg of vacuum in a constant temperature chamber. We refer hereafter to the solid and liquid fractions produced by filtration as stearin and olein respectively.

Analytical methods LCT and its fractions generated by crystallization were analyzed for their melting point profile, solid fat content, dropping point and glyceride and fatty acid compositions. Melting curves were performed on a Dupont model 9900 thermal analyzer (Dupont Instruments, Toronto, Ontario) by the method of Timms (Timms, 1980). The samples were initially cooled to -43 °C and held for 5 minutes and then heated at a rate of 5°C/min to 67 °C.

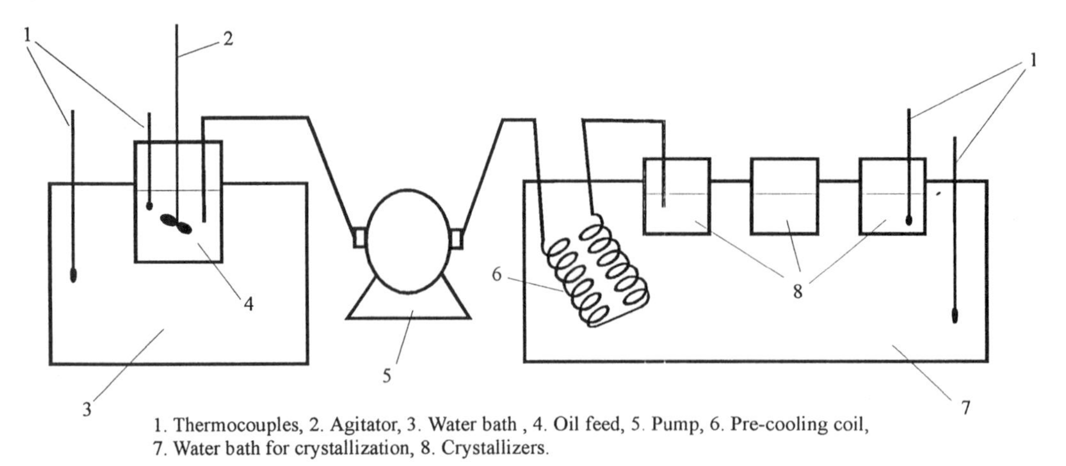

Fig. 1 Schematic diagram of experimental setup

Solid fat content was determined based on the melting curve by the method of Mortensen (Mortensen, 1981).

Melting point of LCT and dropping points of LCT and its stearin and olein fractions were determined by a Mettler Thermo- System FP 800.

Glyceride composition of LCT and its stearin and olein fractions were determined on a GLC equipped with a flame ionization detector (HP5890 Series II, Hewlett Packard, Avondale, PA) using the cold on-column injection techniques (Freeman, 1981; Traitler and Prévot, 1981). The separation was carried out in a J&W Scientific fused silica capillary column (15m×0,25mm i.d.) coated with SE-54 Durabond (0.1 μm) (Chromatographic Specialties, Brockville, ON). Glycerides from C-24 to C-54 were identified from the retention times of a standard mixture of triglycerides (Nu-Chek Prep, Inc., Elysian, MN). Samples were diluted to 10 mg/ml in distilled decane (Caledon, Georgetown, ON), and ca. 0,3 μl was injected onto the column. The detector temperature was kept at 355 °C. The initial oven temperature was held at 200 °C for 2 minutes, subsequently increased at a rate of 10 °C/min to 250 °C, and then programmed at a rate of 5°C to 350 °C and held at this temperature for 10 minutes. The carrier gas was hydrogen and the column head pressure was maintained at a constant pressure of 55 kPa.

Fatty acids from LCT and its fractions were converted into their methyl esters by the classical method (Luddy *et al.*, 1968). The same GLC with hydrogen as carrier gas was used as for triglyceride analyses. The separation was carried out in a J&W Scientific fused silica capillary column (30m×0,25 mm i.d.) coated with OV-225 Durabond (0,2 μm) (Chromatographic Specialties, Brockville, ON). About 1 μl of sample solution was injected automatically by the split technique (50 ml/min of the split flow). The column head pressure was maintained at 55 kPa. The inlet temperature was 215 °C and the detector temperature was 230 °C. The initial oven temperature was 35 °C and increased at a rate of 20 °C/min to 180 °C and held for 3 minutes, and then programmed at a rate of 3 °C/min to 200 °C and held for 13.08 minutes. Relative response factors were determined by analysis of a standard mixture of fatty acid methyl esters (Nu-Chek-Prep, Inc., Elysian, MN).

Results and Discussion

Cooling curves Fig.2 shows the cooling curves of LCT without agitation at a cooling rate of 0,1 °C/min and 0,5 °C/min respectively. Temperature in molten LCT decreased to a certain level, then increased at the start of nucleation due to release of crystallization heat, and then slowly levelled off to the water bath temperature. However, different cooling rates resulted in different nucleation temperatures and nucleation lag time. With slower cooling rate, the nucleation lag time was longer and the degree of supercooling for crystallization was smaller. However, Schaap and Rotten (1976) reported that for crystallizing whole milk fat the cooling rate had no substantial effect on the cooling curve.

DSC melting curve and solid fat content DSC can be used to monitor melt crystallization by examining the peaks registered (Deffense, 1993). Stearin and olein fractions produced under various crystallization conditions were characterized by their differential scanning calorimetry and solid fat content. DSC curves and solid fat contents of stearin fractions produced at various flow rates are compared in Fig. 3 and Fig. 4, respectively. At the flow rate of 30 ml/min, besides the main peak, the DSC curve registered the smallest second peak, compared with other flow rates. This indicates that at this condition the entrained olein fraction in the stearin fraction was the least. This is also shown in Fig. 4. The solid fat content of stearin produced at this flow rate was also the highest at any temperature in range of 20-45°C. Effect of crystallization temperature on the stearin quality was shown in Figs. 5 and 6. Stearin fractions with almost the same quality in terms of DSC melting curve and solid fat content were produced at all temperatures used except at 36 °C.

Dropping point and stearin yield Dropping point of fractions generated by crystallization is also an important physical characteristic for selection of fats in various food-fat applications. Table 1 gives the effect of flow rate on stearin yield and quality of fractions. One-step fractionation of whole milk fat by melt crystallization produces stearin fraction with a dropping point of 40-46 °C (Deffense, 1993) while one-step crystallization of LCT generated a super stearin fraction (D.P. 47,1), though stearin yield was low.

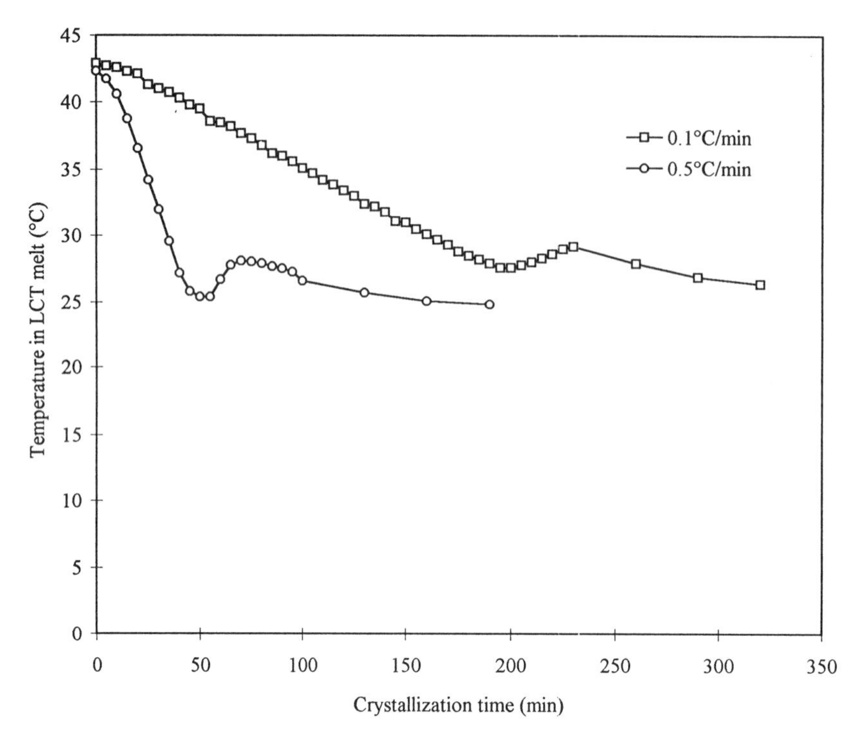

Fig. 2 Cooling curves of crystallizing LCT at different cooling rates

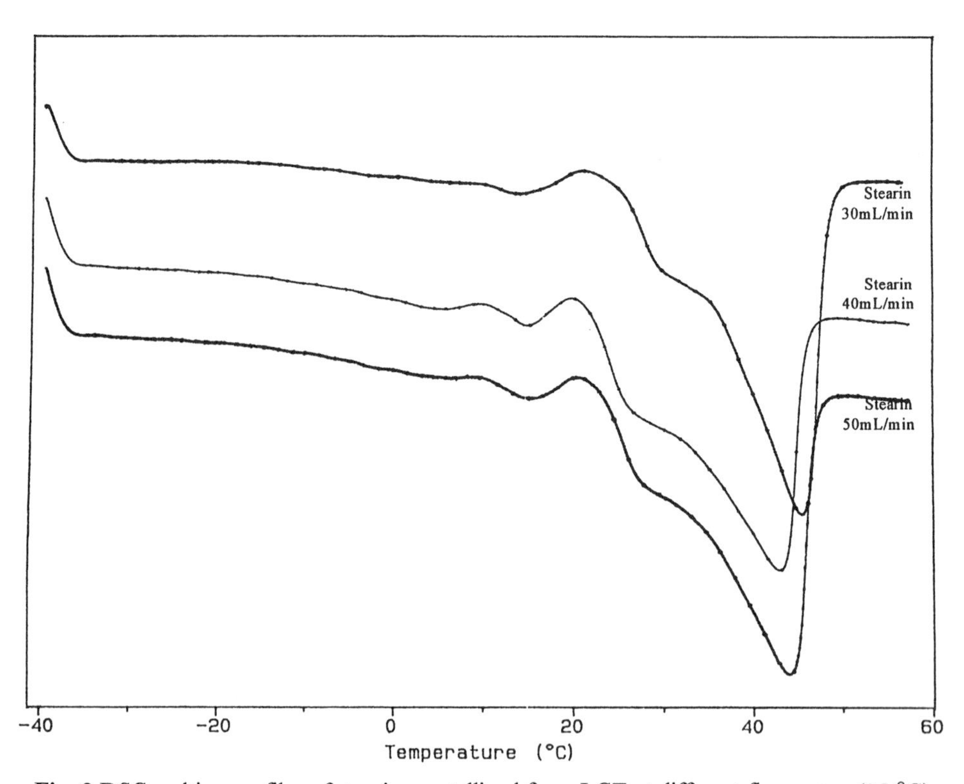

Fig. 3 DSC melting profiles of stearin crystallized from LCT at different flow rates (35 °C)

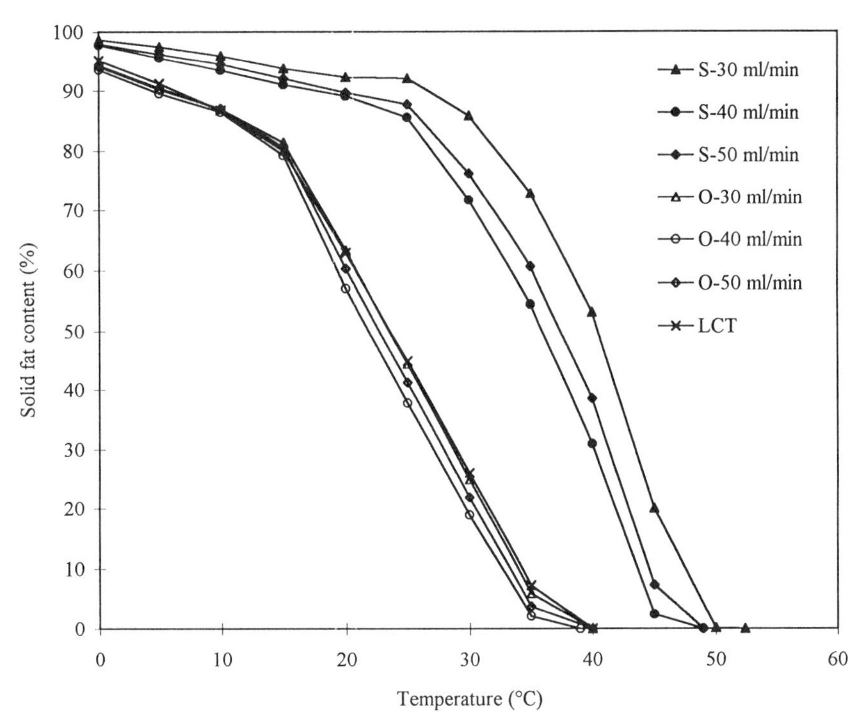

Fig. 4 Solid fat content of LCT and stearin crystallized from LCT at different flow rates (35 °C; S: stearin; O-20: olein)

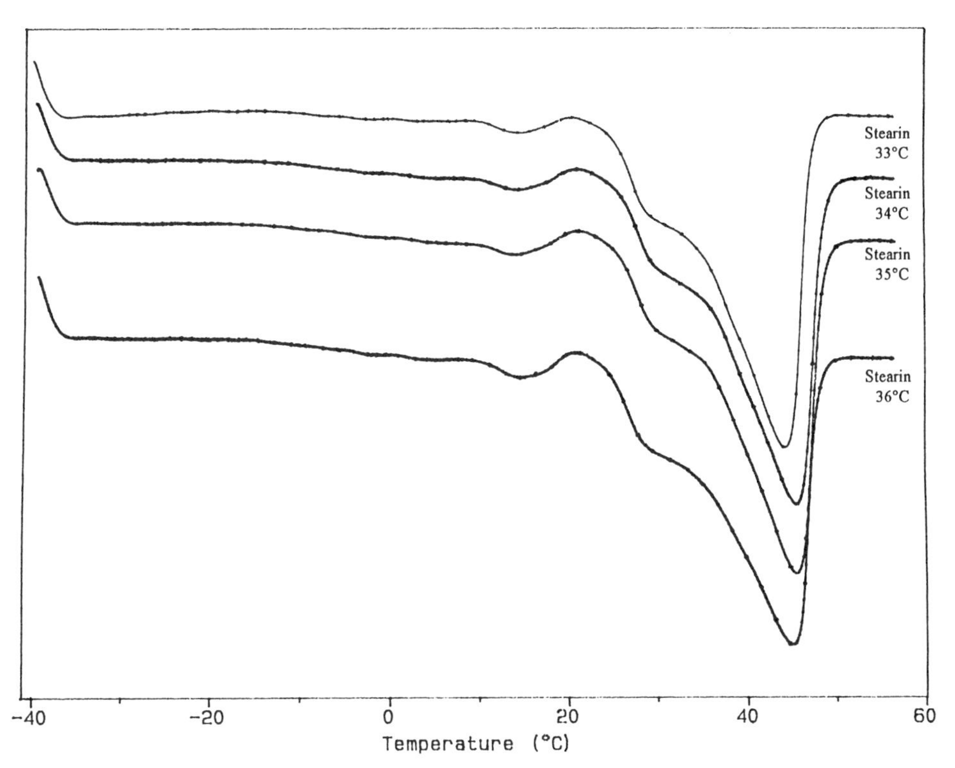

Fig. 5 DSC melting profiles of stearin crystallized from LCT at different temperatures (30 mL/min)

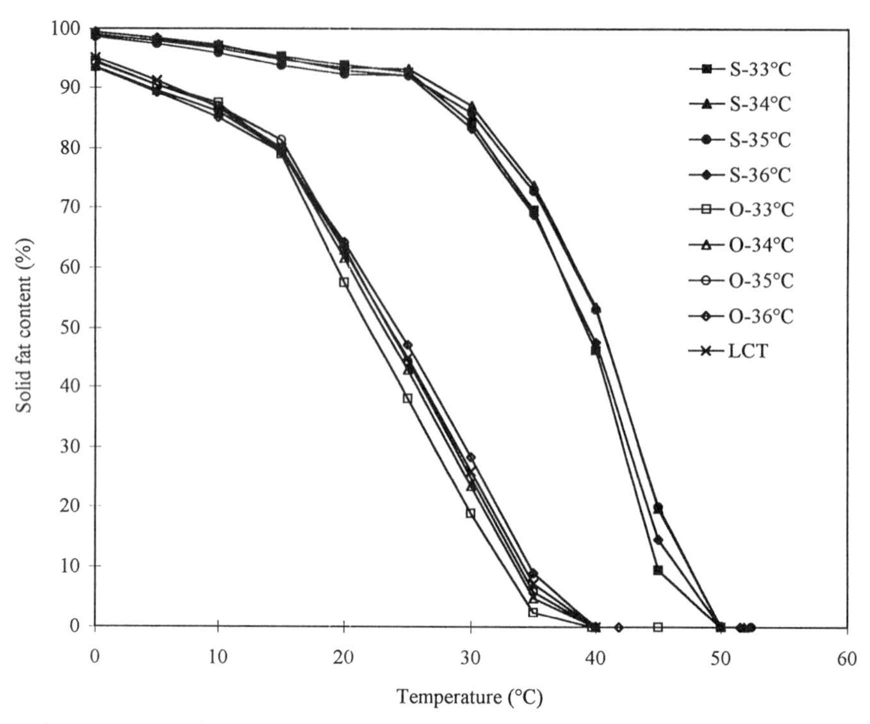

Fig. 6 Solid fat content of LCT and stearin crystallized from LCT at different temperatures (30 mL/min; S: stearin; O: olein)

Table 1. Stearin yield and dropping point of stearin and olein fractions crystallized from LCT at different flow rates (35 °C).

Flow rate (ml/min)	30	40	50
Stearin yield (%)	8,2	18,3	13,9
D.P. of stearin (°C)	47,1	44,4	45,4
D.P. of olein (°C)	34,8	33,1	33,9

Table 2. Stearin yield and dropping point of stearin and olein fractions crystallized from LCT at different temperatures (flow rate: 30ml/min)

Temperature (°C)	33	34	35	36
Stearin yield (%)	12,7	7,9	8,2	4,7
D.P. of stearin (°C)	46,3	47	47,1	46,5
D.P. of olein (°C)	33,7	34,6	34,8	35,7

Effect of temperature on stearin yield and dropping point of stearin is given in Table 2. It can be seen that the lower the crystallization temperature, the higher the stearin yield, though difference in stearin yield was marginal except for 36°C. Stearin fractions with almost the same quality in terms of dropping point was produced as the difference in crystallization temperatures were small (1°C apart).

Effect of agitation on stearin yield and dropping point of stearin is shown in Table 3. Although stearin fractions crystallized from LCT with and

Table 3. Effect of agitation on stearin yield and dropping point of stearin and olein fractions (temperature: 35°C; flow rate: 30 ml/min)

Agitation speed	0 rpm	100 rpm	200 rpm
Stearin yield (%)	8,2	15,8	18,3
D.P. of stearin (°C)	47,1	45,8	45,2
D.P. of olein (°C)	34,8	33,2	32,7

without agitation were about the same quality in terms of dropping point, stearin yield was doubled when agitation was applied during crystallization. Similar results were reported for whole milk fat (Fouad et al., 1990). This is most likely due to secondary nucleation as explained by Grall and Hartel (1992).

Crystallization kinetics Melting point and dropping point of LCT were 35,7°C and 36°C respectively. Supercooling level is defined here as difference between melting point of LCT and crystallization temperature. Fig. 7 shows the effect of crystallization temperature on crystal growth rate and the final average crystal size in the case of no agitation. Crystallization temperature is an important parameter because it is the driving force for crystallization. The largest final average crystal diameter of about 160 μm was obtained at 33 °C after only 4 hours of crystallization of LCT fraction. Shi *et al.* (1991) reported similar results for whole milk fat. Fig. 8 demonstrates the effect of flow rate on crystal growth rate and the final average crystal size. The final average crystal diameter was the largest (150 μm) at 30 ml/min. The reason that flow rate comes to play is that turbulent flow in the coil promotes the primary nucleation while it also governs the cooling rate. The final crystal size is crucial for filtration of crystal slurry. Large uniform crystals make filtration easier and less olein entrapped in stearin. Therefore, in terms of filterability, optimal operation parameters for crystallization of LCT seemed to be 33 °C , i.e., supercooling level of 3 °C, and 30 ml/min.

Effect of agitation on nucleation and average crystal size is shown in Fig. 9 and Fig. 10, respectively. When agitation was applied, crystal number per unit volume increased rapidly during the first hour of crystallization and maintained on a high level, while without agitation crystal number per unit volume essentially was unchanged during the entire course of crystallization. This clearly indicates that secondary nucleation was induced by agitation. From Fig. 10, it can be seen that agitation shortened crystal growth time and decreased growth rate, thus resulted in smaller final average crystal size. This implies that as a result of secondary nucleation, too many nuclei were produced so that growth of individual crystals was suppressed.

Fatty acid composition Fatty acid distribution in

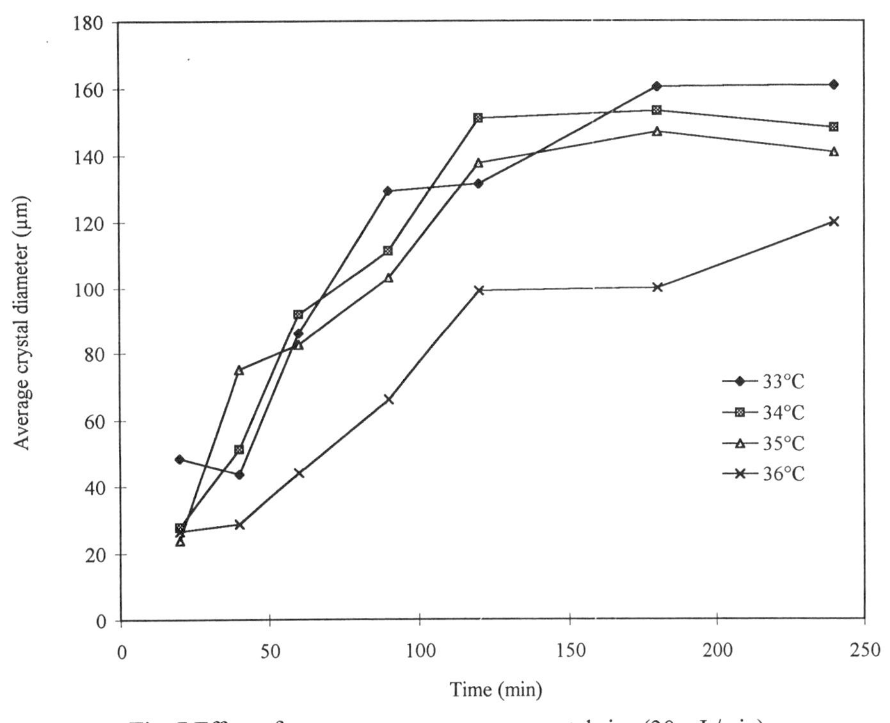

Fig. 7 Effect of temperature on average crystal size (30 mL/min)

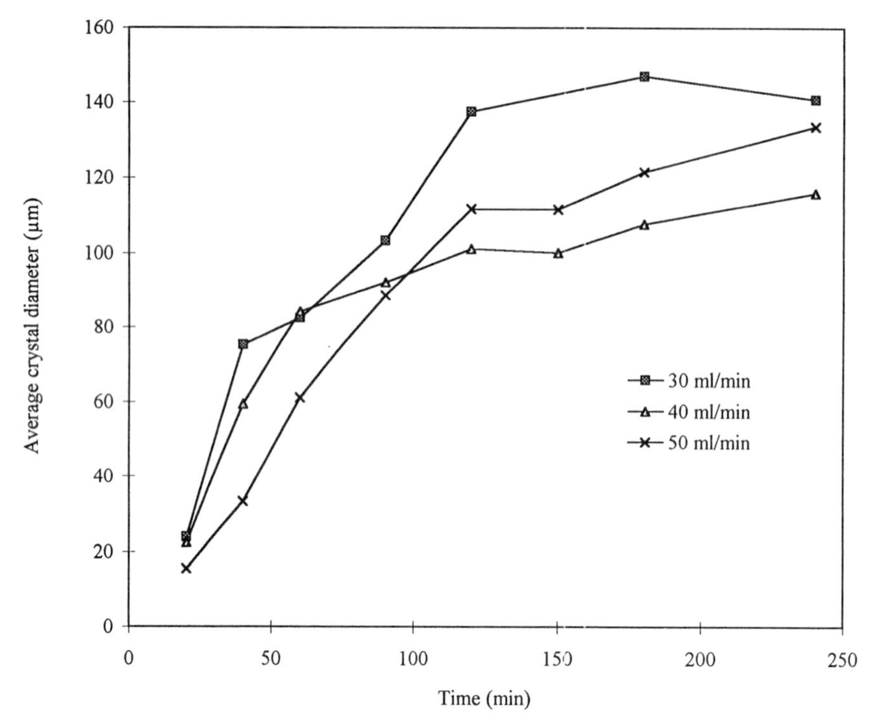

Fig. 8 Effect of flow rate on average crystal size (35 °C)

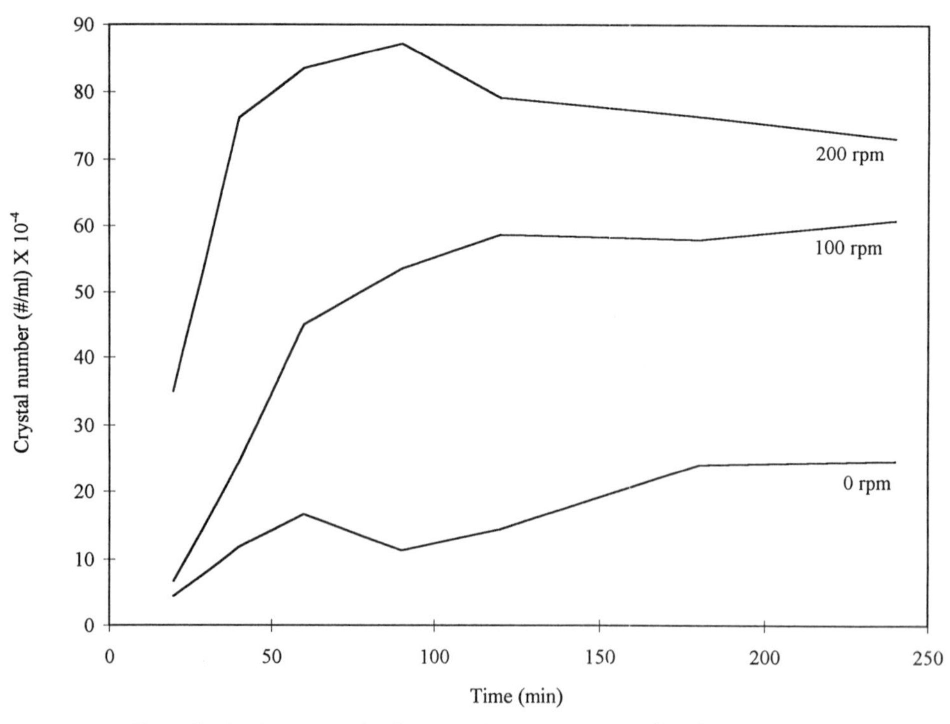

Fig. 9 Effect of agitation on nucleation rate (temperature: 35 °C; flow rate: 30 mL/min)

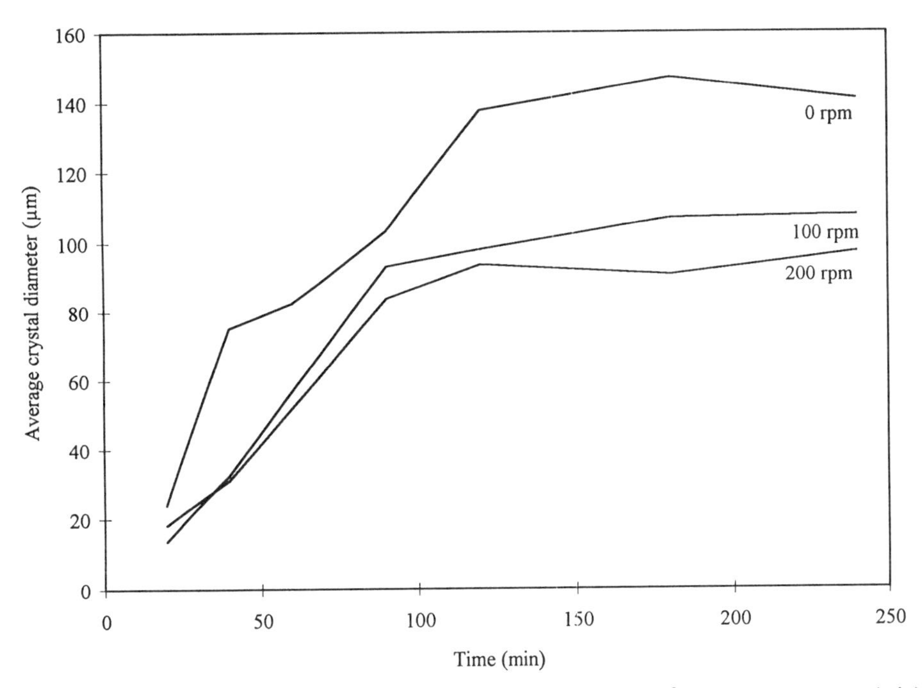

Fig. 10 Effect of agitation on average crystal size (temperature: 35 °C; flow rate: 30 mL/min)

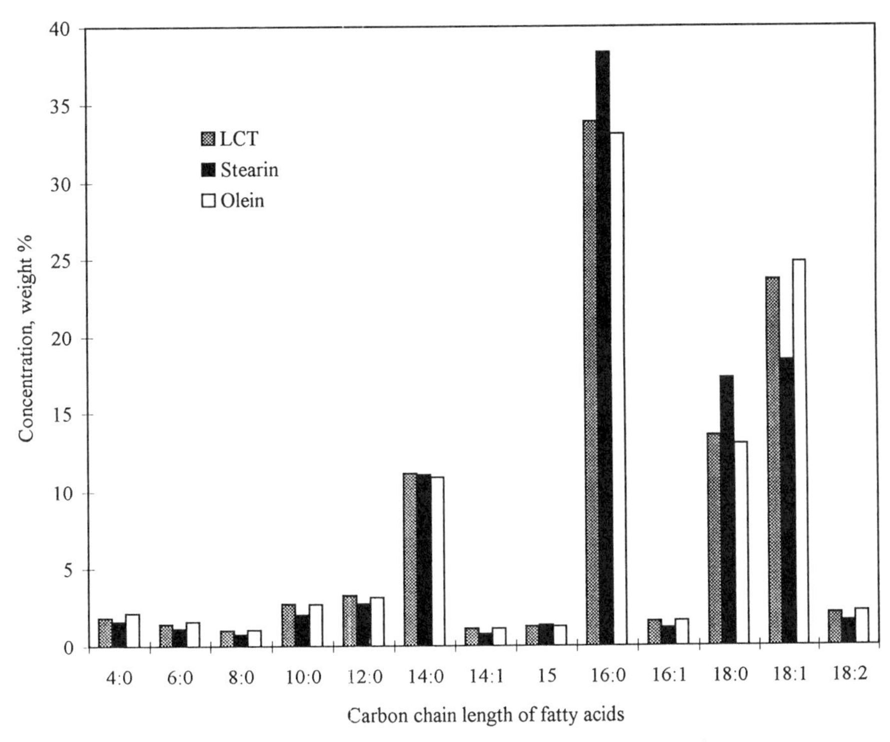

Fig. 11 Fatty acid composition of LCT and stearin and olein fractions crystallized from LCT (temperature: 35 °C; flow rate: 30mL/min)

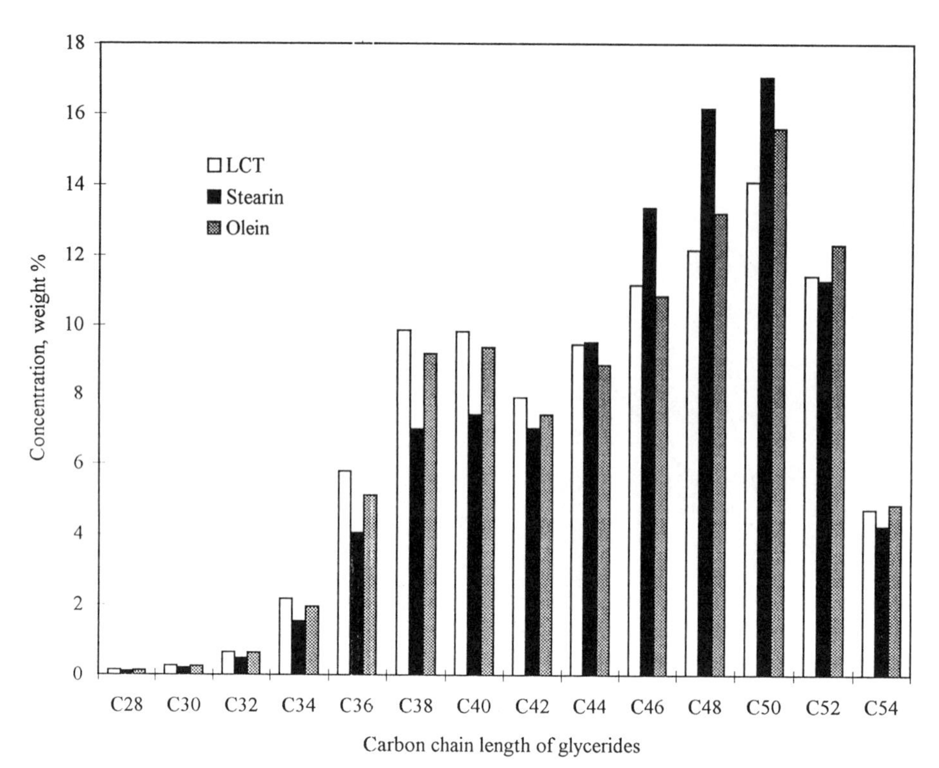

Fig. 12 Glyceride composition of LCT and stearin and olein fractions crystallized from LCT (temperature: 35 °C; flow rate: 30mL/min)

LCT and stearin and olein fractions is shown in Fig. 11. Similar to the results in the literature (Deffense, 1987; Grall and Hartel, 1992), no much separation was evident in terms of fatty acid composition. The general trend was that palmitic acid and stearic acid were concentrated in stearin fraction while palmitoleic acid and oleic acid were enriched in olein fraction.

Glyceride compositions Glyceride analytical results are shown in Fig. 12. Glycerides from C32 to C42 and C52 and C54 were concentrated in olein fraction while glycerides from C44 to C50 were concentrated in stearin fraction. Changes in triglyceride composition after crystallization were more evident than that in fatty acid composition.

In conclusion, melt crystallization of LCT fraction offers only minor advantages over the crystallization of whole milk fat in terms of crystallization kinetics, technical properties such as dropping point and plasticity, and almost no advantage in terms of fatty acid composition.

Acknowledgement

The authors thank Agropur Cooperative agro-alimentaire, Natural Sciences and Engineering Research Council of Canada, Dairy Farmers of Canada, and Université Laval for their financial support.

References

Alex, T.; Degen, P.J.; Dehn, J.W., UK Patent, GB 2270925, 1994.

Amer, M.A.; Kupranycz, D.B.; Baker, B.E. *JAOCS*, **1985**, 62, 1551.

Arul, J.; Boudreau, A.; Makhlouf, J.; Tardif, R.; Sahasrabudhe, M.R. *J. Food Sci.* **1987**, 52, 1231.

Arul, J.; Boudreau, A.; Makhlouf, J.; Tardif, R.; Bellavia, T. *JAOCS* **1988**, 65,1642.

Badings, H.T.; Schaap, J.E.; De Jong, C.; Hagedoorn, H.G. *Milchwissenschaft* **1983**, 38, 95.

Black, R.G. *Aust. J. Dairy Technol.* **1975**, 30, 153.

Boudreau, A.; Arul, J. , *J. Dairy Sci.* **1993**, *76*, 1772.

Breeding, C.J.; Marshall, R.T. *JAOCS* **1995**, 72(4), 449.

Deffense, E. *Fat Sci. Techn.* **1987**, 89, 502.

Deffense, E. *JAOCS* **1993**,70, 1193.

deMan, J.M. *Can. Inst. Food Techn. J.* **1968**, 1, 90.

Fouad, F.M.; van de Voort, F.R.; Marshall, W.D.; Farrell, P.G. *JAOCS*, **1990**, 67,981.

Foley, J.; Brady, J.P. *J. Dairy Res.* **1984**, 51, 579.

Grall, D.H.; Hartel, R.W. *JAOCS* **1992**, 69, 741.

Grönlund, B.; Heikonen, M.; Moisio, T. *Milchwissenschaft* **1988**, 43, 424.

Luddy, F.E.; Barford, R.A.; Herb, S.F.; Magidman, P. *JAOCS.* **1968**, *45*, 549.

Makhlouf, J.; Arul, J.; Boudreau, A.; Verret, P.; Sahasrabudhe, R. *Can. Inst. Food Sci. Technol. J.* **1987**, 20, 236.

Mortensen, B. K. *IMV Document*, F-doc 84, **1981**.

Mulder, H.; Walstra, P. *The Milk Fat Globule: Emulsion Science as Applied to Milk Products and Comparable Foods*, The Universities Press, Belfast, 1974.

Schaap, J.E.; Rotten, G.A.O. *Netherlands Milk Dairy Journal* **1976**, 30, 197.

Shi, Y.; Lang, B.; Boudreau, A.; Arul, J. *AOCS (Canadian section) Annual Meeting*, **1991**.

Timms, R.R. *Aust. J. Dairy Technol.* **1980**, *35*, 47.

Tirtiaux, A. *JAOCS*, **1983**, 60, 425.

Traitler, H.; Prévot, A. *Rev. fse Corps Gas* **1981**, *6*, 263.

Voss, E.; Beyerlein, U.; Schmanke, E. *Milchwissenschaft* **1971**, 26, 605.

The Effect of Physical State and Glass Transition on Crystallization in Starch

Kirsi Jouppila and **Yrjö H. Roos**, Department of Food Technology, P.O.Box 27 (Viikki B), FIN-00014 University of Helsinki, Helsinki, Finland

Crystallization of amorphous compounds is related to diffusion and affected by molecular mobility and viscosity. The present study investigated the effects of physical state and glass transition on crystallization and melting behavior. Mixtures of amorphous starch and distilled water in sealed glass ampoules were stored at temperatures ranging from 10 to 90°C. Crystallinity was determined from melting enthalpy using differential scanning calorimetry. Crystallization was observed to depend on storage temperature. The extent of crystallization and melting behavior were also governed by temperature difference between storage temperature and glass transition temperature. The crystallization data allows evaluation and prediction of starch crystallization and stability in various materials containing starch.

Native starch exists as a semicrystalline mixture of amylose and amylopectin in starch granules. Heating of native starch in excess water results in gelatinization and the loss of crystallinity is well-known. Crystallization of gelatinized starch in various materials may occur during storage causing quality defects. Crystallization phenomena of gelatinized starch are often studied in gels by observation of the effects of limited gelatinization on crystallization behavior. The rate and extent of crystallization may vary depending on the degree of gelatinization (Eliasson, 1985). Usually the linear amylose component crystallizes rapidly but crystallization of the amylopectin fraction occurs slowly. Zeleznak and Hoseney (1986) found that crystallization was only slightly affected by water present during gelatinization and it was primarily controlled by water content after gelatinization during storage.

Storage temperature has been shown to affect starch crystallization. The extent of crystallization was found to be highest in starch gels containing 50% solids (w/w) and stored at 4°C (Longton and LeGrys, 1981; Eliasson, 1985). However, the extent of crystallization in a starch gel at the same water content at 12 and 25°C was observed to be more extensive than in gels stored at 4 and 40°C (Zeleznak and Hoseney, 1987a).

Crystallization of amorphous polymers is affected by their physical state and molecular mobility (Flory, 1953; Roos, 1995). The physical state of biopolymers including starch is strongly dependent on water content and their plasticization by water has been well-documented (Slade and Levine, 1991). The physical state and glass transition of starch at several water contents have been reported in several studies (*e.g.* Zeleznak and Hoseney, 1987b; Kalichevsky *et al.*, 1992; Laine and Roos, 1994; Jouppila *et al.*, 1995). It has been suggested that crystallization of starch is kinetically controlled by T_g (Slade and Levine, 1991). Crystallization ceases below T_g, but it may occur above T_g at a rate depending on temperature difference between storage temperature and glass transition temperature, $T - T_g$. Melting enthalpy of crystallized starch was found (Laine and Roos, 1994) to increase with increasing $T - T_g$.

The aim of the present study was to investigate the effects of the physical state and glass transition on crystallization in amorphous starch.

Experimental

Corn starch (Sigma Chemical Co., water content 10.8%) suspensions (5 %, w/w, in distilled water) were gelatinized under continuous stirring by heating until the boiling temperature was achieved. Gelatinization of starch granules was confirmed visually using polarized light microscopy. The gels were frozen, freeze-dried (for at least 72 h at pressure <0.1 mbar), powdered and stored in vacuum desiccators over P_2O_5 at 24°C until a constant weight was achieved (at least 5 days). Mixtures of anhydrous, amorphous starch and distilled water were prepared in glass

1054–7487/96/0196$12.00/0

ampoules. Starch contents of the mixtures were adjusted to 60, 70, and 80% of solids. Glass ampoules were sealed in acetylene flame and sterilized in an autoclave at 121°C for 20 min to avoid microbial spoilage during storage.

Autoclaved ampoules were stored for 2 or 4 weeks at various temperatures ranging from 10 to 90°C with respective $T - T_g$ values ranging from 20 to 110°C. The T_g values used were calculated with the Gordon and Taylor equation (Gordon and Taylor, 1952) where the T_g of -135°C (Johari *et al.*, 1987) was used for amorphous water, the predicted T_g of 243°C (Roos and Karel, 1991) was used for anhydrous starch, and the value for a constant, k, was 5.2 (Laine and Roos, 1994). The predicted T_g values of mixtures with solids contents of 60, 70, and 80% were -50, -18, and 29°C, respectively.

The melting behavior and extent of crystallization after storage were determined from the location and size of a melting endotherm (Fig. 1). A differential scanning calorimeter (DSC) (Mettler TA 4000 analysis system with DSC-30 low temperature cell, a TC10A TA processor, and GraphWare TA72AT.2 thermal analysis software) was used. The DSC was calibrated for temperature using *n*-hexane (melting point, -95.0°C), distilled water (melting point, 0.0°C), and indium (melting point, 156.6°C), and for heat flow using indium (latent heat of melting, 28.5 J/g). The water content of the samples was adjusted to 80% before DSC runs to ensure full dissolution of the crystalline fractions at the melting temperature of the samples stored at various water content and temperature conditions. Duplicate samples were analyzed for crystallinity and melting behavior at intervals of 2 or 3 days. Gelatinization enthalpy for the native corn starch with 80% water was determined to be 12.4 J/g of solids.

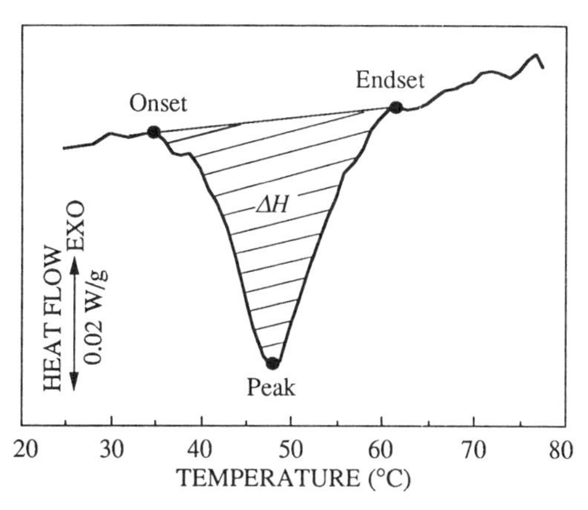

Fig. 1. Determination of onset, peak, and endset temperatures and enthalpy of melting (ΔH) using DSC at 5°C/min, showed for a sample containing 60% solids and stored at 20°C for 20 days.

Results and Discussion

Starch was observed to crystallize during storage at temperatures above T_g (Fig. 2). Melting enthalpies of the crystallized samples leveled off after 4 to 12 days of storage depending on storage temperature. Time needed for the leveling off increased at the lower storage temperatures and $T - T_g$ values. Crystallization occurred most slowly in samples with 60 and 70% of solids stored at 10 and 20°C.

The extent of crystallization in starch was observed to be dependent on storage temperature (Fig. 3). The extent of crystallization in samples

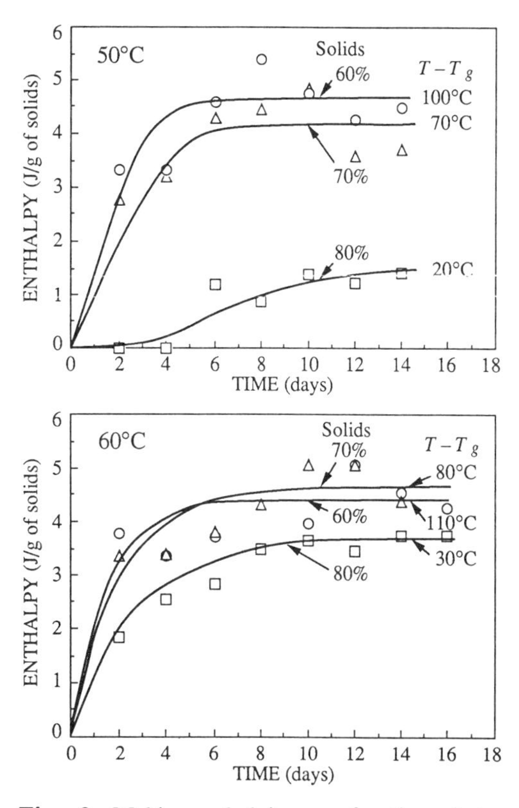

Fig. 2. Melting enthalpies as a function of storage time for samples with various solids contents and $T - T_g$ values stored at 50 and 60°C.

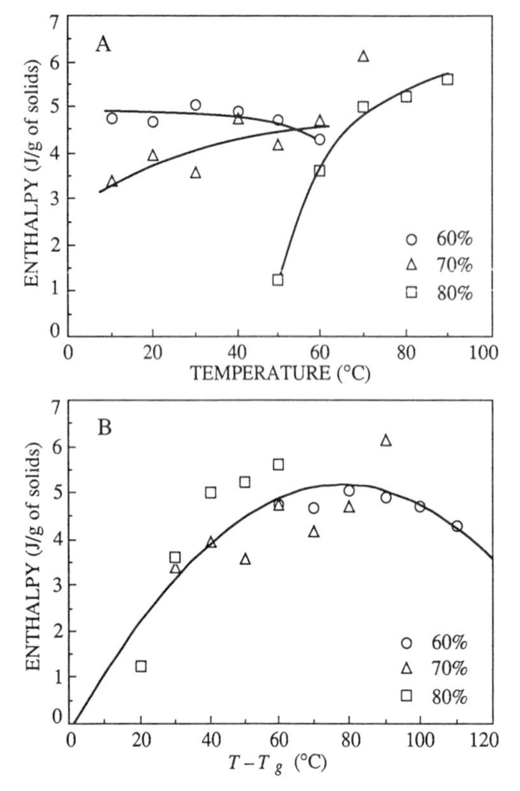

Fig. 3. Relationships between melting enthalpy and storage temperature (A) and melting enthalpy and $T - T_g$ (B). A second order polynomial was fitted to the experimental data (ΔH = -0.00089 $(T - T_g)^2$ + 0.14 $(T - T_g)$ - 0.21, R^2 = 0.64). Values of melting enthalpy are average melting enthalpies determined after leveling off during storage at various temperatures.

containing 80% solids increased with increasing storage temperature, as was also reported earlier (Laine and Roos, 1994). An increasing extent of crystallization with increasing storage temperature was also observed to occur in samples containing 70% solids. The extent of crystallization decreased slightly with increasing storage temperature in samples containing 60% solids. This crystallization behavior was probably due to changes in the physical state and T_g. Plotting of the melting enthalpy against $T - T_g$ suggested low crystallinity after storage either at low or high $T - T_g$ conditions (Fig. 3). Such behavior is typical of crystallization of amorphous compounds (Slade and Levine, 1991; Roos, 1995). At low $T - T_g$ values nucleation is fast but crystal growth is kinetically restricted and slow. At high $T - T_g$ conditions crystal growth is fast but nucleation occurs slowly. Thus the relationship was observed to be parabolic, as suggested by Slade and Levine (1991). It is obvious that at low $T - T_g$ mobility of molecules was low. Also according to Flory (1953), the proportion of crystalline material varies depending on conditions of crystallization. The degree of perfection of the crystallites may be quite low if proper annealing conditions have not been employed, *i.e.*, at temperatures close to T_g.

The melting behavior of the crystallized starch was determined from the location of the melting endotherm (Fig. 4). Onset temperature of melting was below 40°C in samples stored at temperatures below 50°C and containing less than 70% solids or at temperatures below 80°C when the solids content was 80%. Storage at higher temperatures caused a sharp increase in the onset temperature of the melting endotherm. Also endset temperatures increased with increasing storage temperature. The average temperature range of melting was 30°C. Melting occurred between 50 and 70°C in samples containing 80% solids, as reported earlier (Laine and Roos, 1994). The melting range of crystallized starch was narrower in starch gels stored at higher temperatures than in starch gels stored at lower temperatures (Longton and LeGrys, 1981). Such behavior could not be confirmed in the present study.

Peak temperatures of melting endotherm also increased with increasing storage temperature. The peak temperature of the melting endotherm of

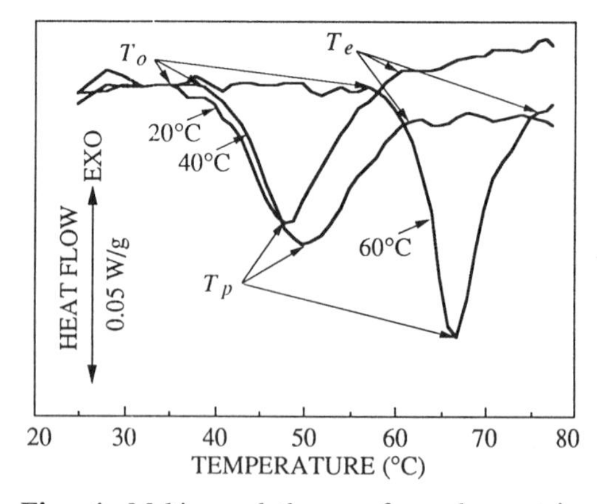

Fig. 4. Melting endotherms of samples containing 60% solids stored at 20, 40, and 60°C for 20, 14, and 8 days, respectively, showing the effect of storage temperature on the onset, peak, and endset temperatures.

crystallized starch has been reported to increase with increasing storage temperature (Zeleznak and Hoseney, 1987a). It was suggested that the crystal structure of the crystallized starch may change due to annealing during storage. However, it was observed using X-ray diffractometry that the crystal structure was independent on storage temperature. Storage temperatures closer to melting temperature of crystals were proposed to allow formation of a more perfect crystalline structure. This is also a well-known phenomenon occurring in synthetic polymers (Flory, 1953).

The results of the present study suggest that at low $T - T_g$ values less perfect crystalline regions are formed that have a lower melting temperature. After the initial crystallization the crystalline regions formed may act as barriers for molecular rearrangement and the extent of crystallization becomes limited. As the $T - T_g$ is raised to above 60°C with an increase in storage temperature or water content more perfect crystallites with a melting enthalpy of about 5 J/g of solids are formed. This behavior agrees with the theory (Flory, 1953; Slade and Levine, 1991; Roos, 1995).

Conclusions

The temperature difference between storage temperature and glass transition temperature in gelatinized starch governs both the extent of crystallization and the melting behavior of the crystallized starch. Crystallization at low $T - T_g$ values is less perfect than well above T_g. The crystallization dependence on T_g allows evaluation and prediction of starch crystallization and the stability of products containing starch.

References

Eliasson, A.-C. In *New Approaches to Research on Cereal Carbohydrates*; Hill, R.D.; Munck, L., Ed.; Elsevier Science Publishers B.V.: Amsterdam, The Netherlands, 1985; pp 93-98.

Flory, P.J. *Principles of Polymer Chemistry*, Cornell University Press: Ithaca, NY, 1953.

Gordon, M.; Taylor, J.S. *J. Applied Chem.* **1952**, *2*, 493-500.

Johari, G.P.; Hallbrucker, A.; Mayer, E. *Nature (Lond.)* **1987**, *330*, 552-553.

Jouppila, K.; Ahonen, T.; Roos, Y. In *Food Macromolecules and Colloids*; Dickinson, E.; Lorient, D., Ed.; The Royal Society of Chemistry: Cambridge, UK, 1995; pp 556-559.

Kalichevsky, M.T.; Jaroszkiewicz, E.M.; Ablett, S.; Blanshard, J.M.V.; Lillford, P.J. *Carbohydr. Polym.* **1992**, *18*, 77-88.

Laine, M.J.K.; Roos, Y. *Proceedings of the Poster Session. International Symposium on the Properties of Water. Practicum II.* **1994**, pp 109-112.

Longton, J.; LeGrys, G.A. *Starch/Stärke* **1981**, *33*, 410-414.

Roos, Y.H. *Phase Transitions in Foods*; Academic Press, Inc.: San Diego, CA, 1995.

Roos, Y.; Karel, M. *J. Food Sci.* **1991**, *56*, 1676-1681.

Slade, L.; Levine, H. *Crit. Rev. Food Sci. Nutr.* **1991**, *30*, 115-360.

Zeleznak, K.J.; Hoseney, R.C. *Cereal Chem.* **1986**, *63*, 407-411.

Zeleznak, K.J.; Hoseney, R.C. *Starch/Stärke* **1987**a, *39*, 231-233.

Zeleznak, K.J.; Hoseney, R.C. *Cereal Chem.* **1987**b, *64*, 121-124.

Effect of Impurities on Solubility, Growth Behaviour and Shape of Sucrose Crystals: Experimental and Modelling

Masashi Momonaga, Manufacturing & Technology Laboratories, Fujisawa Pharm. Co. Ltd. 2-1-6, Kashima, Yodogawaku, Osaka 532, Japan
Silke Niehörster, Sabine Henning, Joachim Ulrich, Universität Bremen, Verfahrenstechnik/FB4, Postfach 330440, D-28334 Bremen, Germany

The effect of various impurities (fructose, maltose, glucose, ethanol) on solubility, growth behaviour and shape of sucrose crystals is described. Additionally, a comparison between experiments and modelling is presented. Maltose and glucose show that the solubility of sucrose is decreased. Furthermore, the growth behaviour is modified by both additives. The growth of faces (110) is retarded. In the presence of ethanol and fructose the (001) face growth rate is reduced. The experimentally found results agree with the modelling.

As most crystallization processes are not carried out in pure solutions, the effects of impurities on the kinetics of crystal growth are important. Especially, the growth and shape of crystals are apparently affected by present impurities. It is well known that sucrose crystals can have various shapes due to different combination of forms. Occurrences of particular forms depend on the conditions of nucleation and growth. There are many studies in literature (e.g. Aquilano, 1984; Bubnik, 1992; Chu, 1989; Heffels, 1986; Kraus, 1994; Saska, 1983; Smythe, 1967a,b) about impurity effects on sucrose crystallization.

In the first part of this study the effect of supersaturation, temperature of solubility, growth rate and crystal shape is investigated. Pure and impure sucrose seed crystals were grown from pure and impure solutions. In the second part the comparison is discussed between experiments and modelling.

1. Experimental

1.1 Materials

All solutions were prepared using distilled water and analytical grade sucrose. Fructose, maltose, glucose and ethanol were used as impurities. Impure crystals of sucrose were prepared by crystallization in solution with different impurity contents. The impurity concentrations in the crystals were 1, 5 and 10wt% for fructose, maltose and glucose.

1.2 Method

The solubility was evaluated by measuring the dissolution rate of sucrose crystals in pure and impure solutions. The growth rates were determined at a given temperature, supersaturation and impurity content by the following technique. The length of each face (c, b) was measured as shown in Fig. 1. Measuring time was 3h after maintaining a constant temperature. The temperatures were 303K and 323K. Supersaturations were varied between 0.25 and 1.0 [g/g-H_2O]. Impurity contents ranged from 1 to 10wt% corresponding to an impurity/water ratio of 3 to 30 [g/100g-H_2O]. A stagnant cell with a heating jacket (Kruse, 1993) connected to an image analyser was used to observe the crystals.

2. Results

2.1 Effect of impurities on dissolution rates of pure crystals

The effect of impurities on the dissolution rates of pure sucrose crystals are shown in Fig.

1054–7487/96/0200$12.00/0

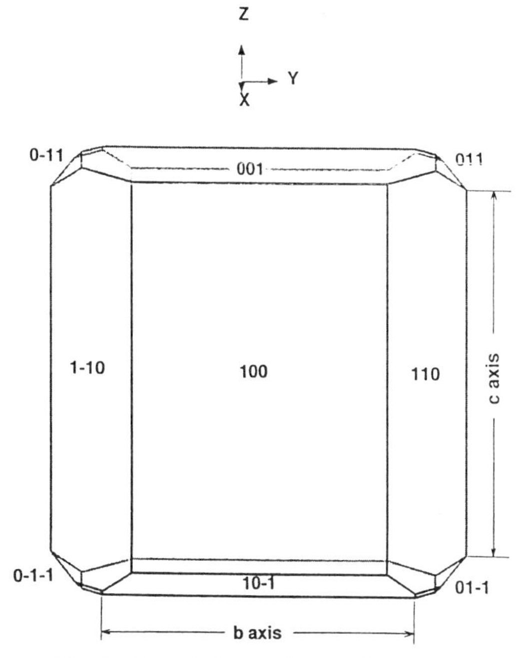

Fig.1 Crystal shape of normal sucrose.

2. Sucrose's dissolution rates are decreasing with increasing impurity concentration. For instance, at an undersaturation of 0.5 [g/g-H_2O] the dissolution rate in the presence of 10wt% glucose is approximately three times lower compared to pure solution. Ethanol shows no significant effect on the dissolution rates.

2.2 Effects of impurities on growth rates of pure crystals and crystal shape

Impurities were added to sucrose solutions in various concentrations. The crystal growth rates were measured at a temperature of 303K and different supersaturations. The impurity effects on the growth rates are shown in Fig. 3. At low supersaturations the growth rate increased in the presence of glucose (10wt%), maltose (5, 10wt%) and ethanol (10wt%), because the solubility decreased. At higher supersaturations an increasing impurity concentration is reducing the growth rate in comparison to pure crystals grown in pure solution. The growth rate decreased of about 20-50% by fructose, 20-40% by maltose and 10-30% by glucose. Over the whole range of used supersaturations ethanol (10wt%) is increasing the growth rate due to a decreasing viscosity and an increasing diffusion coefficient of the sucrose solution.

Impurity effects on the growth rates of (001) and (110) faces are shown in Fig. 4 and Fig. 5. The growth rate of face (001) or accurate lythe growth of its length b is decreasing with increasing impurity concentration in cases of fructose, maltose and glucose. Between these impurities the differences are small. In the presence of ethanol the length b of face (001) is growing faster. All impurities changed (110) face growth rates. With increasing impurity concentration the (110) face growth rate reduction increased. When using glucose the reduction of the (110) face growth rate is small.

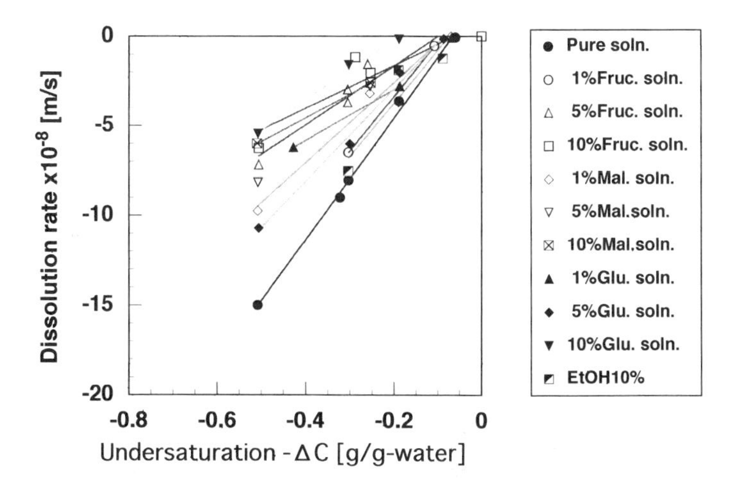

Fig.2 Effect of impurities on dissolution rate for pure crystals.

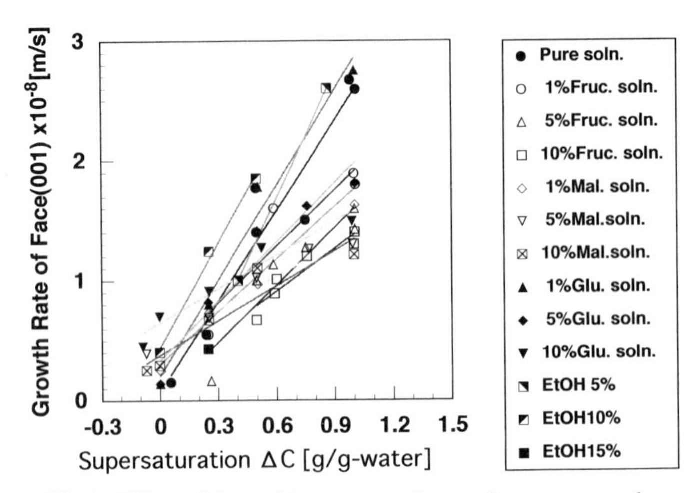

Fig.3 Effect of impurities on growth rate for pure crystals.

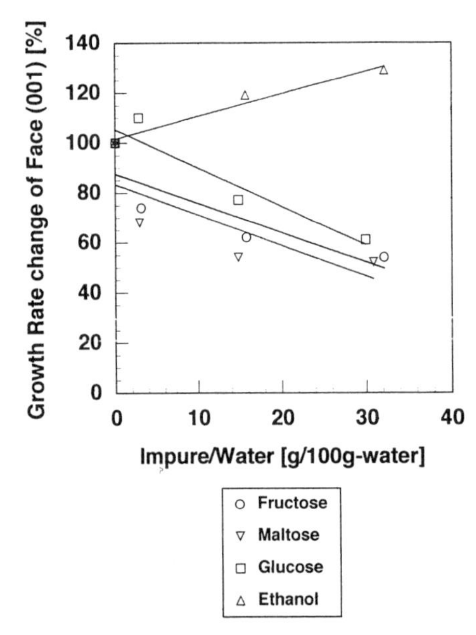

Fig.4 Effect of impurities on reduction of growth rate of face (001).

Since the growth rate of face (110) is measured as length c the results can be explained as follows. The impurities fructose, maltose and ethanol are reducing the length growth c more than glucose. This means that the morphological importance of face (110) is decreasing when using fructose, maltose or ethanol and is increasing when using glucose.

If the growth behaviour of sucrose is changed by impurities the crystal shape can be modified. The left end of the crystal differs from the right

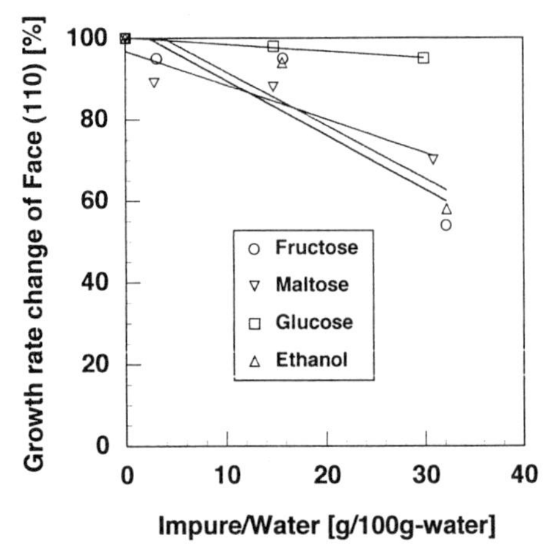

Fig.5 Effect of impurities on redcuction of growth rate of face (110).

pole, because the b-axis resp. the twofold axis of symmetry is asymmetrical in length (Smythe, 1967b). Therefore, the c/b left pole ratio is plotted versus the c/b right pole ratio in Fig. 6. In literature (Bubnik, 1992) a c/b ratio of about 0.7 is given for pure sucrose crystals. The experimental result for pure sucrose is 0.73. Changes in crystal shape will be achieved if the ratio c/b is changed. All impurities altered the ratio. Maltose as impurity increased the c/b ratio to 1.0. In the presence of glucose as impurity the c/b ratio is 1.2. A bigger value than 0.7 means that the (110) and (1-10) face growth rate is suppressed concerning face (001) and (10-1). Thus, the crystal is elongated in c-direction. In the case of ethanol as impurity, the c-axis is

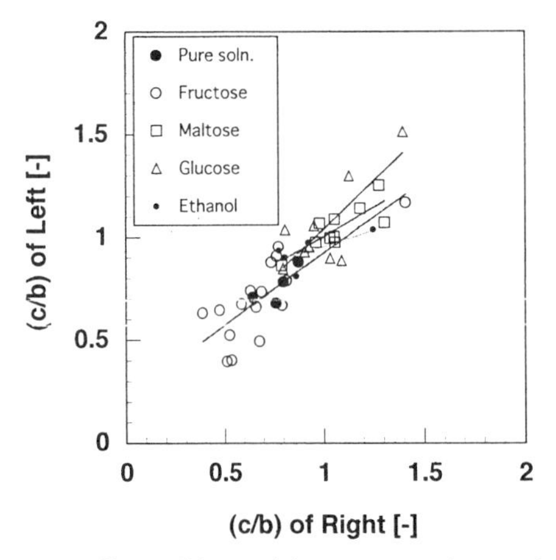

Fig.6 Effect of impurities on growth rate for impure crystals.

growing faster, too. It was observed that the growth rate reduction of face (110) was the biggest (see also Fig. 5). The c/b ratio is 0.6 for fructose as impurity. Comparing the (110) and (1-10) faces growth rates with those of faces (001) and (10-1) the latter ones are growing slower.

The experimentally achieved crystal shapes are shown in Photos1-5. These photos show that fructose (Photo2) did not change the shape clearly in comparison to pure sucrose (Photo1). The corner of faces (011) and (01-1) disappeared in the presence of maltose (see Photo3). In the contrary, the corner of the faces (0-11) and (0-1-1) disappeared using glucose (see Photo4).

2.3 Effects of pure solution on growth rates of impure crystals

Results of impure crystals grown in pure solution are shown in Fig. 7. They are almost the same as those in Fig. 3, where pure crystals are grown from impure solution. With increasing impurity concentration in the crystals the growth rate is decreasing more. However, the growth rate reduction is bigger in impure than in pure solution. Impure crystals change the growth behaviour due to adsorption and inclusion.

2.4 Influence of temperature on the growth rate

The effect of temperature on the crystal growth rate was studied in pure and impure

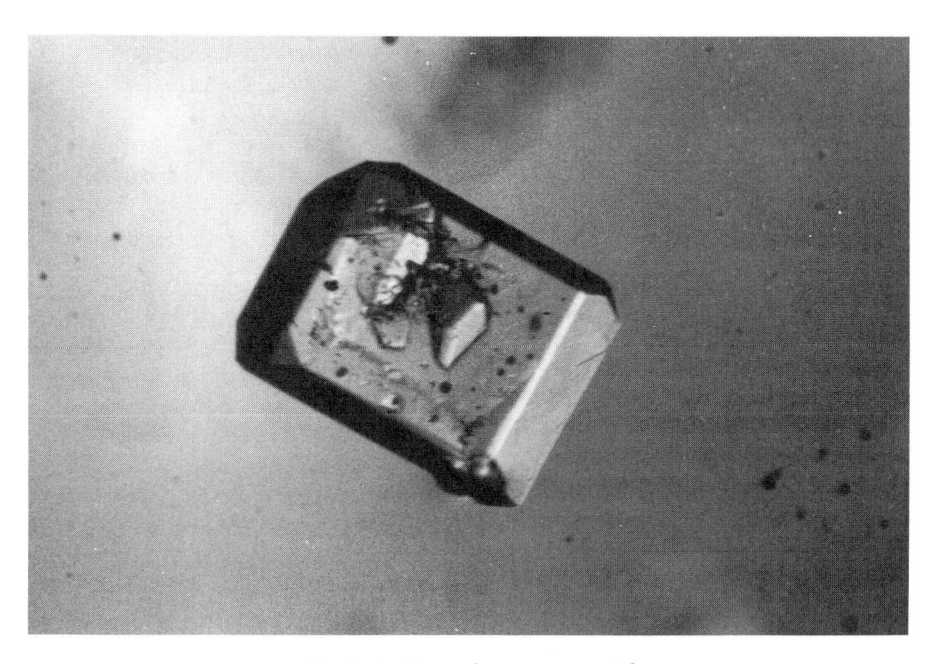

Photo 1 Normal sucrose crystal.

Photo 2 Sucrose crystal in presence of fructose.

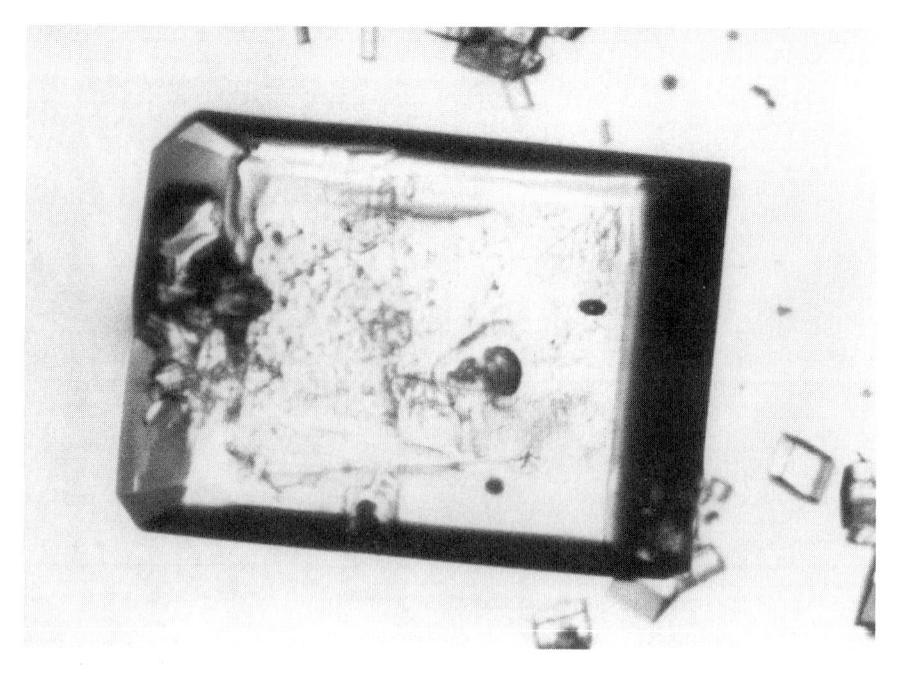

Photo 3 Sucrose crystal in presence of maltose.

Photo 4 Sucrose crystal in presence of glucose.

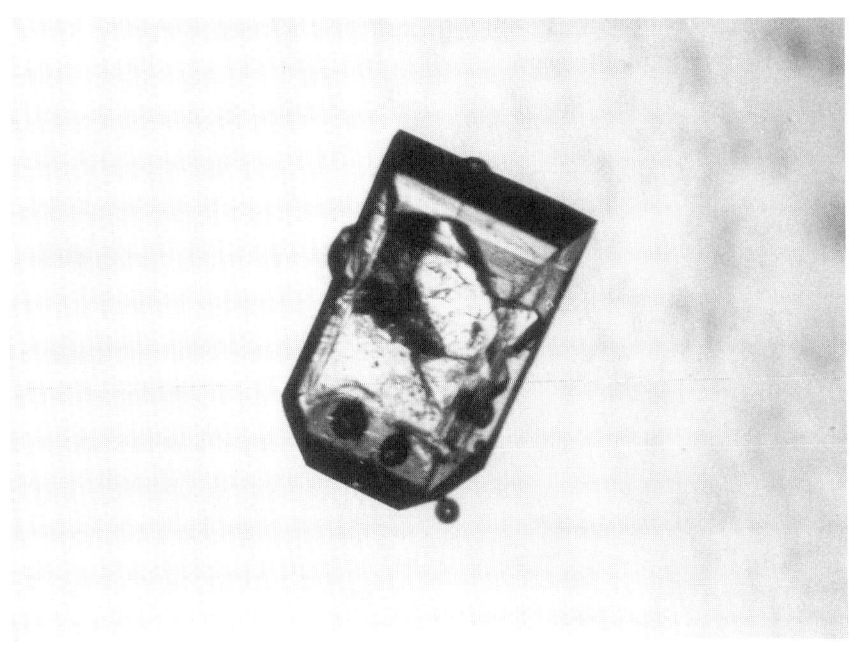

Photo 5 Sucrose crystal in presence of ethanol.

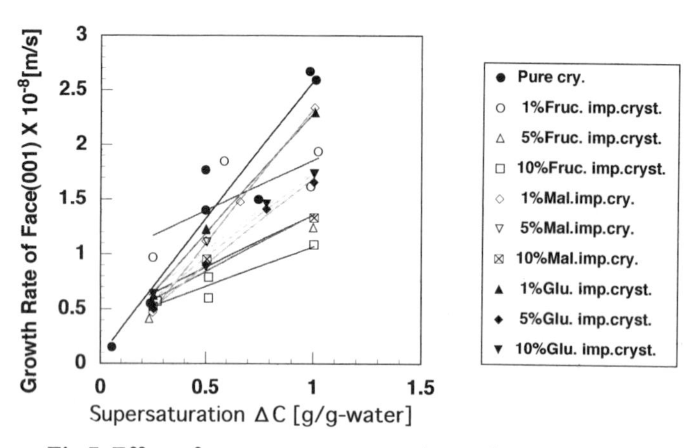

Fig.7 Effect of temperature on growth rate for pure crystals.

solution. Fig. 8 shows the effect of two different fructose concentrations on pure sucrose's growth rate at temperatures of 303K and 323K. The crystal growth rate is in impure solution approximately three times lower compared to pure solution growth for both temperatures.

2.5 Comparison between experiments and modelling

The purpose is to compare the crystal shapes from experiment with the calculated habits influenced by two impurities (glucose and fructose). Calculations were carried out using a commercial software package (CERIUS, 1993). First the habit of pure sucrose has to be calculated. Literature data (Brown, 1973) were used as input. The habit of sucrose was calculated according to the Donnay-Harker law to identify the morphological most important faces ([100], [1-10], [001], [110], [10-1], [0-11], [011], [11-1], [1-1-1]). After an attachment energy calculation using DreidingII force field the binding energy of sucrose on a crystal face was calculated as sum of the slice energy and half the attachment energy.

The influence of the impurities was investigated in the following way. Two molecules are in sucrose's unit cell. Always one host molecule is replaced by an impurity molecule. Then the binding energy of the impurity on a crystal face can be calculated. A binding energy reduction of one kcal/mol changes the attachment energy of about 4.58% for pure sucrose. Therefore, the modified attachment energies in the presence of the impurities can be calculated. Results are shown in Table 1. The ratio of the attachment energies of two faces Eatt,110/Eatt,10-1 was chosen to discuss the influence of the impurities on the crystal shape. The ratio Eatt,110/Eatt,10-1 is 0.74 for pure sucrose. In the presence of glucose the ratio is 0.68 and 1.46 for fructose. The morphological importance of a crystal face decreases with increasing attachment energy. Thus the result of the modelling is that the crystal is elongated in b-direction in the presence of fructose. In the case of glucose the crystal shape is elongated in c-direction.

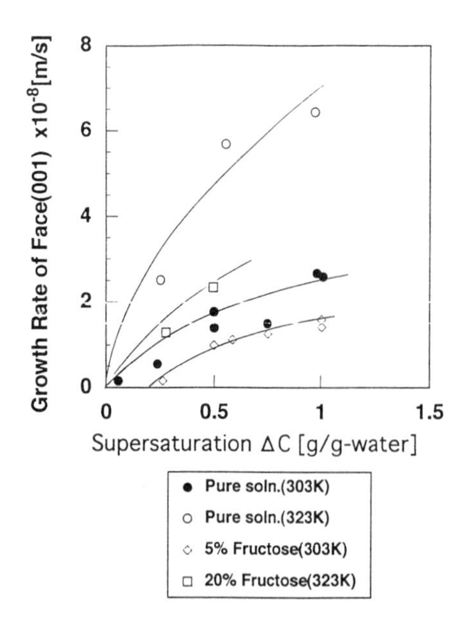

Fig.8 Effect of impurities on axial ratio (c/b).

Table 1: Modelling results

Face	sucrose Eatt [kcal/mol]	+fructose Eatt [kcal/mol]	+glucose Eatt [kcal/mol]
[100]	-43.6	-73.6	-37.3
[1-10]	-72.3	-104.5	-49.8
[110]	-72.3	-100.9	-89.7
[001]	-89.3	-43.0	-136.4
[10-1]	-97.5	-68.8	-132.7
[011]	-119.5	-143.9	-163.8
[0-11]	-119.5	-70.7	-163.7
[1-1-1]	-120.5	-136.4	-135.8
[11-1]	-120.5	-67.6	-129.1
Eatt,110 to Eatt,10-1	0.74	1.46	0.68

3. Discussion

Considering the general effect of the impurities, it is shown that the addition of various impurities to the sucrose solutions decreased the mean growth rate. Growth rate reduction depends on the impurity content and the viscosity of the solution. Decreasing solution's viscosity lead to an increasing growth rate due to an increasing diffusion coefficient. In the case of ethanol as impurity and, e.g. an increasing temperature exactly the effect can be found as mentioned above. The same results can be found in literature (Kelly, 1975; Smythe, 1967a). Other impurities are increasing the solution's viscosity.

Comparing Fig. 4 and Fig. 5 the result is that in the presence of ethanol the crystal shape is elongated in b direction, because the length of face (001) is increasing and (110) is decreasing. The crystal shape is only slightly influenced by fructose and maltose. At a concentration of 30 [g/100g-H_2O] growth rate change is for both faces of about 40% for fructose and 30 to 40% for maltose. Glucose is elongating the crystal shape in c direction, because the length b of face (001) is reduced of about 30% at a concentration of 30 [g/100g-H_2O] compared to only 2% of length c.

The reasons for crystal shape changes are changing growth rates of individual crystal faces. Changes caused by glucose are observed at the crystal's left pole with a disappearing left corner. This result is in good agreement with literature (Bubnik, 1992,1995). Since glucose rings are emerging from (110) faces (Saska, 1983) the growth rates are reduced by incorporating glucose molecules at these faces. Thus the morphological importance of face (110) is increasing. This elongates the crystal in c-direction, which is confirmed by the experimentally achieved c/b ratio of 1.2. The changes of maltose are observed on the right pole of the crystal with a disappearing right corner((011) and (01-1) faces). This is caused by reducing the growth rates of face (110). The ratio c/b is increasing. The resulting crystal shapes are similar when using as impurity either glucose or maltose. Only slight shape changes are caused by fructose. Ethanol as additive is elongating the crystal in c-direction by increasing the length growth of face (001). These results are confirmed by the photos. In comparison to pure sucrose the ratio c/b decreased in the presence of ethanol and fructose. The ratio c/b increased in the presence of glucose and maltose.

The experimentally results agree with the predictions of the modelling. Both ratios, c/b and Eatt,110/Eatt,10-1, give an information concerning the habit change. A ratio of c/b less or Eatt,110/Eatt10-1 bigger than 0.7 (pure sucrose) indicates a crystal elongated in b-direction. This is found for fructose with a c/b value of 0.68 and Eatt,110/Eatt,10-1 of 1.46. For the reverse case a c/b ratio bigger and an attachment energy ratio lower than 0.7 gives a crystal elongated in c-direction, as it is found, e.g. for glucose. The general influence of the impurities on sucrose's crystal shape can be figured out by the presented modelling procedure.

4. Conclusion

Effects of various impurities on the growth behaviour and shape of sucrose crystals were evaluated with following conclusions. Solubilities are changed by glucose, maltose and ethanol. The mean growth rates are decreased by the additives. Little shape changes are observed for each impurity. Glucose and maltose elongated the crystal in c-direction and ethanol and fructose in b-direction. In the case of glucose the corner formed by faces (0-1-1) and (0-11) disappeared. In the case of maltose, the corner formed by faces (011) and (01-1) disappeared. Ethanol is increasing the growth rate due to a viscosity reduction. The agreement between experimental results and modelling is satisfying, but has to be improved.

Acknowledgement

The first author would like to express his thanks to Dr. Yazawa of Fujisawa Pharm. Co. Ltd., who made this research at the University of Bremen possible.

The other authors would like to thank the Senator für Bildung und Wissenschaft der Freien und Hansestadt Bremen for support, who made the other parts of the project possible.

References

Aquilano, D.;Rubo, M.; Vaccari, G.; Mantovani, G.; Sgualdino, G.; *Industrial Crystallization 84*; Ed. Jancic, S.J.; de Jong, E.J.; Elsevier Science Publishers B.V., Amsterdam, **1984**

Brown, G.M.; Levy, H.A.; Acta Cryst. **1973**, B29, 790

Bubnik, Z.; Vaccari, G.; Mantovani, G.; Sgualdino, G.; Kaldec, P.; *Zuckerind.* **1992**, 117, 557

Bubnik, Z.; Kaldec, P.; Proceedings to 11th ICCG, Netherlands, **1995**

CERIUS 3.2, Molecular Simulations Inc. **1993**

Chu Y.D.; Shiau, L.D.; Berglund, K.A.; *J. Cryst. Growth* **1989**, 97, 689

Heffels, S.K.; *PhD Thesis*, TU Delft, **1986**

Kelly, F.H.C.; Keng, M.F.; *The Sucrose Crystal and its Solution*; Singapore University Press, Singapore, **1975**

Kraus, J.; Nyvlt, J.; *Zuckerind.* **1994**, 119, 298

Kruse, M., *PhD Thesis*, Universität Bremen, VDI Verlag, Düsseldorf, **1993**

Saska, M.; Myerson, A.S.; *J. Cryst. Growth* **1983**, 61, 546

Smythe, B.M.; *Aust.J.Chem.* **1967a**, 20, 1097

Smythe, B.M.; *Aust.J.Chem.* **1967b**, 20, 1115

Crystallisation and Polymorphism in Cocoa Butter Fat: In-situ Studies Using Synchrotron Radiation X-ray Diffraction

R.N.M.R. van Gelder, N. Hodgson, K.J. Roberts, A. Rossi, University of Strathclyde, Glasgow, U.K.
M. Wells, M. Polgreen, Cadbury Ltd, Bournville, U.K.
I. Smith, RSS Ltd, Reading, U.K.

We are involved in a research program which seeks to characterise and inter-relate the crystal structure, external morphology and growth kinetics of cocoa butter fat systems. For *in-situ* X-ray studies we have developed a new processing cell which enables the diffraction data to be collected under well defined conditions of temperature and shear. This will allow processing conditions to be optimised on a structural basis. During preliminary experiments we have seen a polymorphic transition taking place at 28.75°C. This probably corresponds to a transformation from Form IV to Form V in cocoa butter. At 33.8°C we observed the melting point of this form, again this is consistent with Form V.

Introduction

In chocolate manufacturing, careful control of the pre-crystallisation or tempering process is very important. This is because it significantly influences the rheological properties of the chocolate and hence its workability during the forming stage. Tempering also affects the physical properties of the end product such a gloss, snap, texture, heat resistance, fat bloom stability, etc. (Murray, 1978; Jewell, 1982; Musser, 1973). These physical properties are related to the polymorphism of the cocoa butter, which comprises the major solid fat in chocolate.

Cocoa butter is a relatively simple fat when compared with other fats of multiple fatty acid composition (Minifie, 1980). Cocoa butter, being the lipid component of the cocoa seed *Thebroma cacao*, is principally made up of three major triglycerides - POP, POS and SOS (S = stearic acid, P = palmitic acid, O = oleic acid), with minor amounts of triglycerides containing linoleic and arachidonic acid (Jurreins, 1968). Cocoa butter can occur in numerous polymorphic crystalline forms, each form having different physical properties such as melting point and crystal morphology.

As part of a wide-ranging research project we are currently characterising and inter-relating crystal structure, external morphology and growth kinetics of the various polymorphic forms of cocoa butter. As part of this we have developed an *in-situ* X-ray processing cell which enables the diffraction data to be collected under well defined conditions of temperature and shear. In this way the structural properties of these materials can be continuously monitored 'on-line' so that fat processing conditions can be optimised. The cell was mounted on the high flux synchrotron radiation beam-line 16.1 at The Daresbury Laboratory (UK). The crystallisation processes which occur during tempering are examined in real time using combined wide and small angle X-ray scattering.

X-ray Studies of Polymorphism in Cocoa Butter

One of the most controversial areas in confectionery science is the discrepancies found between literature data used to classify cocoa butter crystal polymorphs. Since the discovery of polymorphism in cocoa butter, numerous studies have reported different numbers of polymorphs and conflicting melting points for

1054–7487/96/0209$12.00/0

the various crystalline forms (Table 1) (Vaeck, 1951; Vaeck, 1960; Duck, 1964; Willie *et al.*, 1966; Davis *et al.*, 1986; Lovegren et al., 1976).

Today it is generally accepted that cocoa butter can solidify in six crystalline forms, although obvious discrepancies in melting point and differences in nomenclature continue to exist. It is believed that Form V is the preferred polymorph for chocolate production since it gives rise to good demoulding, high gloss, favourable snap and rapid meltdown during eating (Hachiya et al., 1990). Problems of poor texture, lack of contraction and inadequate gloss have been associated with 'lower polymorphs', with no firm data giving the actual polymorph and the amount present. The unsightly 'bloom' formation sometimes seen as a white powdery layer on the surface particularly of dark chocolate is usually associated with the formation of Form VI.

The most widely used method for studying polymorphism is X-ray diffraction. Typical X-ray diffraction patterns of fats exhibit two groups of diffraction lines corresponding to the long and short spacings. The long spacings correspond to the planes formed by the methyl end groups and are dependent on the chain length and the angle of tilt of the component fatty acids of the glyceride molecules. The short spacings refer to the cross sectional packing of the hydrocarbon chain and are independent of the chain length (Chapman, 1962; Jacosberg et al, 1976).

An In-Situ X-ray Cell for On-line Processing

The *in-situ* processing cell, summarised in Fig. 1 enables the synchrotron radiation data to be collected under well defined conditions of temperature and shear so that processing conditions can be optimised on a structural basis. The instrument is based on a typical rheometer design with two aluminium surfaces: a stationery cone and a flat plate which rotates with respect to the cone. The cone angle (4°) ensures uniform shear (shear rate: 0.2 to 22 s^{-1}) throughout the cell and the cone has a 10 mm aperture at a radius of 65 mm. The two parts of the aluminium body have overall dimensions of 200 mm by 240 mm by 40 mm and are separated by 2 mm PVC (polyvinyl chloride) spacers to eliminate heat transfer to the base plate. Three mica (0.025 mm in thickness) sealed arced apertures in the rotating plate are provided to allow transmission of X-rays through the cell for 95% of each revolution. Two rubber 'O'-rings are mounted in grooves on the outer edge of the fixed cone to seal against the inside edge of the bearing. These seals serve to prevent the sample from seeping out of the cell; this is particularly important as the shearing plane is vertical. A window is made in the cone to allow observation of the sample level. The liquid sample is injected into the cell through a valve using a glass syringe. The 12V DC-motor is fitted with a tacho feed-back system for high-accuracy speed control. It drives the rotating plate through a clutch and a 30 to 1 reduction gear box. The cell is mounted on a base plate with two bolts, fitted with springs to ensure the stability of the cell under changing thermal conditions.

The cell design incorporates an optical light probe to measure the turbidity of the sample and thus detect nucleation. The temperature of the cell is controlled with eight Peltier cooling/heating blocks, capable of enabling fast (ca. 10°C/min) and well controlled cooling and heating. The Peltier elements are bonded to brass cooling blocks which are connected via quick release connections to a thermostatically controlled bath recirculating cooling water. A Pt-100 platinum resistance thermometer is mounted in the stationary plate of the cell for an accurate temperature reading.

Instrument control is affected through the use of a modified tempering machine built by Cadbury Ltd, which acts as an interface between the cell and the computer. The computer controls the experiments and collects the data. The control computer is a C.I.L. multifunction instrument model MFI1010 and the computer program is written and compiled using Microsoft Quick Basic.

Materials and Methods

Cocoa butter was provided by Cadbury Ltd (UK). The molten cocoa butter for tempering was injected into the cell through a valve using a

Table 1. Classification of Cocoa Butter Crystalline Forms and their Melting points (°C)

Vaeck (1951)		Vaeck (1960)		Duck (1964)		Wille & Lutton (1966)		Chapman (1971)		Lovergren (1976)	
γ	18.0	γ	17.0	γ	18.0	I	17.3	I	–	VI	13.0
α	23.5	α	21-24	α	23.5	II	23.3	II	–	V	20.0
–	–	–	–	–	–	III	25.5	III	–	IV	23.0
β''	28.0	β'	28.0	β''	28.0	IV	27.5	IV	25.6	III	25.0
β	34.5	β	34-35	β'	33.0	V	33.8	V	30.8	II	30.0
–	–	–	–	β	34.4	VI	36.3	VI	32.2	I	33.5

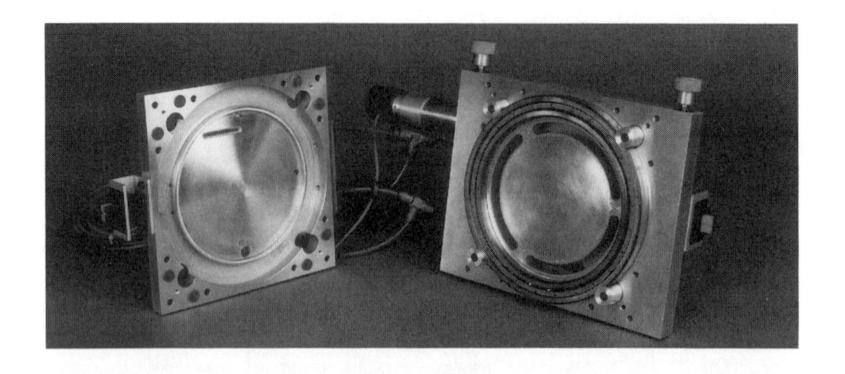

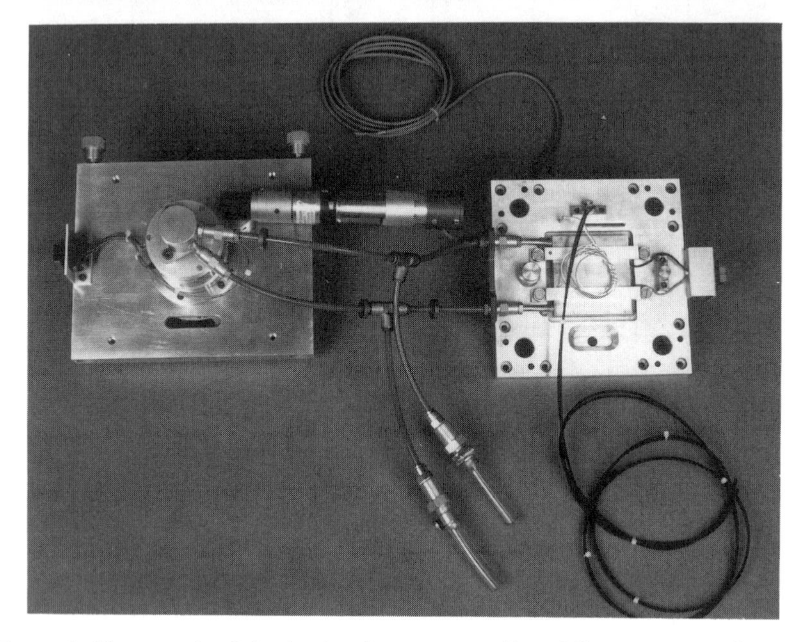

Figure 1. Photograph of the *in-situ* shear stress cell used for *in-situ* experiments.

glass syringe. The sample was then subjected to a cooling/heating process (tempering) which involved, for the data presented here, rapid cooling from 50°C to 15°C and subsequent heating up to 40°C. During the cooling/heating cycle the shear rate was $8s^{-1}$.

For our experiments we made use of the fixed wavelength (1.4Å) X-ray diffraction station 16.1 at the synchrotron radiation source (SRS) in Daresbury (UK) (see *e.g.* Bliss *et al.*, 1995). This station was set-up for combined small and wide angle time-resolved X-ray scattering using two area detectors. The schematic layout is shown in Fig. 2. The SAXS detector was placed at a distance of 1.75m from the sample. The WAXS detector was placed at an angle of 45° above the sample so that the top half of the diffraction rings of the wide angle data could be collected. A helium bag between the WAXS detector and the sample served to minimise air-scattering.

The spectra were analysed using the BSL program (Daresbury Laboratory). BSL is a program that is written specifically for analysing and manipulating 2D data. The SAXS data was calibrated upon comparisons with the d-spacing orders of wet rat tail collagen, while for the WAXS data HDPE (high density poly-ethylene) was used. All data were normalised for beam-decay. Background subtraction was performed

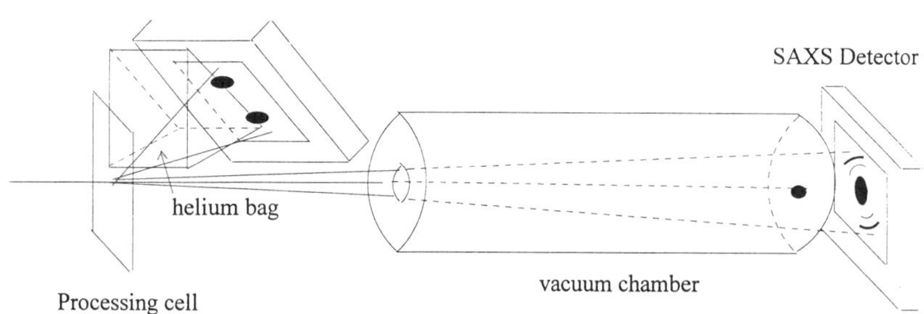

Figure 2. Schematic of Wiggler beamline 16.1 showing the experimental layout for combined small and wide angle X-ray scattering.

using an empty cell for the background pattern and the small angle data was corrected for any inhomogeneities in detector sensitivity. All frames were made with a counting time of 90s.

Results and Discussion

During the cooling phase of the temperature cycle, liquid cocoa butter was cooled down from 50°C to 15°C. The crystallisation process which started to take place at 20.9°C was thus observed *in-situ.* The resulting SAXS spectrum shows one intense ring, which corresponds to a long axis spacing of ca. 50 Å (Fig. 3 (a)), while the WAXS spectrum shows an arc corresponding to a lattice spacing of ca. 4.2 Å (Fig. 3 (a)). The data are consistent with Form III. The intense spots and the arc at 5 Å are due to diffraction from the mica windows of the cell. No polymorphic transitions were observed during this cooling process.

During the heating phase of the temperature cycle a polymorphic transition was observed to take place at about 28.7°C. According to literature data (Willie and Lutton, 1966; Hicklin, Jewell and Heathcock, 1985) this transition probably corresponds to the transition from Form IV to Form V. The SAXS data (Fig. 3 (b)) contains two rings: one at ca. 50 Å and another at 66 Å. During the transition the peak at 50 Å reduces in intensity and disappears as the peak at 66 Å becomes more intense. At 30°C only the latter is still present (Fig. 3 (c)).

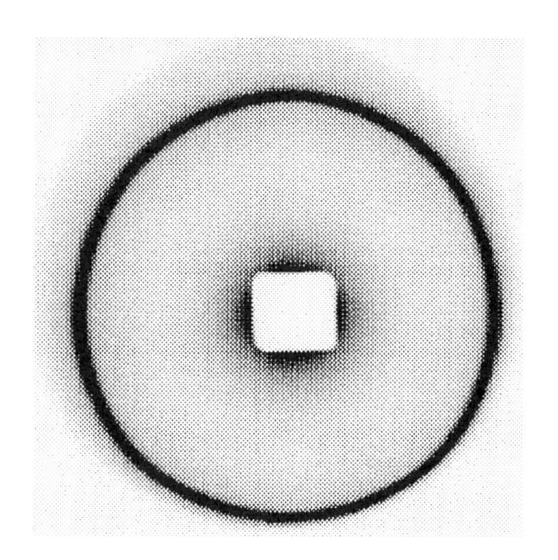

a

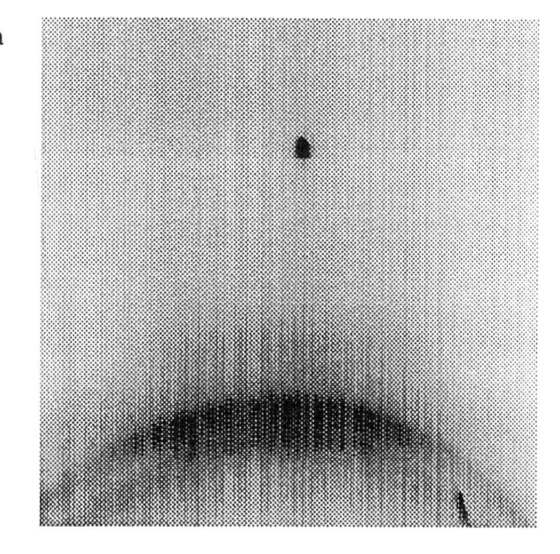

Figure 3. X-ray diffraction data of cocoa butter fat collected *in-situ* on beamline 16.1 at the Daresbury SRS. SAXS (left) and WAXS (right) data for: (a) 20.9°C; (b) 28.7°C; (c) 30°C.

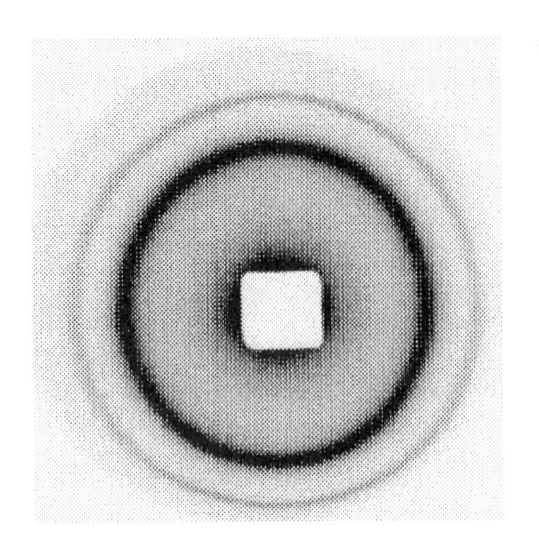

b

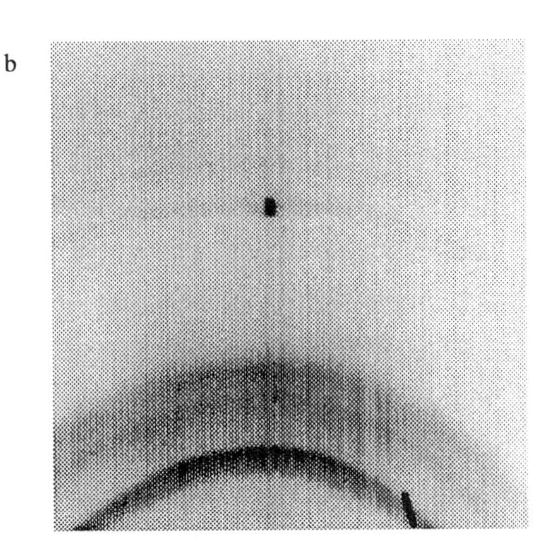

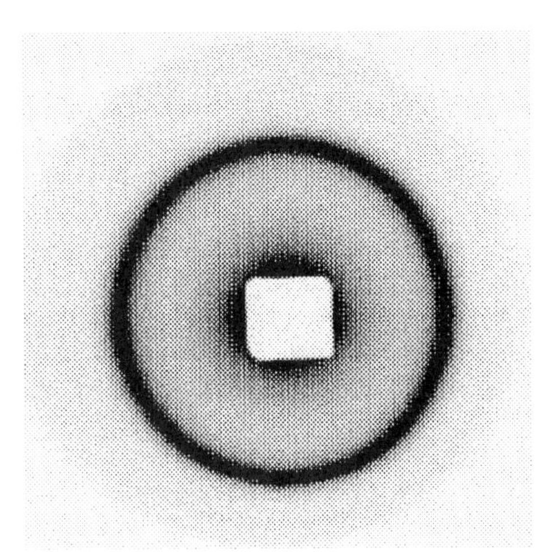

c

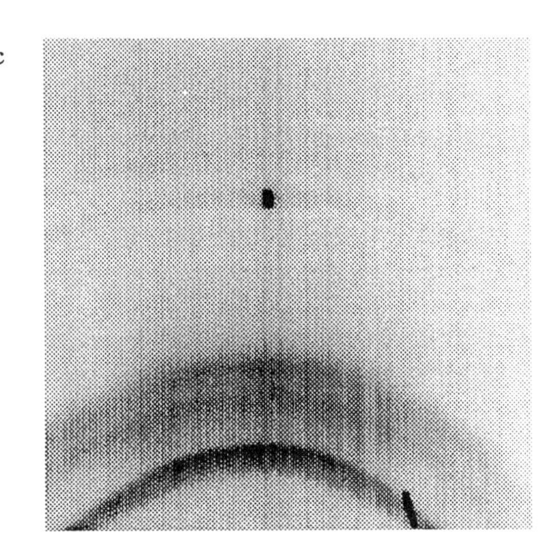

Figure 3. *(Continued).*

In the WAXS spectra (Fig. 3 (b), (c)), during the polymorphic transition, we observed two arcs at ca. 3.91 Å and 3.76 Å, which are also consistent with a transition to form V. At ca. 33.8°C the melting point of the latter form was observed and, according to literature data (Willie and Lutton, 1966), this corresponds to the melting point of Form V.

Conclusions

We have demonstrated the use of a new processing cell which allows us to simulate the tempering process of cocoa butter fat in combination with X-ray diffraction on a real-time basis. During an initial experiment the cocoa butter crystallised at 20.9°C, probably resulting in Form IV. After reheating the sample, at 28.7°C a transition took place to Form V. More experiments are currently underway to quantify the role of processing conditions (temperature and temperature/shear rates) on polymorphic transformations in cocoa butter.

Acknowledgements

This work has been supported by the EPSRC's Soft Solids Processing initiative (Grant

GR/K/42820) in collaboration with Cadbury Ltd and Unilever. One of us (KJR) gratefully acknowledges EPSRC for the current support of a senior fellowship.

We also thank Dr Sasaki for helping with the software development, Dr Hammond for helping with the data analysis, Dr E Towns-Andrews for her co-operation and advice in the use of our cell on station 16.1, Dr T Instone and Dr J. Chambers (Unilever Research) for their interest in the overall project.

References

Bliss, N.; Bordas, J.; Fell, B.D.; Harris, N.W.; Helsby, W.I; Mant, G.R.; Smith, W.; Towns-Andrews, E. *Rev. Sci. Instrum.*, **1995**, 66(2), 1311.

Chapman, D. *Chem. Rev.*, **1962**, 62, 433.

Davis, T.R.; Dimick, P.S. *Proc. PMCA Prod. Conf.*, **1986**, 40, 104.

Duck, W.N. M. Sc. Thesis, Franklin and Marshall College, 1964.

Hachiya, I.; Koyano, T.; Sato, K. *Lipid Technology*, **1990**, 2

Hicklin, Jewell and Heathcock, 1985.

Jacosberg, B. *J. Am. Oil Chem. Soc.*, **1976**, 53, 609.

Jewell, G.G.; *35th P.M.C.A. Production Conference*, **1981**, 63.

Jurreins, G. In *Analysis and Characterisation of oils and fats and fat production*, H.A. Boekenoogen, Ed; London: Interscience, 1968, Vol 2.

Lovegren, N.V.; Gray, M.S.; Feuge, R.O. *J. Am. Oil Chem. Soc.*, **1976**, 53, 108.

Minifie, B. W. In *Chocolate, Cocoa and Confection: Science and Technology*, 2nd edn; AVI Publishing CO., Westport, CT, 1980.

Murray, G.; *Manufacturing Confectioner*, **1978**, 58, 35.

Musser, J.C.; *35th P.M.C.A. Production Conference*, **1973**, 46.

Vaeck, S.V. *Rev. Interm. Choc.*, **1951**, 6, 100; *Manufacturing Confectioner*, **1960**, 40, 35.

Willie, R.L.; Lutton, E.S. *J. Am. Oil Chem. Soc.*, **1966**, 43, 491.

Growth and Dissolution Rate Dispersion of Sucrose Crystals

Jürgen Fabian, Richard W. Hartel, University of Wisconsin - Madison, Dept. of Food Science, USA
Joachim Ulrich, Universität Bremen, Verfahrenstechnik, FR Germany

Experimental results of investigation of the growth and dissolution kinetics of single sucrose crystals in stagnant solution are presented. The data support the constant crystal growth and constant crystal dissolution model. There is no size dependence of sucrose crystals during growth and dissolution. Growth rate dispersion appears at all temperatures and supersaturations studied, as does dissolution rate dispersion.

The Concept of Growth and Dissolution Rate Dispersion

Growth rate dispersion refers to the phenomenon where individual crystals of initially the same size can grow with different rates even though they are all subjected to identical growth environments (temperature, supersaturation, hydrodynamic, etc.) (Mullin 1993).

Review articles by Janse (1976) and Ulrich (1989) summarize and explain the general development and also give information of the model of growth rate dispersion (GRD) for different substances and conditions. Growth rate dispersion in the system sucrose - water has been especially investigated by Shanks (1985), Heffels (1986) and Liang (1987a).

The phenomenon of GRD is primarily explained by differences in surface integration kinetics of different crystals (Valcic 1975; Botsaris, 1982). The dislocation density and structure, varying degrees of stress and strain, as well as differences of amount of surface adsorption of impurities are considered to be the mechanisms causing GRD (Ristic, 1991).

The GRD model is described either by the random-fluctuation-(RF)-model (Wright, 1969; Girolami, 1985; Tavare, 1985 and Garside, 1984) or the constant-crystal-growth-(CCG)-model (Ramanarayanan, 1982 and 1985; Berglund, 1984; Larson, 1985). Both models have been found to describe experimental data depending on the conditions of the specific investigations. The growth rate of single crystals in stagnant solution however, can be best described by the CCG model (Ulrich, 1989).

Dissolution of crystals is generally considered to follow first-order kinetics with respect to undersaturation. This leads to the conclusion that the dissolution process is purely volume diffusion controlled (Mullin, 1969 and 1993; VanHook, 1974). However, it has been shown that some dissolution processes involve a reaction step at the crystal surface (Bovington, 1970; Zhang, 1991; Fabian, 1993a), which can be even more pronounced in the presence of additives (Sears, 1958; Nancollas, 1984; Kubota, 1988; Fabian, 1993b).

Most of these publications about dissolution processes deal with inorganic salts and only few investigate organic substances like sucrose. Most do not consider the existence of growth or dissolution rate dispersion. In this study, dissolution of sucrose crystals was investigated, to determine whether sucrose crystals dissolve with a constant rate and if there is evidence for a surface effect on dissolution. If there is a surface "reaction" step involved in the dissolution process, a dissolution rate dispersion can occur.

Methods

Experiments for examination of growth and dissolution kinetics were conducted in a temperature controlled glass cell, placed directly on a microscope stage. Photomicrographs of crystals were taken by a CCD camera and saved as digitized images in an attached computer by using image analyzer software. Images of the observed crystals were taken after distinct time periods and change of their projected area was measured. From the measured area, the diameter of a circle of equivalent area was calculated. The growth rate was expressed as linear growth rate G [m/s].

Experiments were done at four temperatures 30, 40, 50 and 60 °C. At each temperature, three under- and three supersaturations were used. The crystals used were selected from extra pure sucrose in the size range between 500 and 1300 µm, based on the diameter of circle of equivalent projected area.

1054–7487/96/0216$12.00/0

All the crystals were chosen to have the same history (same sample, quality), were of perfect shape and had smooth surfaces and sharp edges.

Results and Discussion

a) Constant Crystal Dissolution (CCD)

The model of constant crystal growth has already been justified for sucrose crystals as mentioned above. The results of this work show that the observed single sucrose crystals dissolved with a constant rate at all experimental conditions. **Figure 1** shows the change of diameter of single sucrose crystals versus time over the experimental duration of 120 minutes. This dissolution behavior was observed for all temperatures and undersaturations investigated.

b) Growth and Dissolution Rate Dispersion (GRD and DRD)

The growth and dissolution rates obtained from experiments at 30 °C are plotted in **Figure 2** versus the % total solids. In **Figure 3**, the results for the experiments at 40°C are plotted, while **Figures 4** and **5** present the data for 50 °C and 60 °C, respectively.

Both GRD and DRD occur at all temperatures and all concentration differences. This indicates that individual crystals have different growth and dissolution rates even if they are subjected to the same environmental conditions. Size and width of the dispersion varies depending on temperature and concentration. Generally, the width of the dispersion increased with increasing under- or supersaturation and hence, rate of dissolution or growth. The dissolution rate appears to be always bigger than the growth rate at comparable

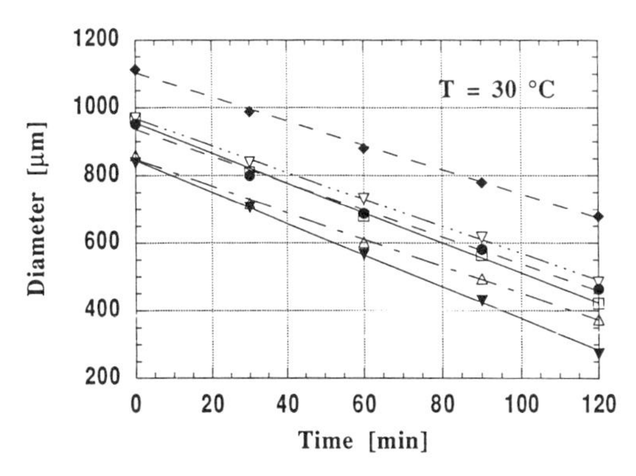

Fig. 1: Constant Crystal Dissolution for six sucrose crystals in stagnant solution (T = 30°C, 65.19 % TS)

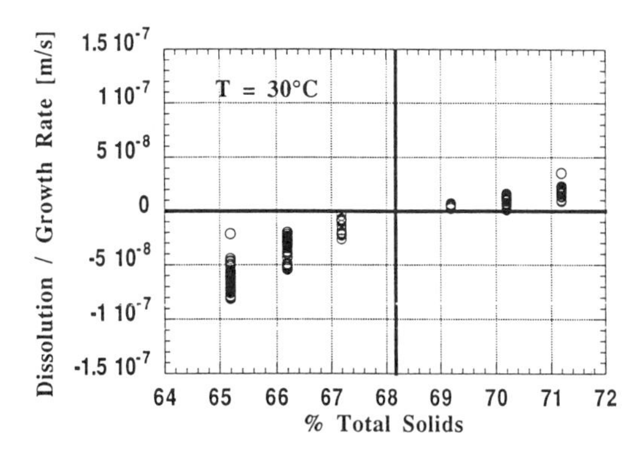

Fig. 2: Growth and dissolution rate dispersion of single sucrose crystals in stagnant solution versus % total solids (T = 30°C)

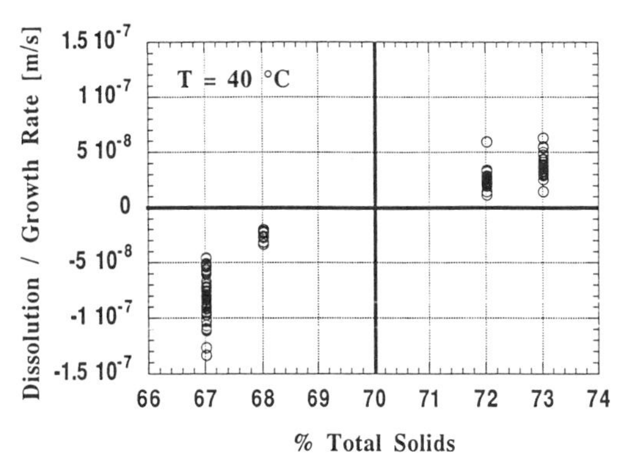

Fig. 3: Growth and dissolution rate dispersion of single sucrose crystals in stagnant solution versus % total solids (T = 40°C)

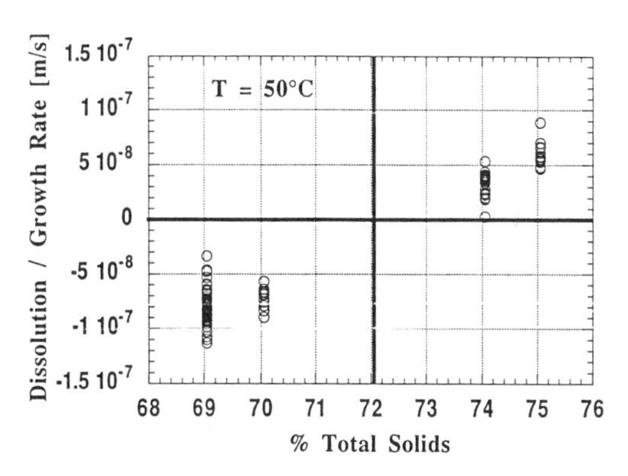

Fig. 4: Growth and dissolution rate dispersion of single sucrose crystals in stagnant solution versus % total solids (T = 50°C)

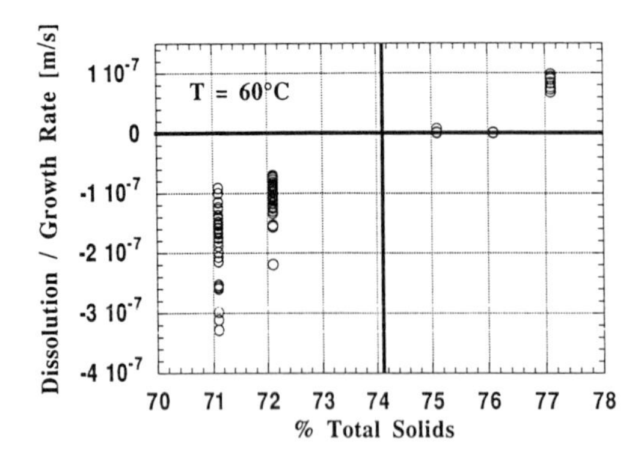

Fig. 5: Growth and dissolution rate dispersion of single sucrose crystals in stagnant solution versus % total solids (T = 60°C)

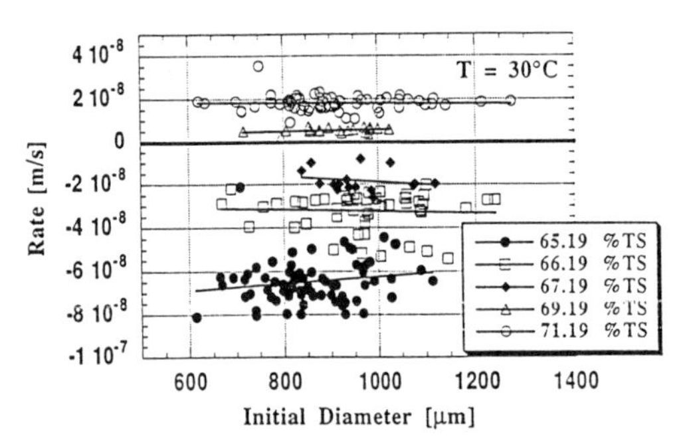

Fig. 6: Growth and dissolution rates of single sucrose crystals in stagnant solution versus initial crystal size (T = 30 °C)

concentration differences. However, the difference depends on the temperature.

c) Size Independent Growth and Dissolution

Previous research (Shanks, 1985; Liang, 1987b; Ulrich, 1990) shows that crystal growth is not size dependent. The results obtained in this research also show that growth as well as dissolution is independent of the initial size of the sucrose crystals. **Figure 6** shows the growth and dissolution rates for the temperature of 30 °C for three different under- and two super-saturations versus the initial size of crystals. Lines through the measured points were obtained by linear regression.

Acknowledgments

The authors appreciate the support of the Deutscher Akademischer Austauschdienst (DAAD) which made this research possible.

References

Berglund, K.A.; Larson, M.A., *AIChE J.* 30, **1984**, 280-287

Botsaris, G.D., In: *Industrial Crystallization 81*, S.J. Jancic and E.J. deJong (Eds.), North-Holland, Amsterdam, **1982**, 109-116

Bovington, C.H.; Jones, A.L., *Transactions of the Faraday Society*, 66, **1970**, 2088-2091

Fabian, J., Ulrich, J., *12th Symposium on Industrial Crystallization*, Z.H. Rojkowski (Ed.), Warsaw, Poland, Zaklad Poligraficzny Wlodarski, Warszawa, 21.-23.Sept.,**1993a**, Volume 2, p4-041 to 4-046

Fabian, J., Ulrich, J., *Delft and Bremen International Workshop for Industrial Crystallization* (DIWIC and BIWIC), J. Ulrich, O.S.L. Bruinsma (Eds.) 13.-16.Sept., **1993b**, 26-33

Garside, J.; Tavare, N.S., *Inst. Chem. Eng. Sympos.* Ser. 87, **1984**, 767-782

Girolami, M.W.; Rousseau, R.W., *AIChE J.* 31, **1985**, 1821-1828

Heffels, St.-K., Dissertation, Technische Universitaet Delft, **1986**

Janse, A.H.; deJong, E.J., *Industrial Crystallization,* J.W. Mullin (Ed.), Plenum Press **1976**, 145-154

Kubota, N.; Uchiyama, I.; Nakai, K.; Shimiza, K.; Mullin, J.W., *Industrial and Engineering Chemistry Research*, 27, **1980**, 930-

Larson, M.A.; White, E.T.; Ramanarayanan, K.A.; Berglund, K.A., *AIChE J.* 31, **1985**, 90-94

Liang, B.M.; Hartel, R.W.; Berglund, K.A. *AIChE J.*, 33, **1987a**, 12, 2077-2079

Liang, B.M.; Hartel, R.W.; Berglund, K.A. Chem. Eng. Sci., 42, **1987b**, 2723-2727

Mullin, J.W.; Gaska, J., *Canadian Journal of Chemical Engineering*, 47, **1969**, 483-489

Mullin, J.W., *Crystallization,* Butterworth-Heinemann, Oxford, Third Edition **1993**

Nancollas, G.H.; Zawacki, S.J., In: *Industrial Crystallization 84* (9th Symposium, The Hague), S.J. Jancic and E.J. deJong (Eds.), Elsevier, Amsterdam, **1984**, 51-60

Ramanarayanan, K.A., *Growth rate dispersion in batch and continuous crystallizers*, PhD-Thesis, Iowa State University of Science and Technology, Ames, **1982**

Ramanarayanan, K.A.; Berglund, K.A.; Larson, M.A, *Chem. Eng. Sci.* 40, **1985**, 1604-1608

Ristic, R.I.; Sherwood, J.N.; Shripathi, T., In: *Advances in Industrial Crystallization* , J. Garside, R.J. Davey, A.G. Jones (Eds.) Butterworth-Heinemann, Oxford, **1991**, 77-91

Sears. G.W., In: *Growth and Perfection of Crystals* (proceedings), R.H. Doremus,

B.W. Roberts, D. Turnbull (Eds.), International Conference on Crystal Growth, Cooperstown, N.Y., **1958**, 441-445
Shanks, B.H.; Berglund, K.A., *AIChE J.*, 31, **1985**, 5, 152-154
Tavare, N.S., *The Canadian J. of Chem. Eng.* 63, **1985**, 436-442
Ulrich, J., *Crystal Research and Technology*, 24, **1989**, 3, 249-257
Ulrich, J.; Kruse, M., In: *Proceedings of the 11th Symposium on Industrial Crystallization*, A. Mersmann (Ed.), Garmisch-Partenkirchen, Germany, 18-20 Sept., **1990**, 361-366
Valcic, A.V., *J. Cryst. Growth*, 30, **1975**, 129-136
VanHook, A., *Crystallization: Theory and Practice*, Reinhold, New York **1961**
VanHook, A., *International Sugar Journal* 6, **1974**, 905, 131-134
Wright, P.G.; White, E.T., In: *Proceedings of the Queensland Society in Sugar Cane Technology,* 36th conference, Brisbane, **1969**, 299-309
Zhang, J.; Nancollas, G.H., In: *Advances in Industrial Crystallization* , J. Garside, R.J. Davey, A.G. Jones (Eds.), Butterworth-Heinemann, Oxford, **1991**, 47-62

General Papers

Study of Organic Supersaturated Solutions: Theory and Experiment

Allan S. Myerson and Alexander F. Izmailov, Department of Chemical Engineering, Polytechnic University, Six MetroTech Center, Brooklyn, New York 11201

Hydrodynamic equations describing crystal growth from solutions are supplemented by the thermodynamics of metastability. The main reason for accounting of thermodynamics of metastability is due to the non-trivial dependences of solution density, shear viscosity and solute mass diffusivity on solute concentration in supersaturated solutions. Combination of hydrodynamic and thermodynamic equations allows understanding of conditions which determine effectiveness of crystal growth. The containerless electrodynamic levitation of supersaturated solution microdroplets is suggested and discussed as the main experiment for the study of thermodynamics of metastability.

The problem of crystal growth of organic materials from supersaturated solutions is well-known. However, until recent there has been no appropriate formalism which accounts both the hydrodynamic and thermodynamic aspects of the problem. Development of such a formalism is of paramount importance since such characteristics of crystal growth from solutions as the solute mass diffusivity D, growth rate η, thickness δ_{diff} of the Diffusion Boundary Layer (DBL) are strongly dependent on solute concentration. This dependence which was first experimentally studied in [1-9] becomes more significant with deeper penetration into a region of metastability. Recent experimental and theoretical studies [10,11] demonstrated that the dependence of any process, occurring in a metastable state, on solute concentration c becomes crucial in the supersaturated region. In papers [10,11] the first attempt to combine together the hydrodynamics of crystal growth from solutions and thermodynamics of metastability was developed and applied for the study of crystal growth from supersaturated solutions of ADP ($NH_4H_2PO_4$), KDP($(C_3H_5NO_2)_3H_2SO_4$),TGS ($(C_3H_5NO_2)_3H_2SO_3$) and glycine. The results obtained for the DBL thickness δ_{diff}, two-dimensional solute concentration profile $c(x_1,x_2)$ and solute diffusional flux $j_{diff}(x_1,x_2)$ demonstrated the non-trivial dependence on supersaturation and thermodynamics of metastability. Therefore, the further understanding of the thermodynamics of solution metastable state is important.

The study of bulk supersaturated solutions is very limited because of high odds for a heterogeneous nucleation to occur at low supersaturations. This nucleation can be triggered by dust impurities, dirt, and most likely, container walls. However, in order to properly investigate thermodynamics of supersaturated solutions it is necessary to have an opportunity to study these solutions at high supersaturations. To the best of our knowledge the only experimental technique which allows such investigation to be performed is an electrodynamic levitation of electrically charged microdroplets of supersaturated solutions. This technique has already been successfully applied for the study of highly supersaturated organic and non-organic solutions [12-20]. The main advantage of the technique is the containerless levitation of microdroplets which contain almost no dust impurities and dirt. The deepest penetrations (up to *45* molal) into the metastable region, approaching almost spinodal region (region of absolute instability), were recently reported [18-20] in the study of supersaturated aqueous solution of glycine at *298K*.

In *Section 1* we describe the general scheme of accounting for thermodynamics of metastability in the Navier-Stocks equations describing the gravity induced steady laminar motion of supersaturated solution within and beyond the diffusion boundary layer. Impact of concentration dependence of solute mass diffusivity, solution shear viscosity and density on crystal growth is discussed in this *Section* as well. In *Section 2* the description of the Spherical Void Electrodynamic Lavitator Trap (SVELT), which allows thermodynamic investigations at high supersaturations, is given.

1054–7487/96/0222$12.00/0

1. The Role of Concentration Dependence of Solution Shear Viscosity, Density and Solute Mass Diffusivity in Crystal Growth from Solutions

Current analytical approaches for the descriptions of crystal growth from supersaturated solutions [21-23 and references therein] neglect almost all the significant particularities of the solution metastable state. These include the non-trivial dependence of the solution density ρ, shear viscosity η and solute mass diffusivity D on the solute concentration c. The usual practice [21-23] has been to assume that these physical characteristics are constants which are independent of solute concentration. However, recent experimental studies [1-9] of supersaturated aqueous solutions of inorganic and organic salts have demonstrated that the solution density ρ, shear viscosity η, and solute mass diffusivity D are dependent on the solute concentration c: $\rho = \rho(c)$, $\eta = \eta(c)$, and $D = D(c)$. This dependence becomes more significant with deeper penetration into metastable region. For example, the diffusivity $D(c)$ declines to zero at a spinodal line which separates metastable and unstable states. These facts have necessitated the development of an appropriate formalism [10,11] to describe crystal growth from supersaturated solution taking into account the dependence of physical properties of supersaturated solutions on solute concentration.

Let us consider the situation where a supersaturated solution is mixed in a result of natural convection. Because solution has a finite viscosity and sticks to a crystal solid interface, an unmixed boundary layer is formed. Within this layer, adjacent to the growing crystal surface, solution can be assumed to be stationary and the solute mass transfer is achieved by means of ordinary diffusion. In the present paper we assume that crystal surface and solution have the same temperatures and, thus, there is no heat transfer.

Solution metastability implies that its density, shear and bulk viscosities, and solute mass diffusivity are some functions of solute concentration. Due to the crystal growth process this concentration is different far from the growing surface than in proximity to it. It is usually assumed that the entire solute concentration change occurs only within the DBL and beyond this layer it is virtually a constant: $c = c_\infty$, although it fluctuates in time and space. Thus, within the DBL, for the case when the solution density $\rho(c)$, shear viscosity $\eta(c)$ and solute mass diffusivity $D(c)$ are weak functions of the solute concentration c one can write the following Taylor expansions:

$$\rho(c) = \rho(c_\infty) + \frac{\partial \rho(c)}{\partial c}\Big|_{c=c_\infty} \Delta c \, [1 + O(1)], \qquad (1)$$

$$\eta(c) = \eta(c_\infty) + \frac{\partial \eta(c)}{\partial c}\Big|_{c=c_\infty} \Delta c \, [1 + O(1)], \qquad (2)$$

$$D(c) = D(c_\infty) + \frac{\partial D(c)}{\partial c}\Big|_{c=c_\infty} \Delta c \, [1 + O(1)]. \qquad (3)$$

Let us take $x_2 = 0$ as the surface (010) of the growing crystal plate, directing x_2-axis into a solution and x_1-axis vertically upward. The lower ledge of the plate corresponds to the value $x_1 = 0$. As a supersaturated solution flows parallel to the crystal plate, the thickness δ_{diff} of the DBL increases with the distance x_1 from the lower ledge where a flow meets a plate. Therefore, the thickness δ_{diff} can be considered as an increasing function of x_1: $\delta_{diff} = \delta_{diff}(x_1)$. Thus, the general Navier-Stokes equations describing the gravity-induced two-dimensional steady laminar motion of an incompressible supersaturated solution within ($x_2 \leq \delta_{diff}(x_1)$) and beyond ($x_2 > \delta_{diff}(x_1)$) the DBL acquire the form:

$$\sum_{j=1}^{2} \frac{\partial \Pi_{ij}}{\partial x_j} = \delta_{i,1} g [\rho(c) - \rho(c_\infty)] \theta[\delta_{diff}(x_1) - x_2], \qquad (4.A)$$

$$\frac{\partial v_1}{\partial x_1} + \frac{\partial v_2}{\partial x_2} = 0, \qquad (4.B)$$

where

$$\Pi_{ij} = -p\,\delta_{ij} + \rho(c) v_i v_j + \eta(c)\left(\frac{\partial v_i}{\partial x_j} + \frac{\partial v_j}{\partial x_i}\right), \quad i,j = 1,2.$$

In the above expressions Π_{ij} is the density tensor of the solution momentum flow, p is the solution pressure, v_1 and v_2 are the solution velocity components, g is the gravity acceleration, and $\theta[\delta_{diff}(x_1) - x_2]$ is the symmetrical unit-step function. The continuity equation (4.B) assumes that supersaturated solution is an incompressible fluid. Utilizing the fact that the DBL thickness $\delta_{diff}(x_1)$ is very small compared to the length L along a crystal plate one can significantly simplify equations (4.A). Thus, in the particular case where $\delta_{diff}(x_1)/L << 1$ the equation describing solution motion along the x_1-axis acquires the form:

$$\rho(c)\left(\frac{\partial v_1}{\partial x_1}v_1 + \frac{\partial v_1}{\partial x_2}v_2\right) \approx -\frac{\partial}{\partial x_2}\left[\eta(c)\frac{\partial v_1}{\partial x_2}\right] +$$

$$+ g[\rho(c) - \rho(c_\infty)]\theta[\delta_{diff}(x_1) - x_2]. \quad \textbf{(5)}$$

This form implies the following assumptions imposed on solution motion: (A) viscosity exerts a significant influence only within the DBL, (B) flow velocity must become zero at the solution-crystal interface, (C) the solution retardation in the DBL is caused by viscous forces alone, (D) the pressure change in the DBL is determined by its change outside and in the natural convection case is negligible since solution outside the DBL is almost stagnant. Substituting into equation (5) expansions (1,2) and keeping only the main terms proportional to Δc and $\partial(\Delta c)/\partial x_2$ one can rewrite equation (5) as follows:

$$\frac{\partial v_1}{\partial x_1}v_1 + \frac{\partial v_1}{\partial x_2}v_2 \approx -\nu_\infty\frac{\partial^2 v_1}{\partial x_2^2} + \omega_\infty(\nu_\infty\frac{\partial^2 v_1}{\partial x_2^2} - g)\Delta c\ \theta[\delta_{diff}(x_1) - x_2] -$$

$$- \nu_\infty\kappa_\infty\frac{\partial}{\partial x_2}\left[\frac{\partial v_1}{\partial x_2}\Delta c\ \theta[\delta_{diff}(x_1) - x_2]\right], \quad \textbf{(6)}$$

where

$$\nu_\infty = \frac{\eta(c_\infty)}{\rho(c_\infty)},\quad \omega_\infty = \frac{\partial \ln[\rho(c)]}{\partial c}\Big|_{c=c_\infty}\quad \kappa_\infty = \frac{\partial \ln[\eta(c)]}{\partial c}\Big|_{c=c_\infty}$$

Boundary conditions for equation (6) have the form:

$$A)\ v_1|_{x_2=0} = v_2|_{x_2=0} = 0, \quad \textbf{(7.A)}$$

$$B)\ v_1|_{x_2=\infty} = v_2|_{x_2=\infty} = 0. \quad \textbf{(7.B)}$$

Condition (7.A) reflects the fact that solution sticks to the crystal solid interface whereas condition (7.B) describes solution as stagnant at an infinite distance from the DBL edge. Thus, equations (4.B and 6) together with boundary conditions (7.A,7.B) describe the hydrodynamic aspects of natural convection corrected by accounting for solution metastability in a system *"supersaturated solution + growing crystal surface"*.

The detailed solution of equations (4.B) and (6) under boundary conditions (7.A) and (7.B) is given in [10,11]. It allows the following expression for the DBL thickness $\delta_{diff}(x_l)$:

$$\delta_{diff}(x_1) = \left[\frac{72\nu_\infty^2(2\kappa_\infty - \omega_\infty)x_1}{|g|\omega_\infty}\right]^{\frac{1}{4}}. \quad \textbf{(8)}$$

Thus, by means of result (8) we have derived how hydrodynamics of natural convection is related to thermodynamic metastability of supersaturated solutions. In expression (8) metastability effects are taken into account through the dependence of the solution density $\rho(c_\infty)$ and shear viscosity $\eta(c_\infty)$ on the bulk solute concentration c_∞.

Analysis of expression (8) obtained for the DBL thickness $\delta_{diff}(x_1)$ allows with the following conclusions. First, we have obtained the well-known result that $\delta_{diff}(x_1)$ grows as $(x_1/|g|)^{1/4}$ [24,25 and references therein]. Second, we have derived how the DBL thickness $\delta_{diff}(x_1)$ depends on the bulk solute concentration c_∞ via such solution static and dynamic characteristics as its density $\rho(c_\infty)$ and viscosity $\eta(c_\infty)$. This dependence provides an opportunity to relate to each other the supersaturated solution static and dynamic characteristics. In particular, expression (8) allows one to understand the relationship between the density $\rho(c_s)$ and the viscosity $\eta(c_s)$ of saturated solutions, where c_s is the saturation concentration at a given temperature and pressure. It is apparent that a boundary layer should vanish at the saturation point since at this point solution and the growing crystal surface are in thermodynamic equilibrium. As it follows from result (8) such a situation is possible only when the following equality is satisfied:

$$\omega_\infty|_{c_\infty=c_s} = 2\kappa_\infty|_{c_\infty=c_s}. \quad \textbf{(9)}$$

Analysis of equation (9) gives that at the saturation point the solution density $\rho(c_s)$ and viscosity $\eta(c_s)$ should be related as follows:

$$\eta(c_s) = cst\ \rho^{\frac{1}{2}}(c_s). \quad \textbf{(10)}$$

This result has been experimentally verified with such inorganic and organic aqueous solutions as NaCl, KCl, Urea, ADP ($NH_4H_2PO_4$), KDP (KH_2PO_4), TGS ($(C_3H_5NO_2)_3H_2SO_4$) and Glycine [1-9] taken at *25° C* and normal pressure. For all these solutions it was found that dependence of their bulk densities $\rho(c_\infty)$ on the bulk solute concentration c_∞ was linear: $\rho(c_\infty) = a_0 + a_1 c_\infty$ (*Table 1* gives coefficients a_0 and a_1 for different solutions). The error of such a linear interpolation of the density experimental data was always within *0.01%*. The experimentally obtained dependence of the solution shear viscosity $\eta(c_\infty)$ on the bulk solute concentration c_∞ for the glycine aqueous solution is presented in *Fig.1* (experimental error of viscosity measurements is within *15%*). In this figure the solid line corresponds to the viscosity experimental data versus solute concentration whereas short dashed line represent the sample function $\eta_{sample}(c_\infty) = cst \cdot \rho^{1/2}(c_\infty)$. It follows from the straightforward comparison between experimental data and the sample function line that their intersection approximately corresponds (error in correspondence is within *15%*) to the saturation concentration c_s for every solution

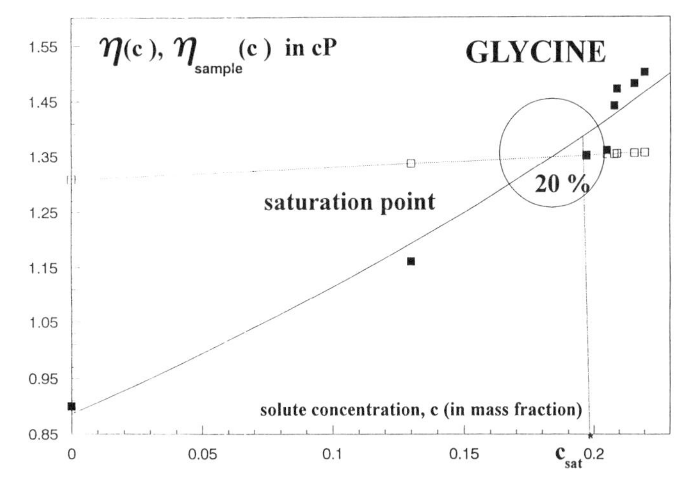

Figure 1. Dependence of the shear solution viscosity $\eta(c_\infty)$ (in cP) on the bulk solute concentration c_∞ (mass fraction) for aqueous solution of glycine. Solid line corresponds to experimental data whereas short dashed line corresponds to interpolation by means of the sample function $\eta_{sample}(c_\infty)$.

Table 1. Dependence of solution density (g/cm^3) on solute concentration (mass fraction). The saturation point is taken at 25° C and normal pressure.

Aqueous Solution of	cst of $\eta_{sample}(c_\infty)$	a_0	a_1	Saturation point at 25° C and normal pressure (in mass fraction)	Intersection point between $\eta(c_\infty)$ and $\eta_{sample}(c_\infty)$ (in mass fraction)
KCl	0.904	1.013	0.571	0.264	0.268
ADP	1.739	0.998	0.564	0.283	0.253
KDP	1.294	0.996	0.744	0.200	0.195
TGS	1.437	1.005	0.367	0.231	0.215
Glycine	1.329	1.000	0.343	0.198	0.183

tested. Therefore, the analytically derived conclusion that at the vicinity of saturation point there is the specific relationship, given by expression (10), between solution viscosity and density is experimentally confirmed with accuracy of *85%*.

To define a complete system of equations describing isothermal solute diffusion in the case of natural convection one has to supplement the general Navier-Stokes and continuity equations (4.A) and (4.B) by the corresponding solute diffusion equation. In the stationary limit the two-dimensional equation for the convective solute diffusion acquires the form:

$$\frac{\partial c}{\partial x_1} v_1 + \frac{\partial c}{\partial x_2} v_2 = \frac{\partial}{\partial x_1}[D(c)\frac{\partial c}{\partial x_1}] + \frac{\partial}{\partial x_2}[D(c)\frac{\partial c}{\partial x_2}]. \quad \textbf{(11)}$$

It is assumed in this equation that the solute mass diffusivity $D(c)$ is dependent on the solute concentration c. In order to solve equation (11) within the DBL it is assumed that there exists expansion (3) for the diffusivity $D(c)$. The following utilization of the fact that the DBL thickness $\delta_{diff}(x_1)$ is very small compared to the characteristic length L of a crystal plate allows one to considerably simplify equation (11):

$$\frac{\partial c}{\partial x_1} v_1 + \frac{\partial c}{\partial x_2} v_2 = D(c_\infty)[\frac{\partial^2 c}{\partial x_2^2} + \gamma_\infty(\frac{\partial c}{\partial x_2})^2 \ \theta[\delta_{diff}(x_1) - x_2]] \ , \quad \textbf{(12)}$$

where

$$\gamma_\infty = \frac{\partial \ln[D(c)]}{\partial c} \Big|_{c=c_\infty}.$$

Solution of this equation can be found in the form [10,11]:

$$c(z) = c_w(z)\ \theta(z_\infty - z) + c_b(z)\ \theta(z - z_\infty). \quad \textbf{(13)}$$

In this expression $c_w(z)$ and $c_b(z)$ are the solutions for the solute concentration profiles within and beyond the DBL, respectively:

$$c_w(z) = c(0) + \frac{1}{\gamma_\infty}\ln[1 + \gamma_\infty z_\infty \beta_0[c(0) - c_s] \int_0^{\frac{z}{z_\infty}} dx\ e^{\frac{9}{4}Scx^4}],$$

$$c_b(z) = c_\infty \ ,$$

where $c(0)$ is the solute concentration on the growing crystal surface ($z = 0$) $c_s \le c(0) \le c_\infty$, the constant $\beta_0 = \beta[c(0)] \ge 0$ is the $c(0)$-dependent coefficient which characterizes the rate of solute exchange between crystal surface and solution and $Sc = \nu_\infty/D_\infty$ is the Schmidt number.

In the crystal growth problem it is essential to know the solute diffusional flux $j_{diff}(x_1,x_2)$ towards the growing crystal surface. This flux is defined as follows:

$$j_{diff}(z) = D(c_w)\frac{\partial c_w(z)}{\partial z} = cst_2^{-1} x_1^{\frac{1}{4}} D(c_w)\frac{\partial c_w(z)}{\partial x_2}. \quad \textbf{(14)}$$

Let us find separately the flux $j_{diff}(z)$ on the DBL edge ($x_2 = \delta_{diff}(x_1)$ or $z = z_\infty$) and on the crystal surface ($x_2 = 0$ or $z = 0$):

$$j_{diff}(x_1, \delta_{diff}(x_1)) = D(c_\infty)\Gamma_\infty, \quad \textbf{(15.A)}$$

$$j_{diff}(x_1, 0) = \delta_{diff}(x_1))\ D[c(0)]\ \beta_0[c(0) - c_s], \quad \textbf{(15.B)}$$

where $D[c(0)]$ is the solute mass diffusivity on crystal surface. Therefore, the ratio $\lambda(\gamma_\infty, \kappa_\infty, Sc) = j_{diff}(x_1,0)/j_{diff}(x_1, \delta_{diff}(x_1))$, which characterizes efficiency of the solute mass transfer towards the growing crystal surface, is given by the following expression:

$$\lambda(\gamma_\infty, \kappa_\infty, Sc) = \lambda(\epsilon, Sc) = \frac{(1+\epsilon)e^{-\frac{9}{4}Sc}}{1 + \epsilon e^{-\frac{9}{4}Sc}\int_0^1 d\xi\, e^{\frac{9}{4}Sc\xi^4}}, \quad \textbf{(16)}$$

where $\epsilon = |\gamma_\infty|/\kappa_\infty$. Analysis of this expression gives that the ratio $\lambda(\epsilon, Sc)$ is the monotonic function of the both variables ϵ and Sc. However, $\lambda(\epsilon, Sc)$ is the increasing function of the variable ϵ $(0 \le \epsilon \le \infty)$ whereas it is the decreasing function of the variable Sc $(0 \le Sc \le \infty)$. For the given Schmidt number Sc the ratio $\lambda(\epsilon, Sc)$ acquires minimum value when $\epsilon = 0$ and maximum when $\epsilon = \infty$:

$$\min[\lambda(x, Sc)|_{Sc=const}] = \lambda(0, Sc) = e^{-\frac{9}{4}Sc},$$

$$\max[\lambda(x, Sc)|_{Sc=const}] = \lambda(\infty, Sc) = [\int_0^1 d\xi\ e^{-\frac{9}{4}Sc\xi^4}]^{-1}.$$

Therefore, the most favorable regime for the solute mass transfer corresponds to such solute concentration regions where the ratio of $dln[D(c)]/dc$ to $dln[\eta(c)]/dc$ acquires a maximum value.

In conclusion let us summarize the results: (A) in the vicinity of saturation concentration c_s there exists the following relationship between the solution shear viscosity and density: $\eta(\rho_s) \propto \rho^{1/2}(c_s)$ and (B) the most favorable regimes for the solute mass transfer towards the growing crystal surface can be achieved when $\epsilon >> 1$. It is well-known that under microgravity conditions one may expect a significant improvement in crystal growth since the DBL thickness increases with decrease of the gravity acceleration constant g (see expression (8)). However, as it follows from expression (16) it is not necessarily the case. For example, at low supersaturations the ratio $(2\kappa_\infty/\omega_\infty - 1)/g$ can be still small even at microgravity. This prevents formation of the appropriate DBL for the crystal growth process to be considerably improved. Thus, to achieve improvement of crystal growth one has to obtain such supersaturation when $2\kappa_\infty/\omega_\infty >> 1$.

2. Spherical Void Electrodynamic Lavitator Trap and Thermodynamic Investigations at High Supersaturations

The theoretical approach to crystal growth developed in the previous *Section* implies knowledge of thermodynamic properties of supersaturated solutions. To study these properties it is necessary: (A) to keep supersaturated solutions in a metastable state for a time enough to accomplish thermodynamic measurements and (B) to achieve high supersaturations. Usually, the bulk experiments with supersaturated solutions allow only fulfillment of the condition (A) at low supersaturations. To achieve high supersaturations it is necessary to avoid heterogeneous nucleation which can be triggered by container walls, dust impurities and dirt. To the best of our knowledge the only experimental technique which allows fulfillment of both conditions (A) and (B) is the electrodynamic suspension which allows a single charged microdroplet to be suspended without a container [12-20]. The combination of a small microdroplet and no container results in the very large supersaturation being achievable since the risk of heterogeneous nucleation is greatly reduced. For example, in our recent studies [18-20] a relative supersaturation of 13.38 was achieved in the suspended glycine-water microdroplets. Normally, in bulk solution values of 1.1-1.3 are the maximum supersaturations achievable. The electrodynamic balance techniques, therefore, provides a significant improvement over other experimental techniques since it allows the preparation of very high supersaturated solutions. The spinodal limit of metastable region was investigated by employing the suspended droplet technique. This technique allows thermodynamic and spectroscopic measurements of high supersaturated levitated microdroplets. In our experiments with levitated microdroplets of supersaturated solutions we employ the developed at Polytechnic University Spherical Void Electrodynamic Levitator Trap (SVELT) [26-28].

For a stationary microdroplet trapped in the SVELT null point its weight mg is balanced by the opposing electrostatic force qV_{dc}:

$$mg = C\frac{qV_{dc}}{z_0},$$

where q is the microdroplet electrical charge, $2z_0$ is the distance between the SVELT spheroid electrodes and C is the SVELT geometrical constant (see *fig.2*). Thus, by means of this equation and under an assumption that the microdroplet charge remains unaltered during an experiment the relative microdroplet mass changes can be easily determined by measuring the balancing *dc*-voltage V_{dc} required to balance the weight mg of the charged microdroplet. This allows to express the microdroplet solute concentration n_0 as follows:

$$n_0 = \frac{1000}{(MW)_{slv}}\left(\frac{V_{dc}^{wet}}{V_{dc}^{dry}} - 1\right)^{-1} \quad (\textit{in molal units}), \qquad (17)$$

where $(MW)_{slv}$ is the solvent molecular weight ($(MW)_{water} = 18$), and V_{dc}^{wet} and V_{dc}^{dry} are the balancing voltages for the solution (wet) and anhydrous (dry) microdroplets, respectively. Therefore, the following two, experimentally justified, assumptions that: (A) solute is nonvolatile, and (B) the solution microdroplet is in equilibrium with its vapor, allow to achieve the desired mean solute concentration n_0 inside

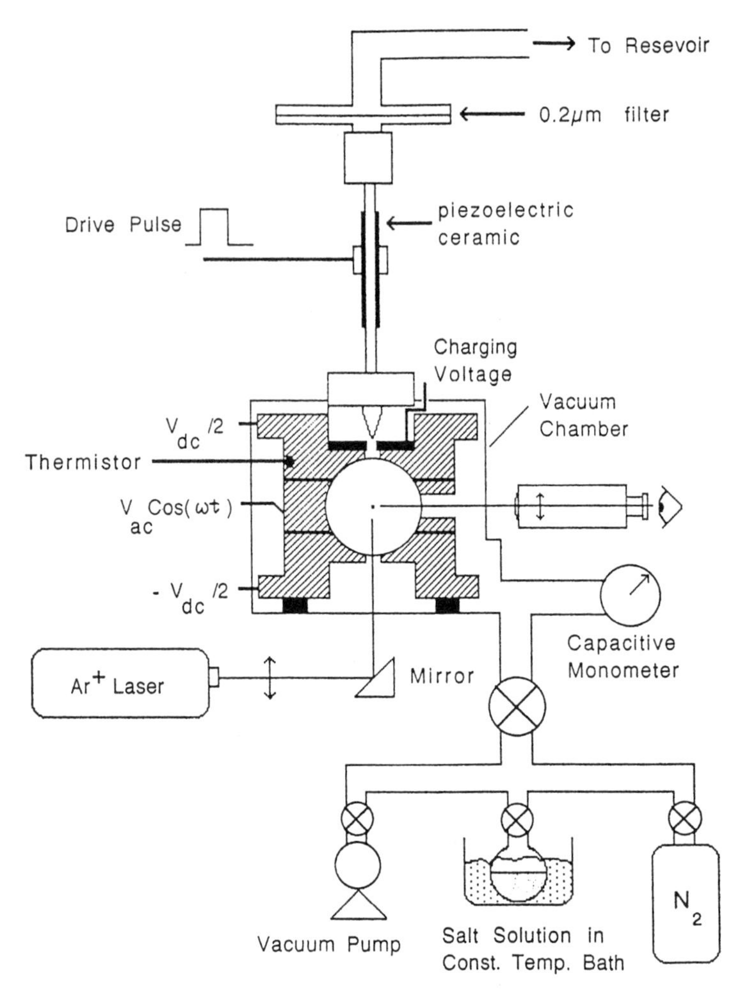

Figure 2. Schematic drawing of the SVELT experimental setup.

of the solution microdroplet by adjusting humidity of the SVELT chamber vapor.

Assumption (B) means that the chemical potentials $\mu_{slv}^{L}(T,n_0)$ and $\mu_{slv}^{G}(T,n_0)$ of solvent in the liquid (solution) and gaseous (vapor) phases are equal. Therefore, the corresponding activities $\lambda_{slv}^{L}(T,n_0) = \lambda_{slv}^{L}(T,0)\, exp[\mu_{slv}^{L}(T,n_0)/kT]$ and $\lambda_{slv}^{G}(T,n_0) = \lambda_{slv}^{G}(T,0)\, exp[\mu_{slv}^{G}(T,n_0)/kT]$ are also equal. Assuming that solvent vapor can be considered as an ideal gas it is straightforward to demonstrate that [29]:

$$\frac{\lambda_{slv}^{L}(T,n_0)}{\lambda_{slv}^{L}(T,0)} = \frac{\lambda_{slv}^{G}(T,n_0)}{\lambda_{slv}^{G}(T,0)} = \frac{P(T,n_0)}{P(T,0)} = \frac{P(T,n_0)}{P_{sat}(T)} = RH(T,n_0).$$

where $P(T,n_0)$ and $RH(T,n_0)$ are the pressure and the relative humidity, respectively, of the solvent vapor which is in equilibrium with the microdroplet of mean solute concentration n_0. $P(T,0) = P_{sat}(T)$ is the pressure of solvent vapor which is in equilibrium with the pure liquid solvent (this pressure is the solvent vapor saturation pressure $P_{sat}(T)$ at the given temperature T). Therefore, the mean solute concentration n_0 and the corresponding solution water activity $\lambda_{slv}^{L}(T,n_0)$ can be easily determined by measuring the balancing *dc*-voltage V_{dc} and the chamber vapor pressure. Once a microdroplet is caught and centered at the SVELT null point, the chamber can be evacuated. After the voltage V_{dc} of the dry microparticle is recorded, solvent vapor above the vapor reservoir is allowed to bleed back into the chamber until the solid microparticle is transformed into the solution microdroplet. The second evacuation is then commenced at a slower rate by adjusting the needle valve. This procedure increases the solute concentration n_0 and, thus leads to deeper penetration into supersaturation(metastability) region. Evacuation is continued at the slower rate until the crystallization point is reached when the balancing *dc*-voltage V_{dc} drops precipitously. After crystallization the evacuation is continued to ensure that there has been no charge loss during the cycle. Therefore, the described experimental procedure allows to continuously record the chamber pressure $P(T,n_0)$ and the voltage V_{dc} . As a result of the solvent vapor evaporation, the balancing *dc*-voltage V_{dc} decreases steadily and the solution microdroplet becomes supersaturated and eventually crystallizes. This measurement is repeated several (usually three) times to ensure the reproducibility of experimental result for the given solution microdroplet. The results obtained in this way for the aqueous glycine solutions are presented in *fig. 3*.

The SVELT experiments with supersaturated microdroplets of electrolyte solutions have allowed to obtain the very high supersaturations. However, deep penetration into the zone of metastable states was accompanied by an increase of the standard deviation $\sigma_{RH}(T,n_0)$ of measurements of the chamber relative humidity. Such a behavior of the standard deviation $\sigma_{RH}(T,n_0)$ with deep penetration into zone of metastability is well understood and discussed [30]. Its increase is conditioned by the corresponding increase of fluctuations of the local solute concentration with approach to the region of unstable states (to

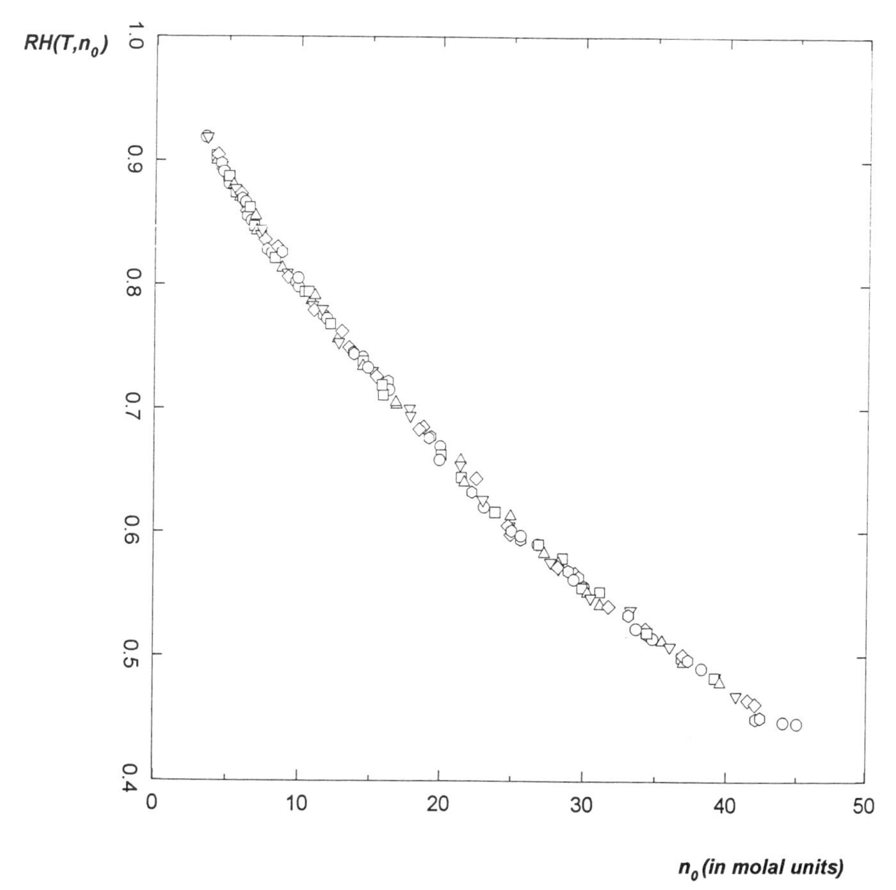

Figure 3. Dependence of the relative humidity $RH(T,n_0)$ of the chamber solvent vapor on the mean solute concentration n_0 of the SVELT confined solution microdroplet. The $RH(T,n_0)$ measurements correspond to $T = 298° K$. Square, triangle and circle points represent experimental data for aqueous solutions of glycine related to various trials.

spinodal line). Therefore, we have retained only such the $RH(T,n_0)$ data the standard deviation of which is within 5%.

In conclusion let us note that the SVELT experimental setup provides unique opportunities in the study of metastable state. The SVELT suspended microdroplets of supersaturated solutions allow: (A) To measure solvent activity of supersaturated solutions in a wide region of solute concentrations including concentrations corresponding to the deeply metastable states close to spinodal line. (B) To measure the solvent activity of supersaturated solutions in a wide temperature region. These measurements combined with the results obtained for various solute concentrations provide an opportunity to obtain phase diagrams (boundary of the metastable states zone in coordinates *temperature + solute concentration*).

Acknowledgements

The authors gratefully acknowledge the support of the National Science Foundation (NSF Grant CTS 9020233) and NASA (NASA Grants NA68-960 and NAG8-1060). The authors thank S.Arnold for his help in creation of the experiment setup.

References:

1. Sorell L.; Myerson A.S., *AIChE J.* **1984,** *28,* 772.
2. Chang Y.C.; Myerson A.S., *AIChE J.* **1984,** *30,* 820.
3. Chang Y.C.; Myerson A.S., *AIChE J.* **1985,** *31,* 890.
4. Chang Y.C.; Myerson A.S., *AIChE J.* **1986,** *32,* 1567.
5. Myerson A.S.; Lo P.Y., *J.Crystal Growth* **1990,** *99,* 1048.
6. Lo P.Y.; Myerson A.S., *J.Crystal Growth* **1990,** *110,* 20.
7. Ginde R.M.; Myerson A.S., in *AIChE Symposium Series 284. Particle Design via Crystallization* **1991,** *87,* 124.
8. Ginde R.M.; Myerson A.S., *J.Crystal Growth* **1992,** *116* , 41.
9. Bohenek M., Ph.D. *Dissertation in Progress,* Polytechnic University.
10. Izmailov A.F.; Myerson A.S., *Phys.Rev.E* **1995,** *52,* 805.
11. Myerson A.S.; Izmailov A.F., *J.Crystal Growth* (in press **1995**).
12. Richardson C.B.; Kurtz C.A., *J.Am.Chem.Soc.* **1984,** *106,* 6615.
13. Kurtz C.A.; Richardson C.B., *Chem.Phys.Lett.* **1984,** *109,* 190.
14. Richardson C.B.; Spann J.F., *J.Aerosol.Sci.* **1984,** *15,* 563.
15. Tang I.N.; Munkelwitz H.R.; Wang N., *J.Colloid Interface Sci.* **1986,** *114,* 409.
16. Cohen M.D.; Flagen R.C.; Seinfeld J.H., *J.Phys.Chem.* **1987,** *91,* 4563.
17. Cohen M.D.; Flagen R.C.; Seinfeld J.H., *J.Phys.Chem.* **1987,** *91,* 4575.
18. Myerson A.S.; Na H.S.; Izmailov A.F.; Arnold S., in *Proceedings of the 12th Symposium on Industrial Crystallization,* Poland: Warsaw 1993, 3-013.
19. Na H.S.; Arnold S.; Myerson A.S., *J.Crystal Growth* **1994,** *139,* 104.
20. Na H.S.; Arnold S.; Myerson A.S., *J.Crystal Growth* **1995,** *149,* 229.
21. Brice J.C. *The Growth of Crystals from Liquids;* North-Holland: Amsterdam, 1973, *vol. XII* in Selected Topics in Solid State Physics.
22. Rosenberger F. *Fundamentals of Crystal Growth I. Macroscopic Equilibrium and Transport Concepts*; Springer-Verlag: Berlin, 1979, *vol. 5* in Springer Series in Solid-State Sciences.
23. Chernov A.A. *Modern Crystallography III. Crystal Growth*; Springer-Verlag: Berlin, 1984, *vol.36* in Springer Series in Solid-State Sciences.
24. Schlichting H., *Boundary Layer Theory;* McGraw-Hill: New York, 1960.
25. Jaluria Y., *Natural Convection. Heat and Mass Transfer;* Pergamon: Oxford, 1980.
26. Arnold S.; Folan L.M., *Rev.Sci.Instr.* **1986,** *57,* 2250.
27. Arnold S.; Folan L.M., *Rev.Sci.Instr.* **1987,** *58,* 1732.
28. Arnold S., *Rev.Sci.Instr.* **1991,** *62,* 3025.
29. GuggenheimE.A., *Thermodynamics;* North-Holland: Amsterdam, 1967.
30. Myerson A.S.; Izmailov A.F., *J.Phys. D: Appl.Phys.* **1993,** *23,* B123.

Growth, Perfection and Defects in C_{60} and C_{70} Single Crystals

K. Kojima, M. Tachibana and Y. Maekawa, Department of Physics, Faculty of Science, Yokohama City University, 22-2 Seto, Kanazawa-ku, Yokohama 236, Japan
H. Sakuma and M. Michiyama, Graduate School of Integrated Science, Yokohama City University, 22-2 Seto, Kanazawa-ku, Yokohama 236, Japan
K. Kikuchi and Y. Achiba, Department of Chemistry, Faculty of Science, Tokyo Metropolitan University, Minami-Ohsawa, Hachioji, Tokyo 192-03, Japan

C_{60} and C_{70} single crystals were grown from vapor by a continuous pulling technique. Large C_{60} crystals up to a size of about 5×3×2 mm^3 and C_{70} crystals of about 1×1×1 mm^3 were obtained. The structural perfection of the grown crystals was examined by X-ray topography and etching method. The grown-in dislocation density in the C_{60} crystals was estimated to be of the order of 10^4 cm^{-2}. The mechanical properties, which are strongly related to the multiplication and motion of dislocations, were investigated using a micro-indentation technique.

The success in efficiently synthesizing fullerenes has generated much interest in the physical properties of this new class of molecular crystals. The growth of single crystals of high quality is prerequisite for studies of intrinsic physical properties of fullerene crystals.

In particular, mechanical properties strongly depend on the quality of crystals. The microhardness of as-grown C_{60} crystals which were grown by solution growth method is one-tenth of that of vacuum annealed C_{60} crystals (Ossipyan *et al.*, 1993; Bobrov *et al.*, 1994). In the solution growth, as-grown crystals contain solvent molecules as impurities so that they have hexagonal close-packed (hcp) structure or contain two kinds of hcp and face-centered-cubic (fcc) structures. When the crystals were annealed in vacuum, the crystal structure changed from hcp to fcc. Therefore, the solution-grown C_{60} crystals show uncertainty of the magnitude of the hardness. On the other hand, it is known that solvent-free C_{60} crystals which were grown by vapor growth method show high perfection with fcc structure. As a result, the vapor-grown C_{60} crystals lead to the high reliability of the hardness (Tachibana *et al.*, 1994). Thus it would be noted that the physical properties of C_{60} crystals strongly depend on the crystal perfection.

C_{60} single crystals with fcc structure have been grown from vapor by various methods, i.e. temperature gradient (Meng *et al.*, 1991), double-temperature gradient (Haluska *et al.*, 1993), vapor transport by inert gas (Liu *et al.*, 1993), periodic oscillation or descending of growth temperature (Li *et al.*, 1994). On the other hand, in the next higher fullerene, C_{70}, which can be isolated in macroscopic quantities, it is difficult to obtain high quality single crystals with hcp structure.

In this work, C_{60} and C_{70} single crystals were grown from vapor by a continuous pulling technique with double-temperature gradient, which is based on that developed by Piper and Polich (1961). The perfection of the grown crystals was examined by X-ray topography and etching method. In addition, the slip system and hardness of C_{60} and C_{70} crystals were investigated using a micro-indentation technique.

Experimental

Crystal Growth

Chromatographically purified C_{60} powder was degassed at 300 °C under a dynamic vacuum of 1–3×10^{-6} Torr for about 24 hours to remove the residual organic solvent. C_{60} single crystals of small size were obtained by sublimation of degassed C_{60} powder at 600°C. The small single crystals were used as the source material for the growth of large single crystals (Tachibana *et al.*, 1994; Li *et al.*, 1994). The source material of about 20 mg was deposited onto the closed end of a long pyrex tube 13 mm in an inner diameter. The tube was evacuated to 1–3×10^{-6} Torr and then sealed off at about length 100 mm. The tube was placed in a horizontal furnace with two sharp temperature gradients oriented oppositely as shown in Fig. 1, where the temperature of the cold middle of the furnace was kept at 530°C, and the temperatures of its hot opposite sides were at 580°C. The tube was advanced towards the source side (the arrow in the figure) at the pulling rate of 1 cm/day in the furnace. In this pulling process, the growth position in the tube is always cleaned by heat treatment at the higher temperature. Large C_{60} crystals were grown at the middle of the furnace for 3–4 days.

C_{70} single crystals were grown in the same procedure as C_{60} crystals. The separation between the source and growth positions was smaller than that for C_{60} crystal growth, since the vapor pressure for C_{70} is lower than that for C_{60}. C_{70} crystals were grown for 7 days.

Assessment of Crystal Perfection

The crystal structures and habit faces of grown crystals were confirmed by Laue method. The defects, especially dislocations, in grown crystals were examined by X-ray topography and etching method. The X-ray topographic observation was carried out by using white beam synchrotron radiation at the Photon Factory of the National Laboratory for High Energy Physics in Tsukuba. As-grown crystals were observed by the synchrotron X-ray topography since it was difficult to cleave, cut or etch them into thin plate specimens. The topographs were recorded on X-ray films (Agfa D2) for short times of a few ten seconds. Etching on the habit faces of the grown C_{60} crystals were carried out by immersing them in toluene for several seconds. The toluene is available for the dislocation etchant of C_{60} crystals (Orlov *et al.*, 1994).

Measurement of Hardness by Micro-Indentation

The plastic deformations of grown crystals were carried out using a micro-hardness tester

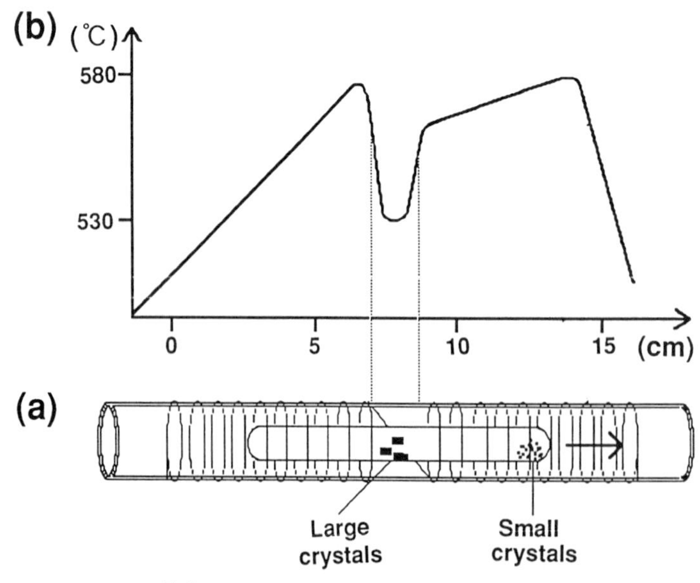

Fig. 1. Experimental setup ((a) furnace and (b) its temperature profile) for crystal growth from vapor by a continuous pulling method.

with a diamond pyramidal indenter. The apparatus was modified for the deformation experiments in a temperature range of about 220–450 K in a vacuum or a gas atmosphere (Tachibana *et al.*, 1994). The indentations were made on {111} and {100} habit faces of C_{60} single crystals and (0001) of C_{70} crystals. The Vickers hardness H_v was estimated through the measurements of the diagonals of the indentations. The indenter was pulled down to be perpendicular to the habit faces at a velocity of 0.01 mm/sec. The contact period of the indenter with the surfaces was 5 seconds. In the measurements of the temperature dependence of the hardness, the load of 2 gf was selected to reduce crack formation around the indentations.

Results and Discussion

Crystal Perfection

C_{60} single crystals up to a size of about $5\times3\times2$ mm^3 were obtained. Most of all grown crystals had a polyhedral shape with smooth and shiny habit faces such as {111} and {100} of fcc structure, which were assigned by Laue method. The synchrotron X-ray topographs of the grown crystals of large size were observed as shown in Fig. 2. The topograph was taken using $1\bar{5}1$ reflection. Some dislocations at A can be observed individually. Their Burgers vectors were identified to be parallel to <110> directions from extinction criterion of dislocation images. Taking into account the disloca-

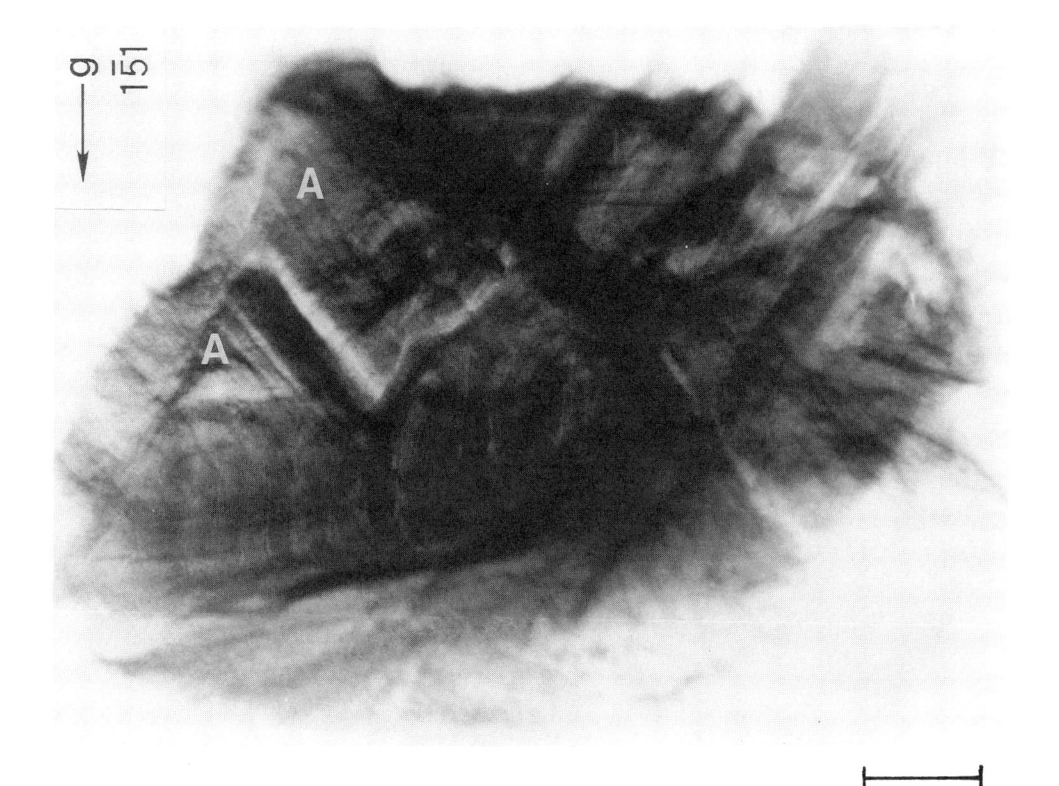

Fig. 2. Synchrotron X-ray topograph of C_{60} single crystal of large size grown from vapor by a continuous pulling method. The arrow indicates the diffraction vector ***g***.

tion theory (Hirth and Lothe, 1982), the Burgers vectors are uniquely defined as $\frac{1}{2}$<110> (10.0 Å). The contrasts of images at crystal edges show the large distortions which are induced by handling of the crystals. These X-ray topographs indicated that the quality of the center region of the grown crystals is high compared with that of crystal edges.

Typical distributions of etch pits on {111} habit faces of as-grown C_{60} crystals are shown in Fig. 3. These etch pits define the positions of the emergent dislocations, which intersect the crystal surface. Most of all shapes of etch pits showed triangle on the surface. The grown crystals are of high perfection except for crystal edges as shown in Fig. 3. Most of all etch pits near the crystal edges distribute along <110> directions. The distributions are parallel to intersections of {111} planes with the surface. The {111} planes correspond to the slip planes of C_{60} crystals with fcc structure as described below. Thus these dislocation distributions at crystal edges were introduced by mechanical deformation during handling of the crystals. Furthermore some isolated etch pits were observed over the grown crystals. These pits correspond to so called grown-in dislocations, which are introduced during crystal growth. The perfections of the as-grown crystals were estimated from the number of the isolated etch pits. Consequently, the grown-in dislocation density was of the order of 10^4 cm^{-2}. These assessments of crystal perfections are consistent with experimental results by X-ray topography.

C_{70} crystals up to a size of about 1×1×1 mm^3 were grown. Most of all grown crystals had a polyhedral shape with habit faces such as (0001) of hcp structure, which were assigned by Laue method. These Laue spots were a little diffusive compared with C_{60}. Thus no clear X-ray topographic images were observed for C_{70} crystals. Furthermore, the grown C_{70} crystals showed the rumpling of the surfaces in Fig. 4. These results indicate that the perfection of C_{70} crystals is much poor compared with C_{60}.

Recently it has been reported that the phase transitions of C_{70} crystals occur not only at around 270 K but also at around 340 K (Vaughan *et al.*, 1991; Verheijen *et al.*, 1992; Van Tendeloo *et al.*, 1993; Mitsuki *et al.*, 1994). This means that the grown C_{70} crystals undergo the phase transition during cooling from growth temperature to room temperature. It seems that the phase transition at around 340

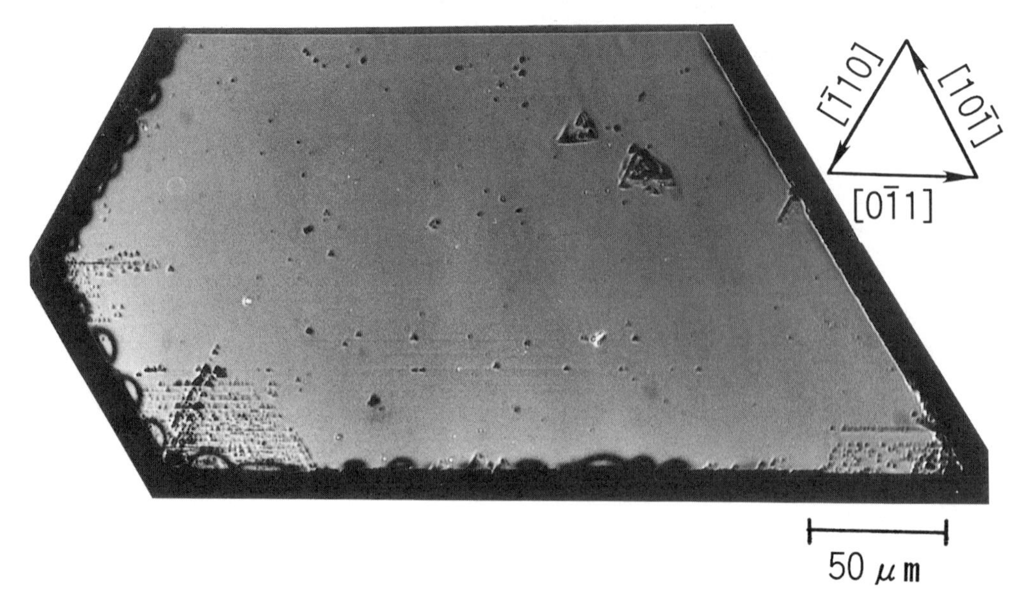

Fig. 3. Distributions of etch pits on (111) habit faces of C_{60} crystals grown from vapor by a continuous pulling technique.

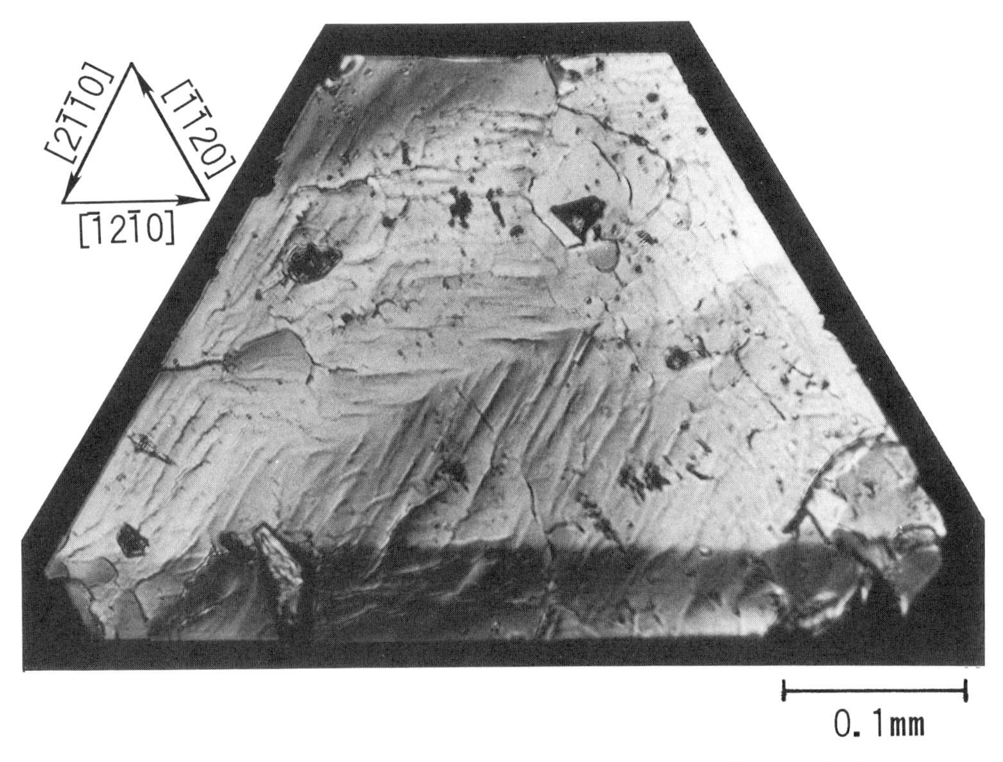

Fig. 4. Optical micrograph of (0001) habit faces of C_{70} single crystals grown from vapor by a continuous pulling method.

K undergoes the discontinuous change in the axial ratio c/a of hcp structure (Verheijen *et al.*, 1992; Mitsuki *et al.*, 1994). Consequently the plastic deformation is induced by the anisotropic thermal expansion of C_{70} crystals since the grown crystals are adhered to the pyrex tube. Thus it is difficult to grow C_{70} crystals of high quality, compared with C_{60} which undergo no such phase transition above room temperature (Weaver and Poirier, 1994).

Slip System by Micro-Indentation

Slip lines around the indentations, which were made on a {111} top surface of C_{60} single crystal at room temperature, were clearly observed. The dislocation etch pits around the indentations showed a typical dislocation rosette pattern in Fig. 5. Both of the symmetry of slip lines and distributions of etch pits correspond to traces of three {111} planes. These results suggest that the slip planes are {111} planes. Thus the distributions of dislocation etch pits show the multiplication and motion of dislocations along the {111} planes. From these etch pit patterns, it is confirmed that the slip systems in C_{60} crystals are {111}<110> systems as fcc crystals (Hirth and Lothe, 1982).

Similar distributions of dislocation etch pits were observed below ~260 K, at which the orientational-ordering-induced fcc to simple cubic (sc) phase transition occurs (Weaver and Poirier, 1994). Moreover, at temperatures higher than room temperature, no drastic change in the dislocation distributions was observed. These results suggest no change in {111}<110> slip system of C_{60} crystals even when the phase transition appear at around 260 K.

In C_{70} crystals, it was difficult to observe slip lines because the surface on the crystals is rumpling and cracking.

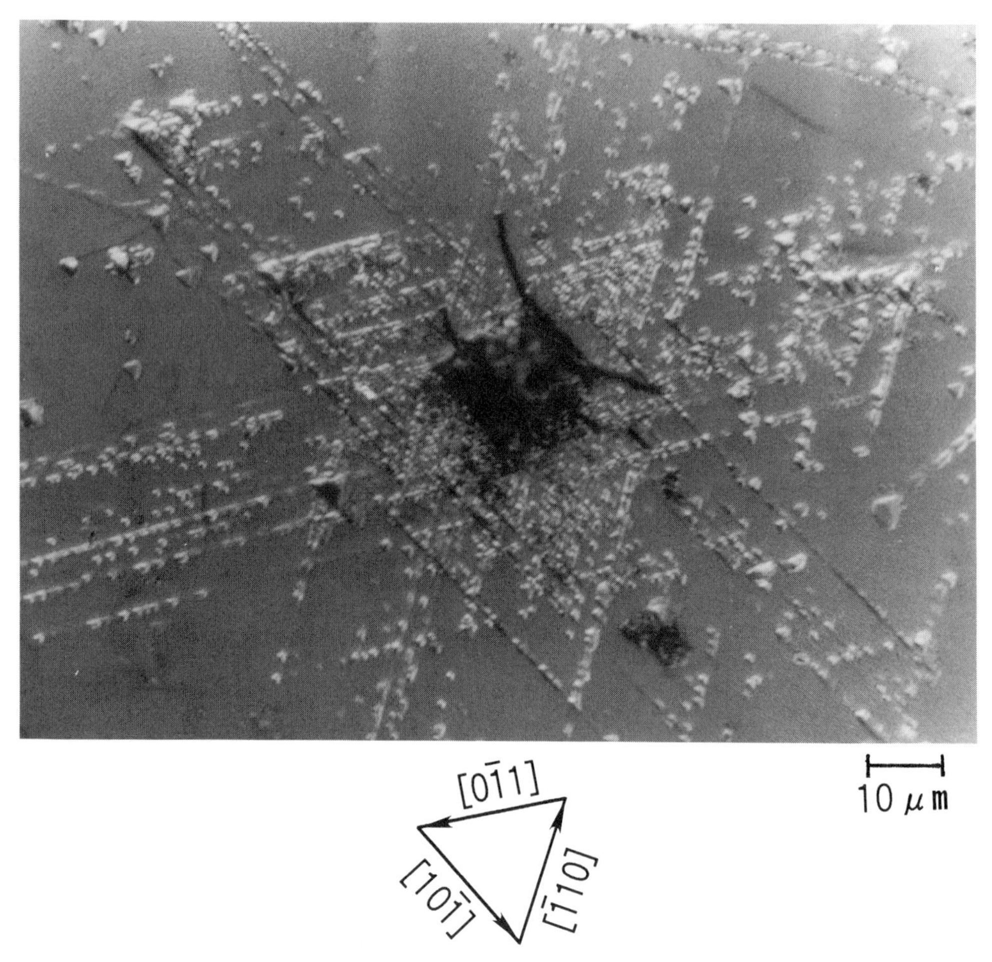

Fig. 5. Distributions of etch pits around the indentations, which were made with the load of 5gf on (111) habit faces of C_{60} single crystals.

Hardness

The Vickers microhardness of C_{60} crystals showed strong temperature dependence. This temperature dependence of the hardness exhibited two interesting behaviors as shown in Fig. 6. One is an abrupt change in the hardness at the phase transition temperature of around 260 K. The other is an anomalous temperature dependence above room temperature: the hardness increased with increasing temperature and reached a maximum at around 370 K. On the other hand, the hardness of C_{70} crystals decreased with increasing temperature above room temperature as shown in

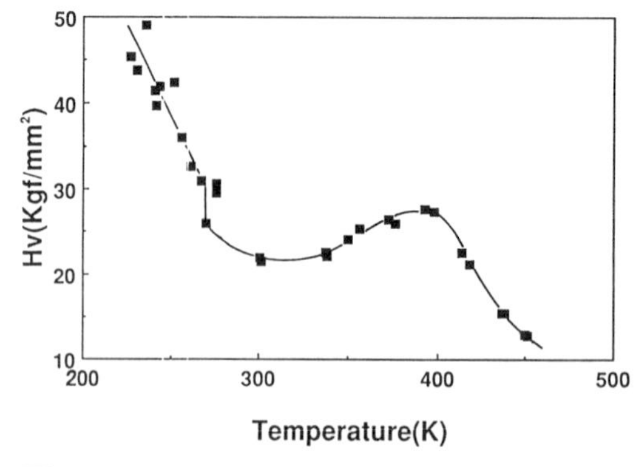

Fig. 6. Temperature dependence of the Vickers hardness H_v on {100} habit faces of C_{60} single crystals in N_2 gas atmosphere.

Fig. 7. No drastic change in the hardness were observed even at the phase transition temperature of around 340 K. Thus the temperature dependence of the hardness on C_{70} crystals is quite weak. Furthermore the magnitude of the hardness of C_{70} crystals was smaller than C_{60} at each temperature. At room temperature, the Vickers hardness of C_{70} crystals was about 10 kg/mm^2, which was about a half of that of C_{60} crystals as shown in Fig. 7.

One of origins of hardness is due to the plastic deformation, i.e. slip, around the indentation as shown in Fig. 5. Thus the hardness is related to the multiplication and motion of slip dislocations. The slip systems in C_{60} crystals were confirmed experimentally to be $\{111\}\langle 110\rangle$ at the temperature range of 240–450 K. On the other hand, the slip systems in C_{70} crystals didn't be identified experimentally, since it was difficult to observe slip line and etch pits.

According to dislocation theory (Hirth and Lothe, 1982), the dislocation elastic energy is proportional to the square of the magnitude of the Burgers vector. This means that the shorter lattice translational vectors are favored as Burgers vectors in crystals. Moreover, Peierls stress is least for the dislocations with the Burgers vectors of the smallest magnitude, and also is lowest, for a given Burgers vector, for the slip planes with the largest lattice spacing (Hirth and Lothe, 1982). Thus, the most possible slip systems in C_{70} crystals with

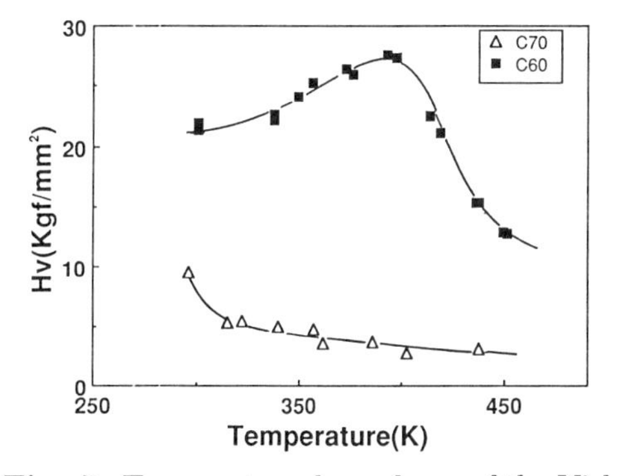

Fig. 7. Temperature dependence of the Vickers hardness H_v on (0001) habit faces of C_{70} single crystals in air. The figure includes the temperature dependence of the hardness on $\{111\}$ faces of C_{60} single crystals.

hcp structure are the $(0001)\langle 11\bar{2}0\rangle$ systems. However, the $(0001)\langle 11\bar{2}0\rangle$ would be unoperated, since the shear stress was applied to be perpendicular to (0001), or parallel to $\{1\bar{1}00\}$ in the deformation experiments. Thus the slip systems in C_{70} crystals in the present work should be the $\{1\bar{1}01\}\langle 11\bar{2}0\rangle$.

The Peierls stress for $\{1\bar{1}01\}\langle 11\bar{2}0\rangle$ slip systems in C_{70} crystals is more than one order as large as that for $(0001)\langle 11\bar{2}0\rangle$, which should approximately correspond to $\{111\}\langle 110\rangle$ slip systems in C_{60} crystals with fcc structures. This means that the hardness of C_{70} crystals should be much large compared with that of C_{60} crystals. However the experimental results showed that the hardness of C_{70} crystals is smaller than that of C_{60} crystals. Thus the smaller hardness of C_{70} crystals would be due to other reasons.

The microhardness of C_{70} crystals shows the quite small values ($\sim$10 kg/mm^2) and the weak temperature dependence. Even in the vapor growth, the perfection of C_{70} crystals is poorer than that of C_{60} crystals. Thus the low quality such as surface rumpling may lead to softening of of the hardness of C_{70} crystals. However, it seems that the lower values of the hardness and the weak temperature dependence of the hardness are characteristics of intrinsic C_{70} crystals since the grown crystals contain no solvents.

Conclusions

Large C_{60} crystals up to a size of about 5×3×2 mm^3 and C_{70} crystals of about 1×1×1 mm^3 were grown from vapor by the continuous pulling technique. The perfection of the grown C_{70} crystals was much poor compared with that of C_{60} crystals, in which the grown-in dislocation density was of the order of 10^4 cm^{-2}. The Vickers hardness of C_{60} crystals, which was measured at temperatures from 220 K to 450 K by the micro-indentation technique, showed strong temperature dependence. In this temperature range, $\{111\}\langle 110\rangle$ slip systems produced by the indentations remained unchanged. On the other hand, no drastic change in the hardness of C_{70} crystals was observed above room temperature, even when the phase transition occurred at around 340 K. The magnitude of the hardness of C_{70} crystals was smaller than that of C_{60} crystals at each temperature.

Acknowledgment

X-ray topographic experiments were partially performed under the auspices of the Photon Factory Program Advisory Committee of the National Laboratory for High Energy Physics (Proposal 95G149). This work (K.K. and M.T.) was supported partially by a Grant-in-Aid for Scientific Research from the Ministry of Education, Science and Culture in Japan.

References

Bobrov, V.S.; Dilanyan, R.A.; Fomenko, L.S.; Lebyodkin, M.A.; Lubenets, S.V.; Orlov, V.I. *Solid State Phenomena* **1994,** *35-36,* 519–526.

Haluska, M.; Kuzmany, H.; Vybornnov, M.; Rogl, P.; Fejdi, P. *Appl. Phys.* **1993,** *A56,* 161–167.

Hirth, J.P.; Lothe, J. *Theory of Dislocations;* 2nd edn, Wiley, New York, 1982; pp 271–287, 436–445.

Li, J.; Mitsuki, T.; Ozawa, M.; Horiuchi, H.; Kishio, K.; Kitazawa, K.; Kikuchi, K.; Achiba, Y. *J. Crystal Growth* **1994,** *143,* 58–65.

Liu, J.Z.; Dykes, J.W.; Lan, M.D.; Klavins, P.; Shelton, R.N.; Olmstead, M.M. *Appl. Phys. Lett.* **1993,** *62,* 531–532.

Meng, R.L.; Ramirez, D.; Jiang, X.; Chow, P.C.; Diaz, C.; Matsuishi, K.; Moss, S.C.; Hor, P.H.; Chu, C.W. *Appl. Phys. Lett.* **1991** *59,* 3402–3403.

Mitsuki, T.; Ono, Y.; Horiuchi, H.; Li, J.; Kino, N.; Kishio, K.; Kitazawa, K. *Jpn. J. Appl. Phys.* **1994,** *33,* 6281–6285.

Orlov, V.I.; Nikitenko, V.I.; Nikolaev, R.K.; Kremenskaya, I.N.; Ossipyan, Y.A. *Pis'ma Zh. Eksp. Teor. Fiz.* **1994,** *59,* 667–670.

Ossipyan, Y.A.; Bobrov, V.S.; Grushko, Y.S.; Dilanyan, R.A.; Zharikov, O.V.; Lebyodkin, M.A.; Sheckhtman, V.S. *Appl. Phys.* **1993,** *A56,* 413–416.

Piper, W.W.; Polich, S.J. *J. Appl. Phys.* **1961,** *32,* 1278–1279.

Tachibana, M.; Michiyama, M.; Kikuchi, K.; Achiba, Y.; Kojima, K. *Phys. Rev.* **1994,** *B49,* 14945–14948.

Tachibana, M.; Michiyama, M.; Kikuchi, K.; Achiba, Y.; Kojima, K. *Trans. Mat. Res. Soc. Jpn.* **1994,** *14B,* 1273–1276.

Van Tendeloo, G.; Amerinckx, S.; De Boer, J.L.; Van Smaalen, S.; Verheijen, M.A.; Meekes, H.; Meijer, G. *Europhys. Lett.* **1993,** *21,* 329–334.

Vaughan, G.B.M.; Heiney, P.A.; Fischer, J.E.; Luzzi, D.E.; Ricketts-Foot, D.A.; McGhie, A.R.; Hui, Y.-W.; Smith, A.L.; Cox, D.E.; Romanow, W.J.; Allen, B.H.; Coustel, N.; McCauley Jr, J.P.; Smith III, A.B. *Science* **1991,** *254,* 1350–1353.

Verheijen, M.A.; Meekes, H.; Meijer, G.; Bennema, P.; De Boer, J.L.; Van Smaalen, S.; Van Tendeloo, G.; Amerinckx, S.; Muto, S.; Van Landuyt, J. *Chem. Phys.* **1992,** *166,* 287–297.

Weaver, J.H.; Poirier, D.M. In *Solid State Physics;* Ehrenreich, H.; Spaepen, F., Eds.; Academic: San Diego, 1994; Vol. 48, pp 18–27, and references therein.

Generation of Chirality in Molecular Crystals Composed of Two Different Achiral Molecules

Hideko Koshima, PRESTO, Research Development Corporation of Japan, and Department of Materials Chemistry, Faculty of Science and Technology, Ryukoku University, Seta, Otsu 520, Japan
Kuiling Ding, Yosuke Chisaka, Naoya Naka and Teruo Matsuura, Department of Materials Chemistry, Faculty of Science and Technology, Ryukoku University, Seta, Otsu 520, Japan

A chiral two-molecule crystal between achiral acridine and achiral diphenylacetic acid was obtained by the spontaneous crystallization from the solution. The source of the chirality is the torsion of two phenyl planes and a carboxyl plane of diphenylacetic acid molecule in the same direction as a three-winged propeller. The crystal of diphenylacetic acid alone is achiral due to the dimer structure of the antipodes.

Chirality is one of the topics in solid-state chemistry (Desiraju, 1989; Ohashi, 1993). Even if a molecule does not have an asymmetric center carbon, i.e. achiral molecule, it can form a noncentrosymmetric crystal by torsional or helical arrangement of the molecules. For instance torsions of flexible molecules such as benzophenone (Banerjee at al., 1938) and benzil (Brown et al., 1965) lead to chiral crystals of right- or left-handed conformation. Such a crystal can be distinguished by knowing whether the space group is chiral or not. In the case of one-molecule crystals, a large number of chiral crystals composed of achiral molecules are already compiled in the JCPDS crystal data series and Cambridge Crystallographic Data Centre. Most frequent chiral space groups are P212121 and P21.

However, chiral two-molecule crystals between two different achiral molecules have been scarecely reported (Elgavi et al., 1973; Fu et al., 1993; Suzuki et al., 1994). In the course of our study of the preparation of two-molecule crystals between different organic molecules and the solid-state photoreaction (Koshima et al., 1994, 1995), we have found that a chiral molecular crystal between acridine and diphenylacetic acid is spontaneously crystallized from the solution. The reason that the chirality generates is discussed in comparison with an achiral crystal of diphenylacetic acid alone and an achiral two-molecule crystal of acridine and diphenylpropionic acid.

Preparation and Structure Analysis of a Chiral Molecular Crystal of Acridine and Diphenylacetic Acid

A chiral molecular crystal (**chiral A**) was prepared by spontaneous crystallization from the equimolar solution of acridine and diphenylacetic acid in acetonitrile. The melting point is 101 °C, which is lower than those of acridine (107 °C) and diphenylacetic acid (148 °C). The crystal structure was determined by X-ray crystallographic analysis to be a typical chiral space group $P2_12_12_1$ (Table 1).

Source of the chirality

We can easily understand that the molecular crystal **chiral A** of acridine and diphenylacetic

1054–7487/96/0239$12.00/0

Table 1. **Crystal Data of X-ray Structure Analysis**

	chiral A	**B**	**C**	**D**
formula	$C_{27}H_{21}NO_2$	$C_{14}H_{12}O_2$	$C_{28}H_{23}NO_2$	$C_{15}H_{14}O_2$
F_w	391.47	212.25	405.50	226.28
space group	P212121	P21/n	P21/n	P21/c
a (Å)	14.93 (1)	12.254 (4)	12.717 (6)	8.556 (1)
b (Å)	25.42 (1)	7.226 (3)	13.061 (9)	12.598 (1)
c (Å)	5.47 (2)	12.737 (2)	13.542 (5)	11.837 (1)
α (deg)	90.0	90.0	90.0	90.0
β (deg)	90.0	90.99	105.57	109.759
γ (deg)	90.0	90.0	90.0	90.0
Z	4	4	4	4
R (%)	5.0	4.7	5.0	6.0

chiral A N · CO_2H C H

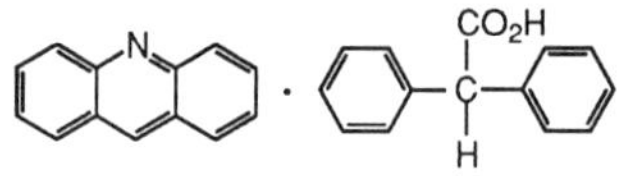

B CO_2H C H

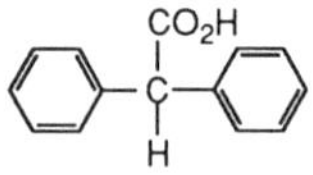

C N · CO_2H C Me

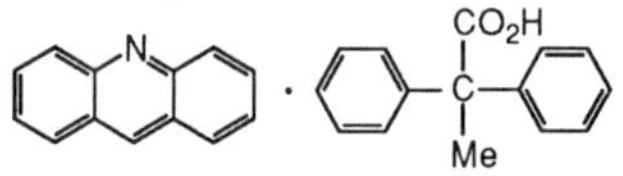

D CO_2H C Me

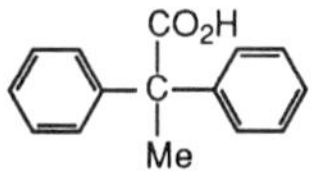

acid has a noncentrosymmetric arrangement in the unit cell (Fig. 1). The molecular pair of acridine and diphenylacetic acid is connected through the hydrogen bonding O–H•••N with the distance H•••N of 1.71 Å as shown in Fig. 2 chiral A. The dihedral angle between acridine plane and carboxyl plane of the molecular pair is 35.5 °. The torsion of two phenyl planes and a carboxyl plane of diphenylacetic acid molecule is the same direction as a three-winged propeller. The torsion angles are listed in Table 2. The four molecular pairs in the unit cell have the same handed conformation. The stereoview in Fig. 3 shows that the molecular pairs stack along c-axis like columns. Conclusively the source of the chirality is the three-winged propeller in the same direction and the column stacking of the molecular pairs.

On the other hand the crystal of diphenylacetic acid alone is not chiral (Table 1). Because the antipodes of diphenylacetic acid molecules form the dimer through the hydrogen bonding O–H•••O with the distance H•••O of 1.69 Å (Fig. 2B). The positive and negative torsion angles are shown in Table 2. Consequently the chirality is offset and does not appear in the crystal.

When all the solution containing equimolar ratio of acridine and diphenylacetic acid is evaporated, initially added acridine and diphenylacetic acid is completely recovered in the form of the crystal **chiral A**. This indicates that the torsion of the three-winged propeller is very flexible in the solution to be changed from two conformations of the antipodes to one conformation of **chiral A** in the crystallization.

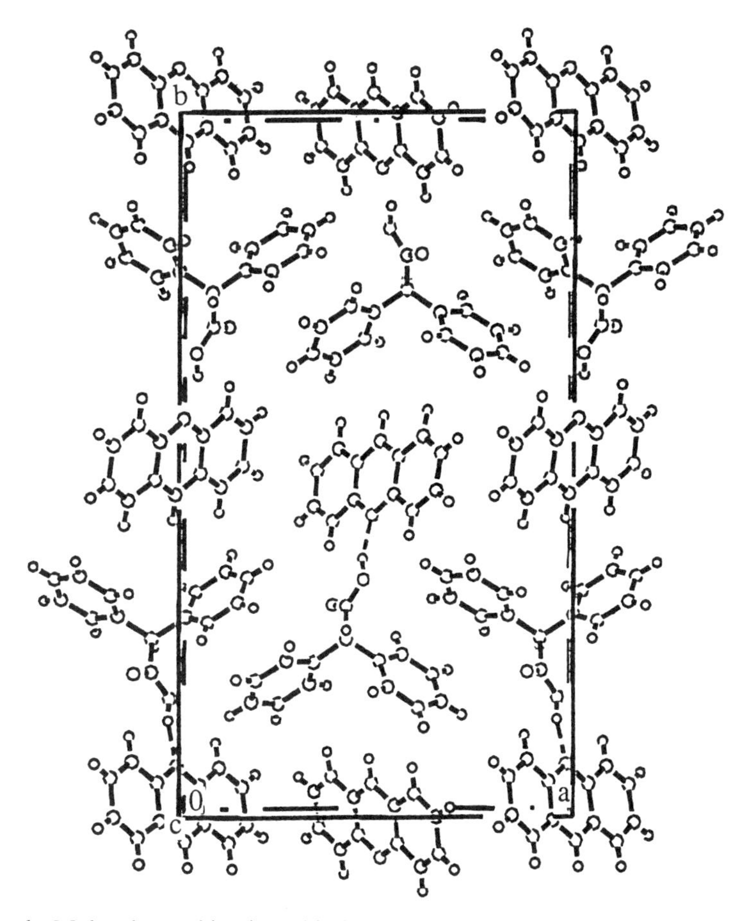

Fig. 1. Molecular packing in a chiral crystal of acridine and diphenylacetic acid.

Although at present we have just obtained only **chiral A** of one-handed (left or right) conformation, another-handed crystal should be existed. It is hoped to succeed in preparing another-handed crystal in future.

We tried to find other chiral molecular crystal based on the same concept. 2,2-Diphenylpropionic acid was selected instead of diphenylacetic acid. A molecular crystal **C** of acridine and 2,2-diphenylpropionic acid was prepared by crystallization from the equimolar solution of the two components in ethanol. The melting point is 139 °C, which is between those of acridine (107 °C) and 2,2-diphenylpropionic acid (178 °C). However **C** crystallized in an achiral space group $P2_1/n$ (Table 1). The stereoview shows the symmetric structure (Fig. 4). The right- and left-handed molecular pairs of acridine and the antipodes of 2,2-diphenylpropionic acid exist in the crystal **C** (Fig. 2C). The positive and negative torsion angles of the three winged-propeller of 2,2-diphenylpropionic acid are listed in Table 2. The distance H•••N of the hydrogen bonding O–H•••N in the molecular pair is 1.70 Å.

One reason that the crystal **C** cannot be crystallized with a chiral space group like the crystal **chiral A** is most probably the steric hindrance of the methyl group of 2,2-diphenylpropionic acid molecule. The C1–C2, C1–H2, C1–H3 and C1–H4 bond lengths of **C** are 1.54 Å, 0.98 Å, 1.08 Å and 0.95 Å, respectively (Fig. 5C). On the other hand, the C1–H1 bond length of **chiral A** is 1.03 Å (Fig. 5 chiral A). Thus the methyl group is too bulky to stack along an axis like **chiral A** (Fig. 3).

Table 2. Torsion angles in the crystals

Torsion angle (deg)	
chiral A	
H1–C1–C2–C3	50.2
H1–C1–C8–C9	13.0
H1–C1–C14–O2	33.6
B	
H1–C1–C2–C3	–53.0
H1–C1–C8–C9	–27.0
H1–C1–C14–O2	–47.2
H1'–C1'–C2'–C3'	53.0
H1'–C1'–C8'–C9'	27.0
H1'–C1'–C14'–O2'	47.2
C	
C1–C2–C3–C4	74.7
C1–C2–C9–C10	17.1
C1–C2–C15–O2	–171.2
C1'–C2'–C3'–C4'	–74.7
C1'–C2'–C9'–C10'	–17.1
C1'–C2'–C15'–O2'	171.2
D	
C1–C2–C3–C4	63.3
C1–C2–C9–C10	9.8
C1–C2–C15–O2	31.9
C1'–C2'–C3'–C4'	–63.3
C1'–C2'–C9'–C10'	–9.8
C1'–C2'–C15'–O2'	–31.9

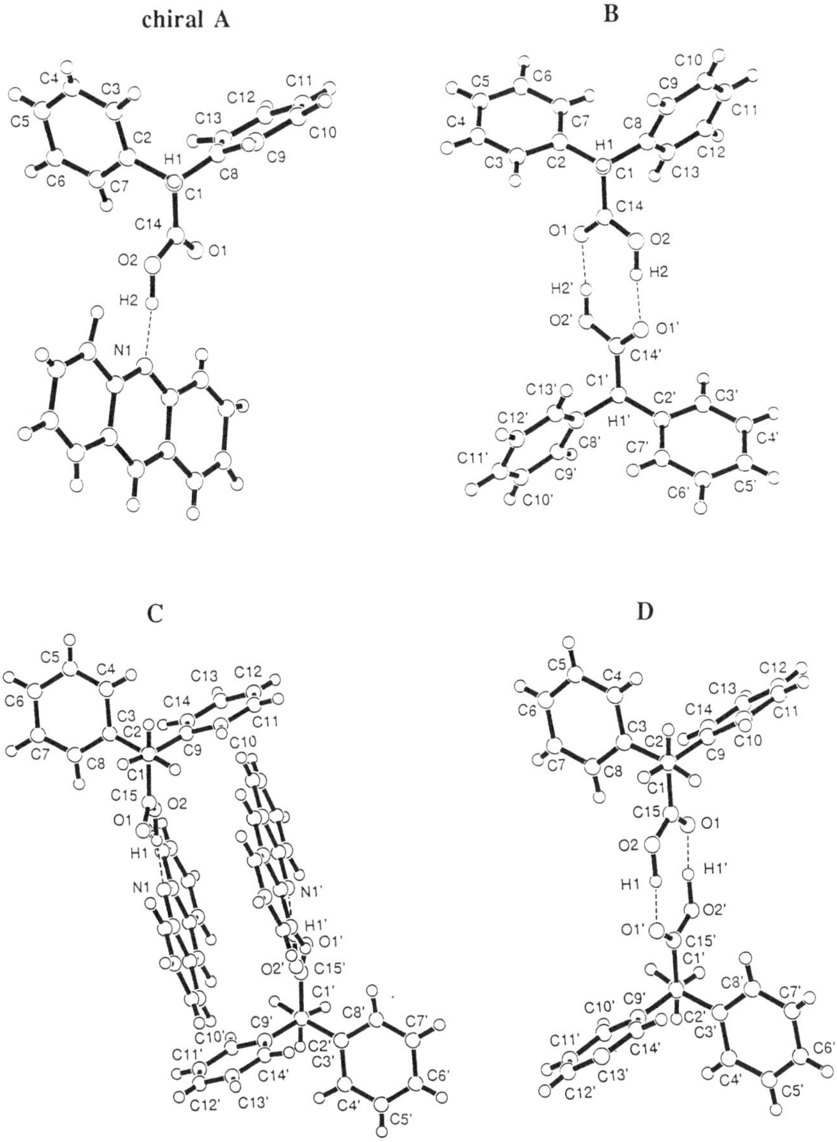

Fig. 2. Molecular pairs in the crystals:
(**chiral A**) Chiral crystal of acridine and diphenylacetic acid
(**B**) Achiral crystal of diphenylacetic acid
(**C**) Achiral crystal of acridine and 2,2-diphenylpropionic acid
(**D**) Achiral crystal of 2,2-diphenylpropionic acid.

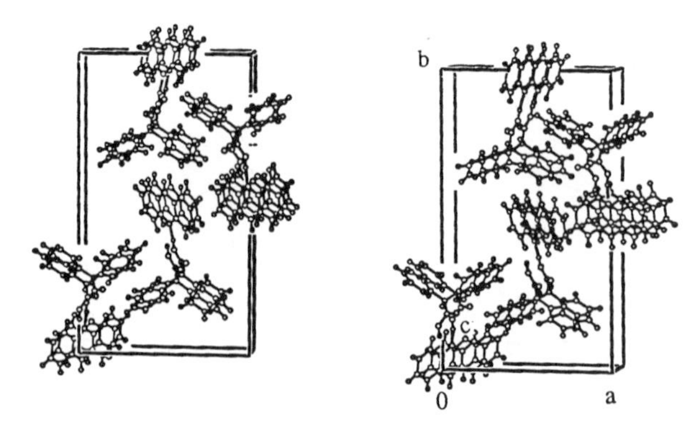

Fig. 3. Stereoview of a chiral molecular crystal of acridine and diphenylacetic acid.

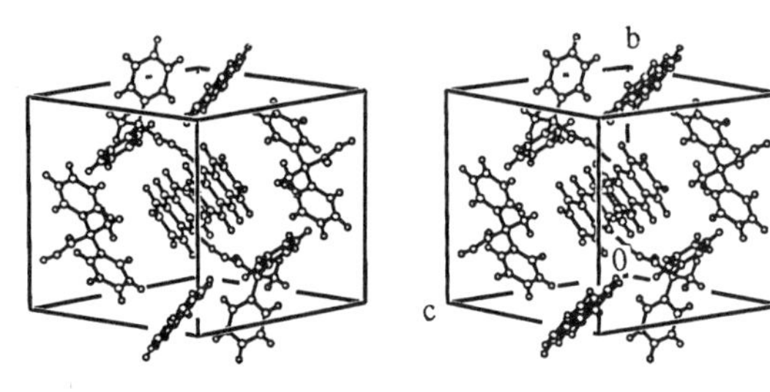

Fig. 4. Stereoview of an achiral molecular crystal of acridine and 2,2-diphenylpropionic acid.

chiral A **C**

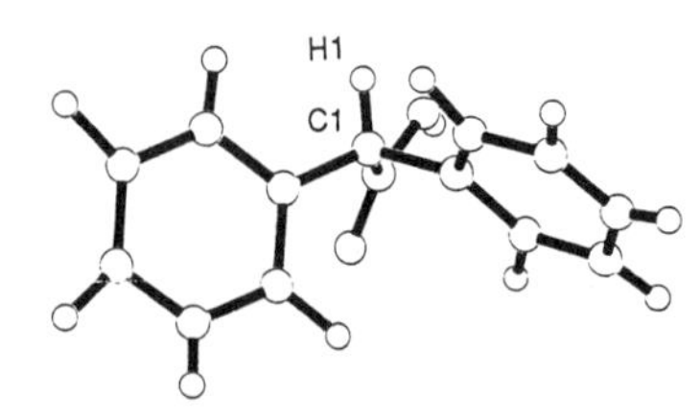

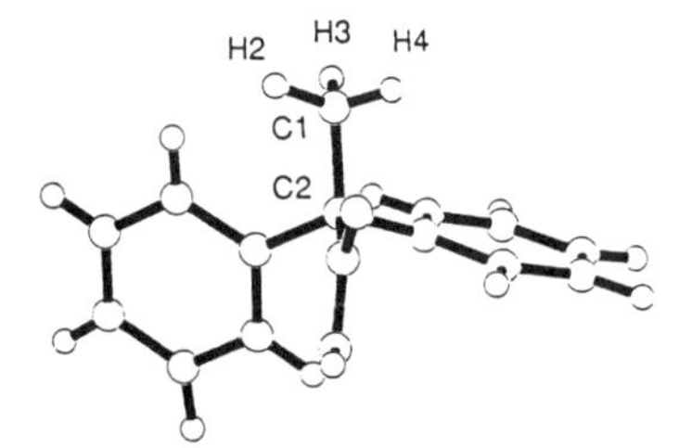

Fig. 5. Diphenylacetic acid and 2,2-diphenylpropionic acid molecules in the molecular crystals **chiral A** and **C.**

References

Banerjee; Haque *Indian J. Phys.* **1938**, *12*, 87.

Brown; Sadanaga, *Acta Cryst.* **1965**, *18*, 158.

Desiraju, G. R. *Crystal Engineering: The Design of Organic solid;* Elsevier: Amsterdam, 1989.

Elgavi, A.; Green, B. S.; Schmidt, G. M. J. *J. Am. Chem. Soc.* **1973**, *95*, 2058-2059.

Fu, T. Y.; Liu, Z.; Scheffer, J. R.; Trotter, J. *J. Am. Chem. Soc.* **1993**, *115*, 12202-12203.

Koshima, H.; Matsuura, T. *Kokagaku* **1995**, *19*, 10-20.

Koshima, H.; Ding, K.; Chisaka, Y.; Matsuura, T. *Tetrahedron Asymmetry* **1995**, *6*, 101-104.

Koshima, H.; Ding, K.; Matsuura, T. *J. Chem. Soc., Chem. Commun.* **1994**, 2053-2054.

Koshima, H.; Maeda, A.; Masuda, N.; Matsuura, T.; Hirotsu, K.; Okada, K.; Mizutani, H.; Ito, Y.; Fu, T. Y.; Scheffer J. R.; Trotter, J. *Tetrahedron Asymmetry* **1994**, *5,* 1415-1415.

Koshima, H.; Yao, X.; Wang, H.; Wang, R.; Matsuura, T. *Tetrahedron Lett.* **1994**, *35,* 4801-4804.

Ohashi, Y., Ed. *Reactivity in Molecular crystals;* VCH: Tokyo, 1993.

Suzuki, T,; Fukushima, T.; Yamashita, Y.; Miyashi, T. *J. Am. Chem. Soc.* **1994**, *116,* 2793-2803.

Role of MeOH in Chiral Combination of Host-Guest Molecules in the Inclusion Crystal

Koichi Tanaka and Fumio Toda, Department of Chemistry, Faculty of Engineering, Ehime University, Matsuyama, Ehime 790, Japan

Ken Hirotsu, Department of Chemistry, Faculty of Science, Osaka City University, Sumiyoshi-ku, Osaka 558, Japan

(*S*, *S*)-(–)-1,4-bis[3-(*o*-chlorophenyl)-3-hydroxy-3-phenyl-1-propyl]benzene (**2**) and (*S*, *S*)-(–)-9,10-bis[3-(*o*-chlorophenyl)-3-hydroxy-3-phenyl]anthracene (**3**) include one enantiomer of chiral guest compound when the complexation is carried out in toluene, but include the other enantiomer and MeOH in a 1:1:1 ratio when the complexation is carried out in MeOH.

We have reported that optically active diol host compounds, (*S*, *S*)-(–)-1,6-di(*o*-chlorophenyl)-1,6-diphenylhexa-2,4-diyne-1,6-diol **1**, (*S*, *S*)-(–)-1,4-bis[3-(*o*-chlorophenyl)-3-hydroxy-3-phenyl-1-propyl]benzene **2** and (*S*, *S*)-(–)-9,10-bis[3-(*o*-chlorophenyl)-3-hydroxy-3-phenyl]anthracene **3** are successfully used for optical resolution of various guest compounds and enantioselective synthesis in their chiral inclusion crystals. We report here a very interesting chiral recognition that **2** and **3** include one enantiomer of racemic guest in the absence of MeOH, but include the other enantiomer together with a MeOH molecule in the presence of MeOH.

For example, when a solution of **3** (0.5 g, 0.76 mmol) and (±)-2-methylpiperidine **4** (0.15 g, 1.52 mmol) in toluene (5 ml) was kept at room temperature for 12 h, a 1:1 inclusion crystal of **3** and (*R*)-(–)-**4** was formed as yellow prisms, which upon distillation at 150°C/20 mmHg gave (*R*)-(–)-**4** of 71% ee [0.05 g, 67% yield, $[\alpha]_D$-3.2° (*c* 0.12, MeOH)]. On the other hand, when the above inclusion complexation was carried out in MeOH, a 1:1:1 complex of **3**, (*S*)-(+)-**4** and MeOH was obtained as yellow prisms, which upon distillation gave (*S*)-(+)-**4** of 62% ee [0.5 g, 67% yield, $[\alpha]_D$ +2.8° (*c* 0.11, MeOH)].

MeOH plays a very interesting role, switching the chirality of the guest compound which is included in the chiral host compound. Similarly, **3** include the (–)-enantiomer of methyl alanate **7** in the absence of MeOH and include its (+)-enantiomer together with MeOH in the presence of MeOH (Table 1). In the case of α-phenylethyl amine **6**, a 1:1:1 complex of **3**, (+)-**6** of 58% ee and MeOH was formed when the complexation was carried out in MeOH.

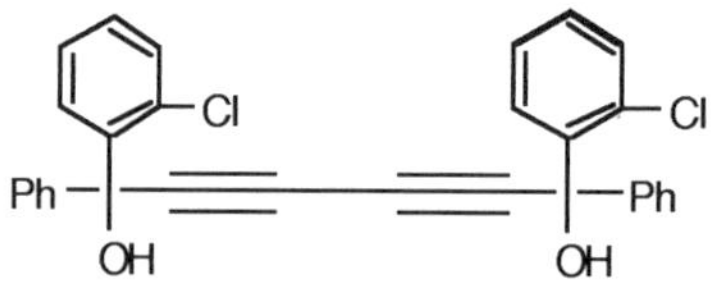

Cl Cl Ph OH OH Ph

(*S*,*S*)-(–)-**1**

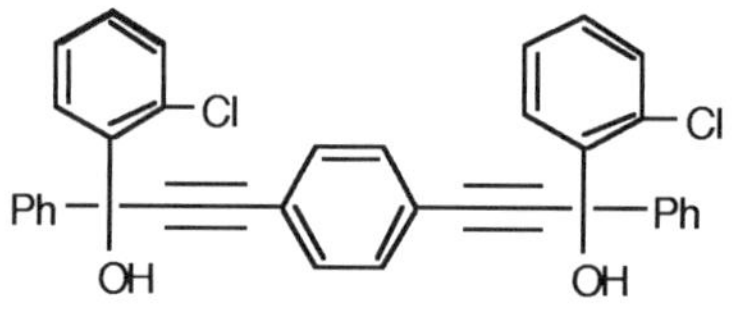

Cl Cl Ph OH OH Ph

(*S*,*S*)-(–)-**2**

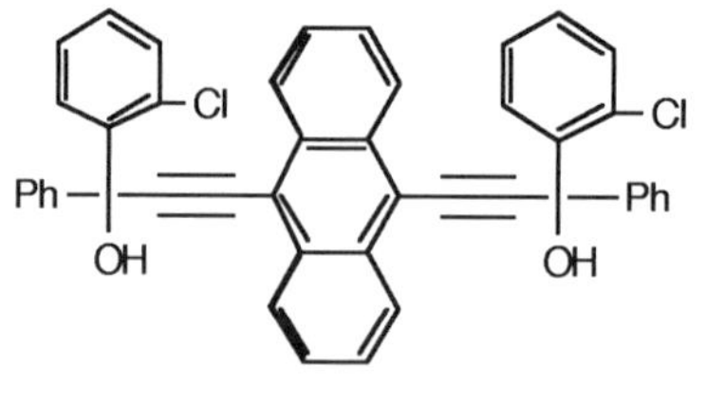

Cl Cl Ph OH OH Ph

(*S*,*S*)-(–)-**3**

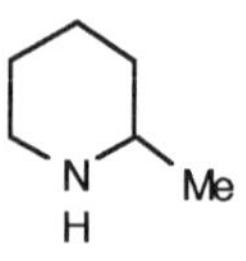

N H Me

4

H N N H Me

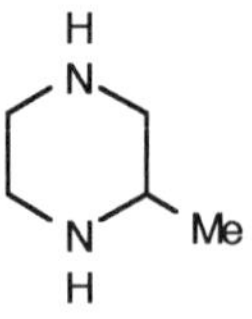

5

1054–7487/96/0246$12.00/0

Table 1 Optical resolution of guest compounds by inclusion complexation with chiral hosts in toluene and MeOH.

Guest	Solvent	Host 1 Complex h:g:s	Host 1 Isolated guest (% ee)	Host 2 Complex h:g:s	Host 2 Isolated guest (% ee)	Host 3 Complex h:g:s	Host 3 Isolated guest (% ee)
4	MeOH	1:1:0	(±)-4 (0)	1:1:0	(–)-4 (32)	1:1:MeOH	(+)-4 (62)
	toluene	1:1:0	(±)-4 (0)	1:1:0	(–)-4 (26)	1:1:0	(–)-4 (71)
5	MeOH	1:1:MeOH	(–)-5 (60)	a		a	
	toluene	1:1:toluene	(–)-5 (10)	a		a	
6	MeOH	a		a		1:1:MeOH	(+)-6 (58)
	toluene	a		1:1:0	(+)-6 (14)	a	
7	MeOH	a		a		1:1:MeOH	(–)-7 (39)
	toluene	a		a		1:1:0	(+)-7 (20)
8	EtOH	a		1:1:EtOH	(–)-8 (78)	a	
	toluene	a		1:2:0	(+)-8 (38)	1:2:0	(±)-8 (0)

a No complexation occured.

Ph-CH(NH2)-Me **6**   Me-CH(NH2)-CO2Me **7**   4-hydroxycyclopent-2-enone **8**

The host **2** also showed similar tendency. Complexation of **2** and (±)-4-hydroxycyclopent-2-enone **8** in toluene gave a 1:2 complex of **2** and (+)-**8**, which upon distillation gave (+)-**8** of 38% ee. When the complexation was carried out in EtOH, a 1:1:1 complex of **2**, (–)-**8** and EtOH was obtained as colorless prisms, from which (–)-**8** of 78% ee was obtained by distillation. EtOH can be replaced by MeOH, but sterically larger alcohols such as PrOH, BuOH did not show the same behavior. In the case of **4**, a 1:1 complex of **2** and (–)-**4** was obtained in both solvents.

However, host **1** showed a different behavior from that of **2** and **3**. Complexations of **1** with **4** both in toluene and in MeOH gave the same 1:1 complex of **1** and *rac*-**4**. In the case of **5**, a 1:1:1 complex of **1**, (–)-**5** of 10 % ee and toluene was formed when the complexation was carried out in toluene, and a 1:1:1 complex of **1**, (–)-**5** of 60% ee and MeOH was obtained from MeOH solution. In this case, MeOH plays an important role for higher enantioselectivity in the inclusion complexation.

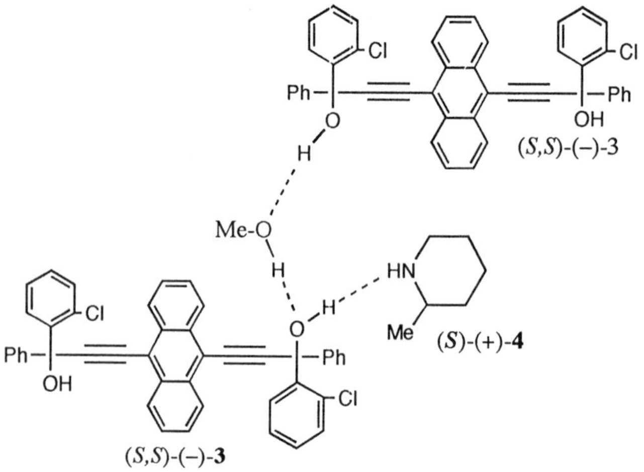

Scheme 1. Schematic illustration of the 1:1:1 inclusion complex of (*S,S*)-(–)-**3**, (*S*)-(+)-**4** and MeOH.

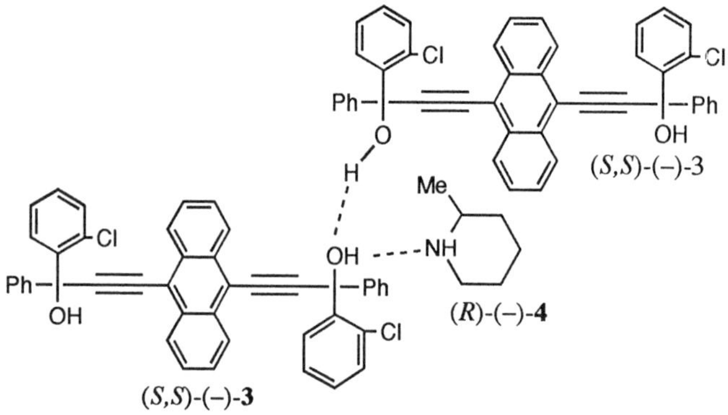

Scheme 2. Schematic illustration of the 1:1 inclusion complex of (*S,S*)-(–)-**3** and (*R*)-(–)-**4**.

In order to clarify the role of MeOH, the X-ray crystal structures of both the 1:1:1 complex of **3**, (*S*)-(+)-**4** and MeOH and the 1:1 complex of **3** and (*R*)-(–)-**4** were analyzed. In the 1:1:1 inclusion crystal of **3**, (*S*)-(+)-**4** and MeOH, a hydrogen bond network, host(1)-OH... MeOH...host(2)-OH...(*S*)-(+)-**4**-NH is constructed (scheme 1). The MeOH plays a space filling role for accommodating (S)-(+)-**4** in a chiral host lattice.

However, in the absence of MeOH, the OH group of host(1) is hydrogen bonded to that of host(2) directly (scheme 2). As the result, the cavity of the complex is more suitable for accommodation of the opposite enantiomer.

References

Toda, F.; Tanaka, K.; Nakamura, K.; Ueda, H.; Oshima, T.; *J. Am. Che. Soc.*, **1983**, *105*, 5151.

Toda, F.; Tanaka, K.; Marks, D.; Goldberg, I.; *J. Org. Chem.*, **1991**, *56*, 7332.

Tanaka, K.; Kakinoki, O.; Toda, F.; *J. Chem. Soc., Perkin Trans.* 1, **1992**, 307.

Tanaka, K.; Toda, F.; *J. Chem. Soc., Perkin Trans. 1*, **1992**, 943.

Tanaka, K.; Ootani, M.; Toda, F.; *Tetrahedron: Asymmetry*, **1992**, *3*, 709.

Toda, F.; Tanaka, K.; Miyahara, I.; Akutsu, S.; Hirotsu, K.; *J. Chem. Soc., Chem. Commun.*, **1994**, 1795.

The Effect of Chromium Ion (III) on the Nucleation of Supersaturated Ammonium Sulfate Solutions

Wei-Ming Sun and Allan S. Myerson, Chemical Engineering Department, Polytechnic University, Brooklyn, 333 Jay st., Brooklyn , NY 11201

The nucleation concentration of an ammonium sulfate solution with and without chromium(III) was measured by levitating micro-sized droplets electrodynamically in a spherical void electrodynamic levitator trap (SVELT) with a water vapor reservoir. The nucleation concentration is a function of chromium (III) concentration. The mechanism of chromium (III) interaction with nucleation will be discussed.

It is well known that the small amounts of an impurity can profoundly effect the nucleation and crystal growth rate. For example, Cooke (1966) found that Pb^{2+} acts as a nucleation agent in a NaCl system but Co^{2+} inhibits the crystal formation of KNO_3. Kitamura et al. measured the effect of K^+(Mullin,1985) and Fe^{3+}(Kitamura,1986) on ammonium-aluminum sulfate crystals and benzoic acid on *o*-chloro-benzoic acid (Kitamura, 1982 , 1983). Kitamura et al.(1992) found that the surface reaction-rate of ammonium sulfate crystal would be affected by the Cr^{3+}. Kubota et al.(1995) found that the adsorbed hydrolysis product of a hydrated chromium(III) suppresses the rate of crystal growth and dissolution.

During the past few years, the electro-dynamic levitator has been used to investigate the properties of aqueous solution at high concentration (Tang, 1984; Cohen, 1987; Na, 1994). In this technique, a single pre-filtered charged droplet is trapped by the superimposed dc and ac electrical field. This technique has the following advantages :

1. The suspended solution particles can be highly supersaturated.
2. Because it is free of foreign surface, the nucleation is homogeneous.
3. The water and solute activity of the system can be determined up to the nucleation point thus allowing evaluation of the true thermodynamics supersaturation.
4. Spectroscopic studies can be performed on the supersaturated droplet.

Because of these advantages, the electrodynamic balance has been proven to be a useful experimental tool for the investigation of a wide variety of problems involving micron-sized particles.

Ginde and Myerson (1993) evaluated the role of additives by measuring their effect on the metastable zone width and on the formation of concentration gradients in vertical columns of supersaturated solution. In presence of the additive, an increase in metastable zone width compared to the pure solution indicate that the additive acts as a nucleation inhibitor. The opposite effect indicates that the impurity is a nucleation enhancer. Using the spherical void electrodynamic levitator trap (SVELT) experiments, we can find the effect of impurity on the metastable zone width without fear of heterogeneous nucleation. We can measure the water activity and obtain solute activity thus allowing us evaluation of any changes in the supersaturation due to the additive.

Experiment

The configuration of experimental apparatus including the vacuum chamber, a particle generator, a water reservoir and the optic detection system is drawn in Fig. 1. A detailed schematic diagram of the SVELT is shown in Fig. 2.

The main experimental apparatus, SVELT, was developed by the Micro particle Photophysics Laboratory at Polytechnic University. The design methods and theoretical details of the SVELT system were described by Arnold (1985,1991).

1054–7487/96/0249$12.00/0

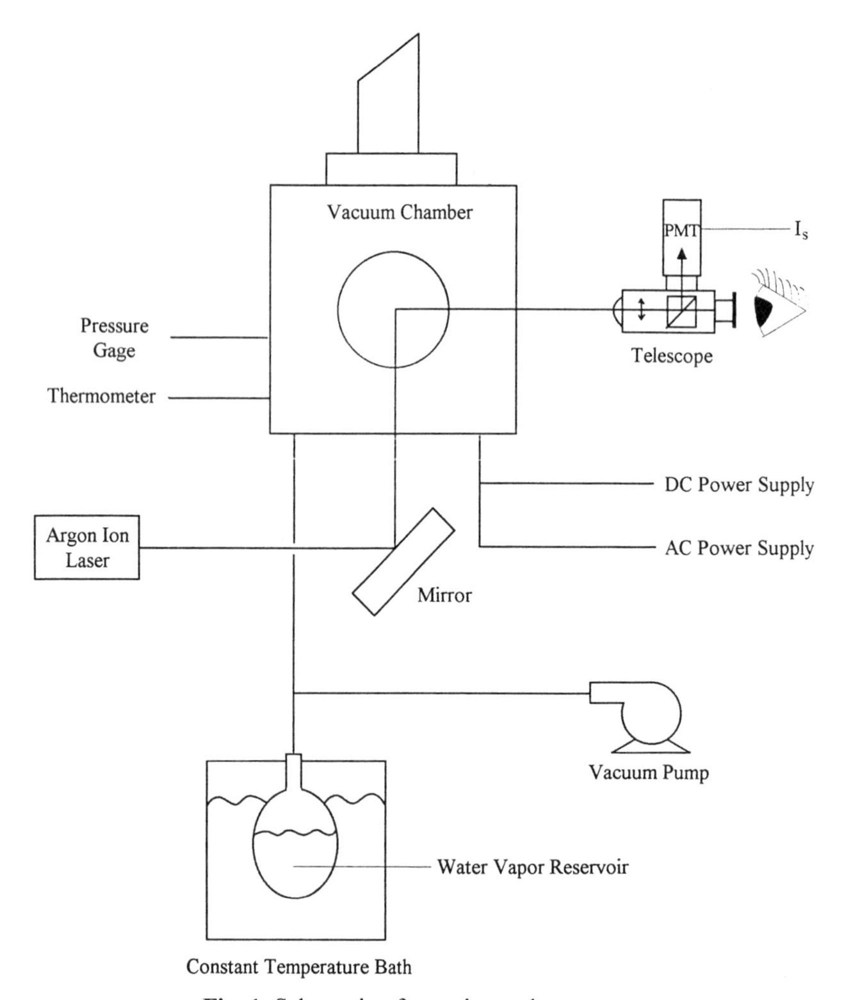

Fig. 1. Schematic of experimental apparatus

Solution particles are charged by going through a ring electrode in front of the particle generator. By adjusting the generator drive voltage, particles as small as 10 microns in diameter have been produced. The charged particles entering the SVELT will be caught in the trap with an AC voltage around 400 volts. Using the DC electrodes between the cap and the bottom, we can let the particle to be pull to the trap center. The weight of this particle, mg, carrying q electrostatic charges is balanced against the gravity force

$$mg = CqV_{DC} / Z_o \qquad (1)$$

where Z_O is characteristic length of the cell and C is a geometric constant. If we assume that the particle charge, q , remains constant during the experiment and the solute is nonvolatile, then , from the equation (1), the relative mass changes of droplet can be determined by measuring the DC voltage change. The molality can be calculated by

$$m = \left(\frac{V_{DC}^{wet}}{V_{DC}^{dry}} - 1\right)^{-1} \frac{1000}{MW} \qquad (2)$$

where MW is the molecular weight of the solute. When a solution particle caught by the SVELT, we can begin to evacuate the chamber to obtain the particle in a dry state. After recording the DC voltage of the dry particle, V_{DC}^{dry}, water vapor from the constant temperature water reservoir is

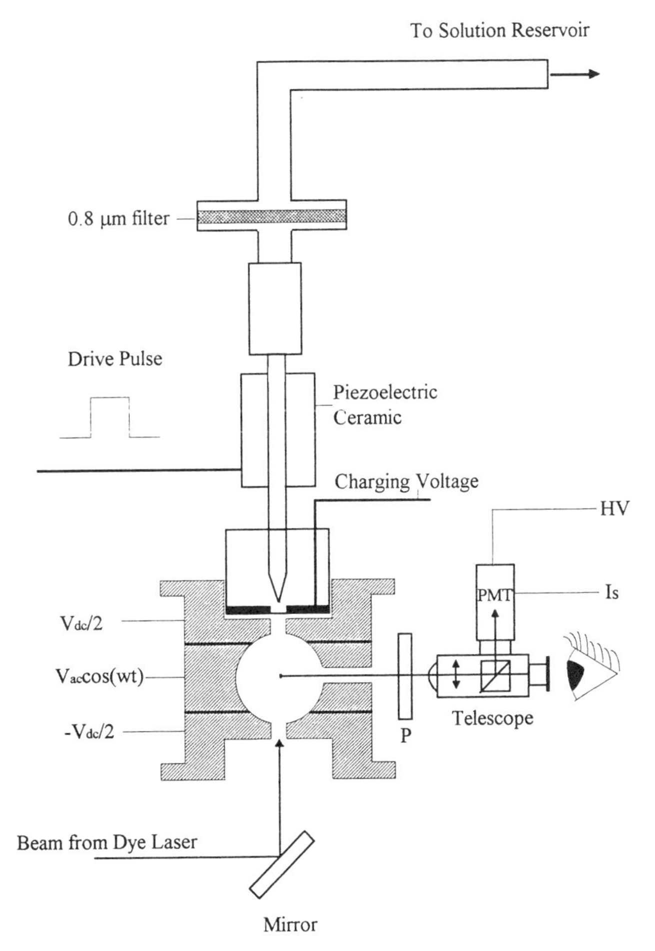

Fig. 2. Schematic of Spherical Void Electrodynamic Levitator Trap (SVELT)

allowed to bleed back slowly to the vacuumed chamber until the solid particle changes back to the solution state. Then another evacuation process is used to evaporate the solvent and to increase the concentration into the supersaturated metastable zone until the crystallization point. As a result of water vapor evaporation, the DC balancing voltage decreases steadily and the solution particle becomes supersaturated. From the balancing DC voltage on this evacuation process, we can get the concentration values of the whole metastable zone until crystallization point. At crystallization there is a sudden drop in the balancing voltage. After crystallization, the evacuation is continued to insure that there has been no change of mass and no charge loss during the cycle. During the entire experiment, the DC voltage is tuned to keep the trapped particle at the center point.

Temperature control on this experiment is very important. Water is circulated from one temperature-controlled water bath through the outwall of water reservoir and the SVELT system. The temperature of the leviator itself is maintained within ±0.1°C and is measured with a thermistor.

This experimental cycle is repeated several times to insure the reproducibility.

Results and Discussion

Kitamura et al.(1992) found that the Cr^{3+} is an inhibitor for the crystal growth of ammonium sulfate during the 1-5 ppm range. Kubota et al. (1995) put the 0.5 - 5 ppm Cr^{3+} into potassium sulfate solution and also found Cr^{3+} to be an inhibitor . The metastable zone width (MZW) of ammonium sulfate should be increased by additive of Cr^{3+}.

Table 1 shows the initial and final chromium (III) level used in the experiments. They are different since water is evaporated during the experiment.

In the Table 2, the nucleation concentration of ammonium sulfate with different amounts of Cr^{3+} is shown. The nucleation concentration of pure ammonium sulfate is 30.097 m that is very close to the value of 29.99 m reported in the literature. (Tang, 1984) At low Cr^{3+} concentration (final concentrations below 2.57 ppm), the additive increased the metastable zone width and acted as an inhibitor. At higher concentrations(24.1 ppm) the additive decreases the metastable zone width.

Fig. 3. shown the effect of Cr^{3+} on the inhibitor zone. Under the 1.5 ppm of the final concentration of Cr^{3+}, the crystallization concentration of ammonium sulfate slightly increases as Cr^{3+} concentration increases until a maximun is obtained and then decreasing.

From Fig. 4., we can find that Cr^{3+} acts differently when the concentration of impurity is higher than 2.577 ppm. The more Cr^{3+}, the lower the crystallization concentration of ammonium sulfate solution. That mean the Cr^{3+} acts as a nucleation enhancer.

The water activity of ammonium sulfate solution with different amount Cr^{3+} is shown in the Table 3 and Fig. 5. The water activity data linearly decreased with increasing molality in the undersaturated region, but some deviation from linearity was observed in the supersaturated region. A detail discussion water activity is found in the previous work. (Na et al., 1994,1995)

From the previous work (Na and Myerson, 1994), the equation of water activity at ammonium sulfate system is :

$$a_w=C0+C1m+C2m^2+C3m^3+C4m^4+C5m^5 \quad (3)$$

where C0=1.21242, C1=-0.090154, C2=0.004529, C3=-1.2121E-4, C4=1.29698E-6, C5=-5.68515E-10.

Using the relationship of osmotic coefficient and water activity as a function of concentration expressed in equations :

$$\ln a/a^* = \ln(m/m^*)+(\phi_m - \phi_m^*)+\int_{m^*}^{m}[\phi_m - 1/m]dm \quad (4)$$

$$vm\phi_m = -1000/M_s * \ln a_w \quad (5)$$

where a_w is water activity; a is solute activity; ϕ_m is osmotic coefficient; M_s is the molecular weight of solvent. The reference state of this equation is chosen on the saturated concentration. The results of water activity are listed in the Table 3.

Electrostatic forces between ions are inversely proportional to the square of the

Table 1. The final concentration (ppm level) of chromium(III) in ammonium sulfate solution

Initial ppm (Cr^{3+})	Final average ppm (Cr^{3+})	standard deviation
0.01	0.259	0.009
0.05	1.384	0.074
0.1	2.574	0.094
1.08	25.36	0.977

Table 2. The crystallization concentration of ammonium sulfate with different amount chromium(III)

Initial ppm (Cr^{3+})	Final ppm (Cr^{3+})	particle 1	particle 2	particle 3	particle 4	particle 5	particle 6	particle 7	average
0.00	0.000	30.07	30.99	30.10	30.08	29.79	29.58	30.08	30.10
0.01	0.259	29.29	29.59	29.38	30.81	30.56	31.29	31.17	30.30
0.05	1.384	31.26	31.28	29.71	29.81	29.80	31.68	31.88	30.77
0.10	2.574	29.07	29.46	28.75	29.61	29.31	29.84	28.70	29.11
1.08	25.36	26.52	24.70	25.46	24.17				25.21

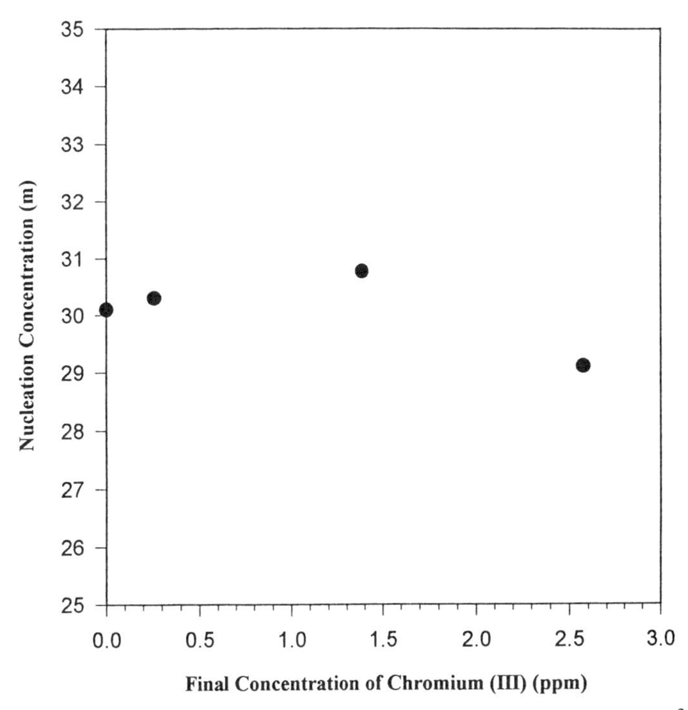

Fig. 3. Nucleation point of $(NH_4)_2SO_4$ solution droplet with trace amount Cr^{3+} at 25°C

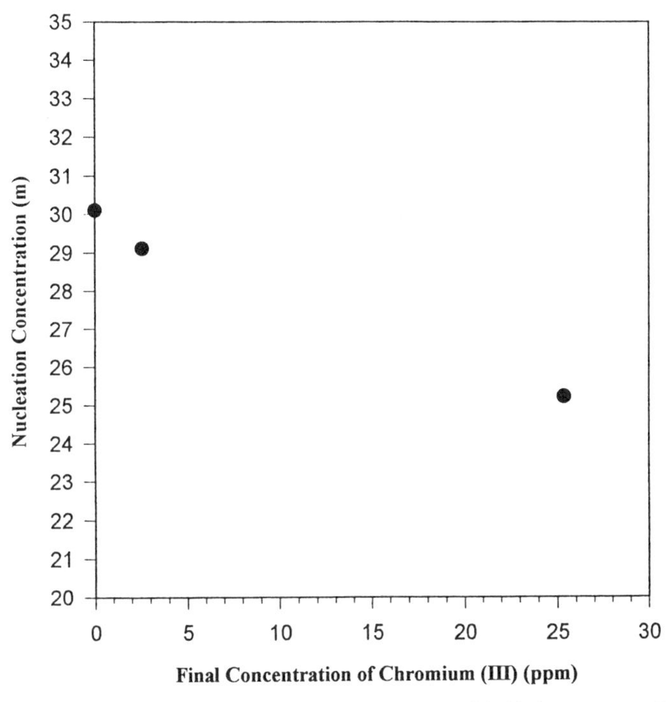

Fig. 4. Nucleation point of $(NH_4)_2SO_4$ solution droplet with higher amount Cr^{3+} at 25°C

Table 3. The water activity of ammonium sulfate on the difference supersaturated concentration with different amount chromium(III)

		water activity of solution (a_w)					average standard deviation
Initial Cr^{3+} conct. (ppm)	Final Cr^{3+} conct. (ppm)	10.0 (m)	15.0 (m)	20.0 (m)	25.0 (m)	30.0 (m)	δ
0.00	0.000	0.649	0.534	0.493	0.419	0.370	0.0236
0.01	0.259	0.712	0.579	0.497	0.434	0.375	0.0274
0.05	1.384	0.685	0.561	0.483	0.426	0.337	0.0616
0.10	2.574	0.662	0.544	0.468	0.408	-	0.0153
1.08	25.36	0.652	0.516	0.429	0.355	-	0.0218
average		0.672	0.547	0.474	0.408	0.361	

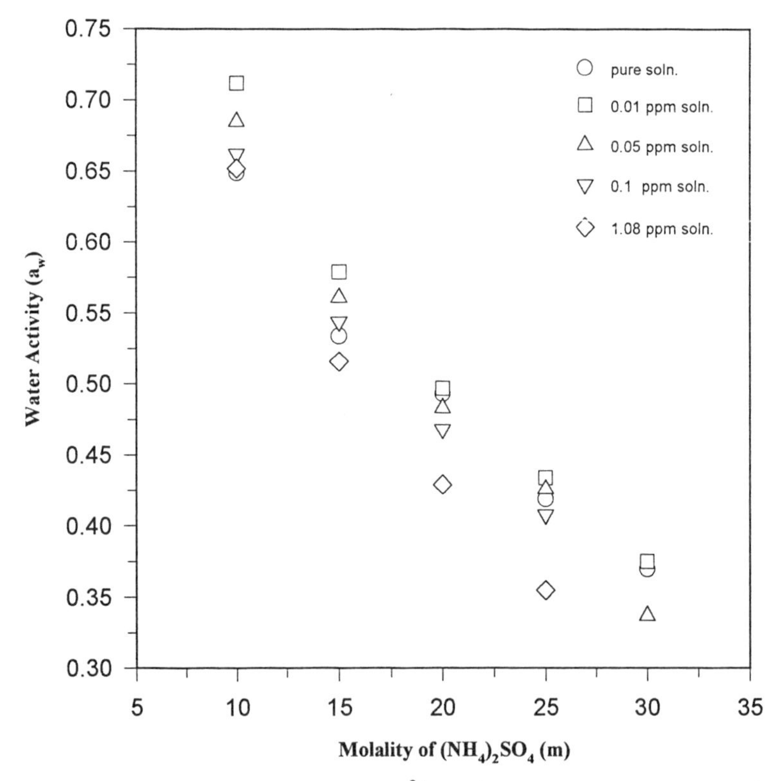

Fig. 5. The effect of water activity by Cr^{3+} on the $(NH_4)_2SO_4$ solution at 25°C

separation distance by Columb's relation (the inverse-square law). They have a much larger range than other intermolecular forces that depend on higher powers of the reciprocal distance. Thus, the solute activity behaviors of electrolyte solutions can be understood by making attempts to find the complicated problems set such as long-range electrostatic forces between ions and short range interactions between ions and solvent molecules. These factors are considered to be effected by ionic charge and size. A hydration-association treatment is useful in understanding the activity in electrolyte solutions. It has been mentioned (Na and Myerson, 1994) that the effect of ionic solvation (hydration) increases activity but that of ion-pair formation counteracts thus effect.

Fig. 5 graphically shows the effect of Cr^{3+} on the water activity at various concentration of ammonium sulfate. Comparing the water activity data at the same concentration of ammonium sulfate, the more Cr^{3+} the lower the water activity value. At low supersaturations (around 10 m), the water activity of pure solution is lower than those with chromium ion. However, the water activity of solution with Cr^{3+} will decrease more quickly than the pure solution when the molality increases. The slope of water activity data of ammonium sulfate increases with increasing amount of Cr^{3+}.

We will investigate this effect further for more fully understand the role of Cr^{3+} in preventing or enhancing the nucleation of ammonium sulfate solutions.

Acknowledgements

The authors gratefully acknowledge the suport of NASA (Grants No. NA68-960 and 6NA8-1060).

References

Arnold, S., *Rev. Sci. Instr.* **1991**,62, 3025

Arnold, s.; Folan, L. M., *Rev.Sci. Instr.* **1985**, 56, 2066

Cohen, M. D.; Flagan, R. C., *J. Phys. Chem.* **1987**, 91, 4583.

Cooke , E. G., *Kristall Tech.* **1966**,1 ,119

Ginde, R. M.; Myerson, A. S. *J. Cryst. Growth* **1993**, 126, 216

Kitamura, M.; Mullin, J. W. *Proc. World Congress III of Chem. Engng.* **1986**, 2, 1048

Kitamura ,M.; Nakai, T. , *Kagaku Kogaku Ronbunshu* **1982**, 8, 442

Kitamura, M. ; Nakai, T. , *J. Chem. Eng., Japan* **1983** ,16, 288

Kitamura, M.; Nakai, T.; Ikemoto, K.; Kawamura,Y. , *International Chem. Eng.*, **1992,** 32, 1, 157

Kubota, N.; Fukazawa J.; Yashiro, H.; Mullin, J. W., *J. Cryst. Growth* **1995**, 149, 113

Mullin, J. W.; Amatavivadhana ,A.; Chakraborty, M., *J.Appl.Chem.* **1970**,20,153

Mullin ,J. W. and Kitamura ,M., *J.Cryst. Growth* **1985**, 71,118

Na, Han-Soo; Arnold, S.; Myerson, A.S., *J. Cryst. Growth,* **1994**, 139, 104

Na, Han-Soo; Arnold, S.; Myerson, A.S., *J. Cryst. Growth,* **1995**, 149, 229

Tang I. N.; Munkelwitz H. R., *J. of Colloid & Inter. Sci.* **1984,** 98, 2, 430

Crystallization of the Molecular Crystals between Nitroanilines and Nitrophenols

Hideko Koshima, PRESTO, Research Development Corporation of Japan (JRDC), Japan
Yang Wang, Teruo Matsuura, Department of Materials Chemistry, Faculty of Science and Technology, Ryukoku University, Seta, Otsu 520, Japan
Hisashi Mizutani, Hiroyuki Isako, Ikuko Miyahara, Ken Hirotsu, Department of Chemistry, Faculty of Science, Osaka City University, Sugimoto, Osaka 558, Japan

Slow evaporation of a 1 : 2 molar solution of p-nitroaniline and p-nitrophenol in dichloromethane-hexane deposited two kinds of single crystals which were mechanically separable, 2 : 1 molecular crystal **A** and 1:3 molecular crystal **B**. Similarly a 1 : 1 molar solution of m-nitroaniline and m-nitrophenol in chloroform-petroleum ether deposited the single crystals of 2 : 1 molecular crystal **C**. The crystal structures of **A, B** and **C** are isomorphic to those of p-nitroaniline, p-nitrophenol (β form) and m-nitroaniline, respectively.

Introduction

Recently we have reported on the preparation of two-component mixed crystals from p-, m- and o-nitroanilines and p-, m- and o-nitrophenols using a melting-resolidification process and on their second harmonic generation (SHG) activity. Among the fifteen mixed crystals prepared, at least eight mixed crystals (p-NA/o-NA,* p-NA/p-NP, m-NA/o-NA, m-NA/m-NP, o-NA/m-NP, o-NA/o-NP, p-NP/m-NP and p-NP/o-NP) were found to form a molecular compound between two components (Koshima, et al., 1995). It is now generally accepted that hydrogen bonding interactions are one of the important factors for the formation of a crystalline molecular compound from two different organic compounds (Koshima, et al., 1994; M. C. Etter and S. M. Reutzel, 1991).

In attempts to obtain the single crystals of molecular compounds for the mixed crystals of nitroaniline-nitrophenol series by the crystallization from a solution of two components, we succeeded in obtaining the single crystals of molecular crystals between p-NA and p-NP and between m-NA and m-NP. This paper describes the isolation of these bimolecular crystals and their X-ray crystallographic analysis.

* NA = nitroaniline; NP = nitrophenol

Results and Discussion

When a 1 : 2 molar solution of p-NA and p-NP in a mixture of dichloromethane and hexane was slowly evaporated at room temperature, two kinds of single crystals deposited which could be mechanically separated: crystal **A**, yellow prisms, m.p. 120-138°C and crystal **B**, yellow needles, m.p. 103-109°C. In a similar manner slow evaporation of a 1 : 1 molar solution of m-NA and m-NP in chloroform and petroleum ether deposited crystal **C**, yellow crystals, m.p. 88-94°C. The elemental analysis of these crystals confirmed that **A, B** and **C** were 2 : 1 p-NA/p-NP, 1 : 3 p-NA/p-NP and 2 ; 1 m-NA/m-NP molecular crystals, respectively. The analytical data were the same within experimental errors for the crystals obtained from several different batches. These results show that each molecular crystal has a definite stoichiometrical ratio, suggesting the formation of a molecular compound, although the possibility of the formation of a solid solution is not eliminated.

The X-ray structural analysis of **A, B** and **C** showed that their amino and hydroxyl groups cannot be discriminated in the crystal structures possibly due to the disorder between two components. Only two amino hydrogens for **A** and **C** appeared on each d-map, while only a hydroxyl hydrogen for **B**

1054–7487/96/0256$12.00/0

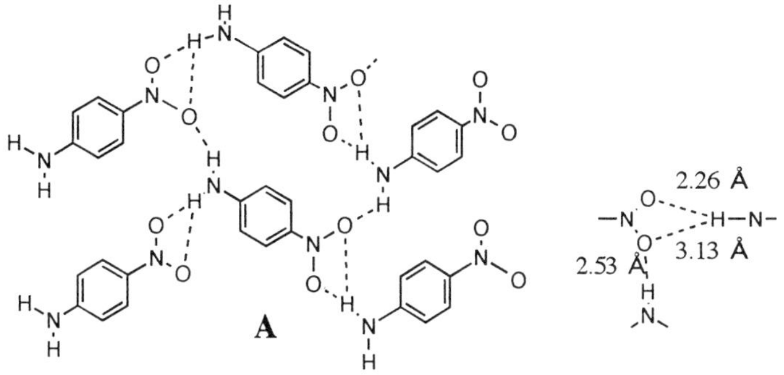

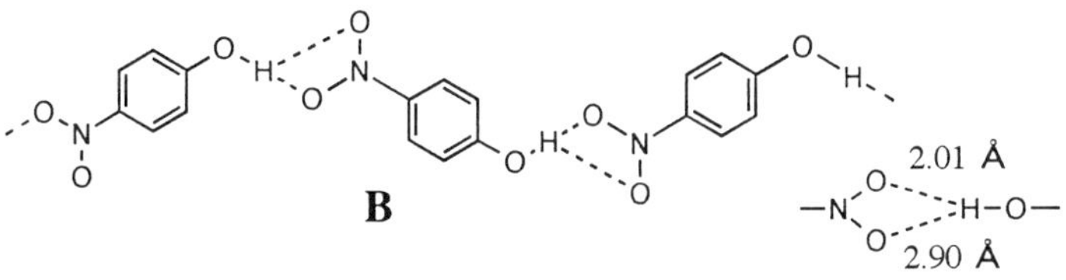

appeared on the d-map. Therefore, the refinements for **A**, **B** and **C** were done with the major component molecules, namely p-NA, p-NP and m-NA, respectively. The R values obtained for **A**, **B** and **C** were 0.060, 0.066 and 0.041, respectively.

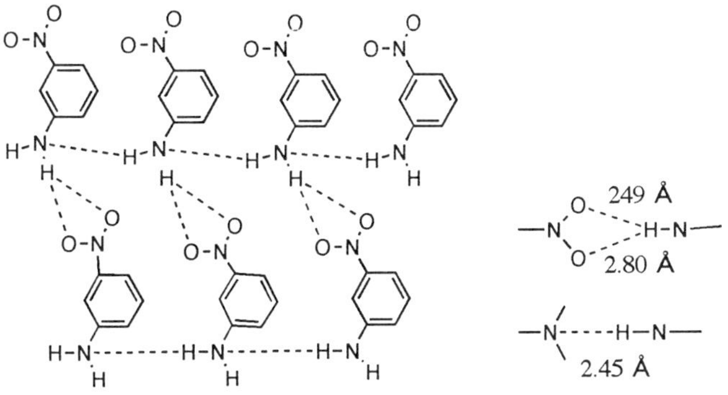

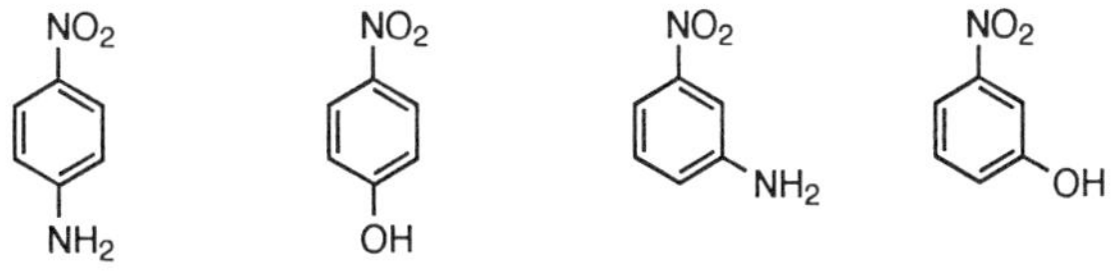

The crystal data of **A**, **B** and **C** are summarized in Table 1 in comparison with those reported for the component compounds, indicating that the crystal structures of **A**, **B** and **C** are isomorphic to those of p-NA, p-NP (β form) and m-NA, respectively. The following figures for **A**, **B** and **C** show the features of the hydrogen bonding in the molecular arrangements of each crystal. (1) In the crystal lattice of **A**, N(O)(O)···H-N hydrogen bonds (2.26, 2.53 and 3.13 Å) connect molecules in a two-dimensional way and the amino group is coplanar to the benzene ring. (2) In the case of **B**, molecules are linearly connected to each other by

Table 1. The crystal data of **A**, **B**, **C**, p-NA, p-NP and m-NP.

Crystal	Cryst. form*	a (Å)	b (Å)	c (Å)	β (°)	Space group	Z	Ref.
A	M	8.508	6.056	12.252	91.73	$P2_1/n$	4	a
B	M	14.990	11.130	3.774	101.58	$P2_1/a$	4	a
C	O	19.347	6.594	4.970	-	$Pca2_1$	4	a
p-NA	M	12.336	6.07	8.592	91.45	$P2_1/n$	4	b
p-NP (α form)	M	11.415	8.780	6.098	103.08	$P2_1/a$	4	c
p-NP (β form)	M	14.743	11.117	3.785	92.65	$P2_1/n$	4	d
m-NA	O	6.501	19.330	5.082	-	$Pbc2_1$	4	e
m-NP (M form)	M	11.240	6.891	8.154	98.05	$P2_1/n$	4	f
m-NP (O form)	O	8.123	11.305	6.777	-	$P2_12_12_1$	4	g

* M = monoclinic, O = orthorhombic.

a) This work.
b) Trueblood and Goldish, 1961.
c) Coppen and Schmidt, 1965.
d) Coppen and Schmidt, 1965a
e) Skapski and Stevenson, 1973.
f) Pandarese, et al., 1975.
g) Coda, et al., 1975.

hydrogen bonds $NO_2 \cdots H\text{-}N$ (2.01 and 2.90 Å). (3) In the crystal lattice of **C**, weaker $\text{-}N \cdots H\text{-}N$ (2.45 Å) and $NO_2 \cdots H\text{-}$ (2.49 and 2.80 Å) hydrogen bonds connect molecules in a two-dimentional way and two amino hydrogens are not coplanar to the benzene ring, since the amino nitrogen atom is near sp^3 hybridization. These lengths of the $N\text{-}O \cdots H\text{-}N$ hydrogen bonds are comparable to those compiled for nitroaniline derivatives (Panunto, et al., 1987).

The available experimental data for the molecular crystals **A**, **B** and **C** cannot discriminate whether they are molecular compounds or solid solutions. In any case, the crystal growth of each molecular crystal might be initiated by the primary nucleation of the major component. In the case of **A** for example, once the primary nucleus of p-nitroaniline is formed, p-nitrophenol molecules are incorporated into the crystal lattice of p-nitroaniline in the definite ratio of 2 : 1.

References

Coda, A.; Fumagalli, M.; Pandarese, F.; Ungaretti, L. Acta Crystallogr., Section A, **1975**, 31, S208.

Coppen, P.; Schmidt, G. M. J. Acta Crystallogr., **1965**, 18, 62-67.

Coppen, P.; Schmidt, G. M. J. Acta Crystallogr., **1965a**, 18, 654-663.

Etter, M. C.; Reutzel, S. M. J. Am. Chem. Soc., **1991**, 113, 2585-2598 and references cited therein

Koshima, H.; Ding, K.; Matsuura, T. J. Chem. Soc., Chem. Commun., **1994**, 2053-2054.

Koshima, H.; Wang, Y.; Matsuura, T.; Mibuka, N.; Imahashi, S. Mol. Cryst. Liq. Cryst., **1995**, in the press.

Pandarese, F.; Ungaretti, L.; Coda, A. Acta Crystallogr., **1975**, B31, 2671-2675.

Panunto, T. W.; Urbanczyk-Lipkowaka, Z.; Johnson, R.; Etter, M. C. J. Am. Chem. Soc., **1987**, 109, 7786-7797.

Skapski, A. C.; Stevenson, J. L. J. Chem. Soc. Perkin Trans. II, **1973**, 1197-1200.

Trueblood, K. N.; Goldish; Donohue, J. Acta Crystallogr., **1961**, 14, 1009-1017.

Gel Growth of the Organomineral Crystal 2-Amino-5-Nitropyridinium Dihydrogen Phosphate and the Quadratic Nonlinear Optical Effect

N.Horiuchi, Advanced Research Center for Science and Engineering, Waseda University, 3-4-1, Okubo, Shinjuku-ku, Tokyo 169, Japan
Y.Uesu, Department of Physics, Waseda University, 3-4-1, Okubo, Shinjuku-ku, Tokyo 169, Japan
F.Lefaucheux and **M.C.Robert**, Laboratoire de Minéralogie-Cristallographie, associé au CNRS et aux Universités Paris 6 & 7, 4 Place Jussieu, 75252 Paris Cedex 05, France

2-amino-5-nitropyridinium dihydrogen phosphate crystals have been grown by sol gel technique to improve the crystalline quality. Optimal growth conditions are determined in terms of the content of tetramethoxysilane and the molar ratio of H_3PO_4 in the growth solution. The quadratic nonlinear optical coefficients have been obtained as d_{15}=13±3pm/V, d_{24}=2.2±0.2pm/V and d_{33}=12±1pm/V by second harmonic generation. It is confirmed that the measured value are significantly small in comparison with the theoretical ones.

A series of 2-amino-5-nitropyridine (2A5NP) whose hyperpolarizability ß is as large as 20 × 10^{-30} esu (Kotler 1992), displays large quadratic nonlinear optical coefficients such as PNP (Sutter 1988) and 2A5NPLT (Zyss 1993). 2-amino-5-nitropyridinium dihydrogen phosphate (2A5NPDP) is an organo-mineral crystal which is proposed by the new approach of the molecular engineering. Since the organic molecule is combined with inorganic anions via hydrogen bonds, mechanical and thermal resistance are improved. This crystal is expected to exhibit large quadratic nonlinear optical coefficients d_{33}=87pm/V (Kotler 1992). However, the measured coefficients of 2A5NPDP, which was grown by free solution technique were considerably smaller than the theoretical ones (Kotler 1992). As dislocations occur easily in organic crystals, it was thought that the poor crystalline quality reduced the nonlinear optical properties.

In this study, 2A5NPDP crystal is grown by sol gel technique. Using the grown crystals with good quality, the quadratic nonlinear optical coefficients are reexamined by SHG and they are compared with the theoretical ones.

Growth Condition

At first, 2A5NPDP is prepared as powder from 2A5NP and H_3PO_4 (Kotler 1992). The 2A5NPDP powder is dissolved in aqueous solution of H_3PO_4 at 50°C and 2A5NPDP is recrystallized by decreasing the temperature to 20°C. Tetramethoxysilane (TMOS) is added to the solution as gelling agent (Lefaucheux 1988). Sol gel technique has mainly two advantages in crystal growth. (1)The gel walls partition the growth solution in small volumes. The convection is suppressed so that the mass transfer is ensured only by diffusion. (2)Nucleation takes place in a confined space so that the number is strongly reduced in comparison with free solution growth technique. In other words, the supersaturation in gel becomes higher than in free solution. For these reasons, polynucleation is strongly inhibited and a dislocation free crystal can be grown.

The molar concentration of H_3PO_4 is increased to dissolve sufficient amount of 2A5NPDP. However, the high acidity is not favorable because the Si–O network in TMOS gel is chemically broken. The content of TMOS can not be increased, because it is mechanically broken when the crystal is growing. Considering these facts, the content of TMOS and the acidity of the aqueous solution of H_3PO_4 are varied (Table 1) to find the optimal growth condition.

Nucleation is difficult and takes place at very high supersaturation (Lefaucheux 1988) leading

Table 1. 2A5NPDP Crystal Growth Condition in TMOS Gel

Growth Condition	A	B
TMOS (Vol. %)	2 – 10	2–3
H_3PO_4 (M)	2 – 6	2–4
Initial Temperature (°C)	50	50
Final Temperature (°C)	20	20
Cooling Rate (°C/day)	2	0.4
Volume of the Solution (cm^3)	10	10

A: Without seed crystal
B: With seed crystal

almost to dendritic growth. So we developed growth on seed in gel with the following procedure: the gel is made as usual with the supersaturated solution and before it becomes too firm, a seed is introduced by letting it drop via its weight until it reaches the middle of the gel column. The gel which is not yet completely formed, is cured after the seeding and growth is initiated by slow cooling. Fig.1 presents a crystal grown by this technique. This crystal is transparent and was cut for nonlinear optical coefficients measurements.

Since 2A5NPDP has a polar structure (Masse 1991), the crystal are elongated along the c-axis. Seeing the size of the as-grown crystals, the optimal growth condition is found as follows: TMOS 2-5%; H_3PO_4 2-4M. Under these conditions, the seeding technique is applied (Table 1). The maximum size of the grown crystal is $5 \times 7 \times 27$ mm^3.

Fig.1. 2A5NPDP crystals grown in TMOS gels. The tube diameters is 20mm. The growth conditions are as follows: TMOS 2%; H_3PO_4 4M.

Quadratic Nonlinear Coefficients

2A5NPDP belongs to the point group *mm2* and has three independent components, i.e., d_{15}, d_{24} and d_{33}, in the transparency region. These values are determined by measuring the second harmonic (SH) intensity. Especially, the coefficients d_{15} and d_{24} can be determined under the phase matching condition (Kotler 1992) and they were also obtained with the crystals which were grown in TMOS gel (Horiuchi 1995). Here, the d_{33} coefficient is obtained by Maker fringe method by comparison with the d_{11} of α-quartz (d_{11}=0.36pm/V) (Singh 1971):

$$\frac{d_{33}^{2A5NPDP}}{d_{11}^{Quartz}} = \frac{n^{\omega,\,2A5NPDP}}{n^{\omega,\,Quartz}} \cdot \frac{\Delta k^{\,2A5NPDP}}{\Delta k^{\,Quartz}} \times \sqrt{\frac{n^{2\omega,2A5NPDP}}{n^{2\omega,Quartz}} \cdot \frac{I^{\,2A5NPDP}}{I^{\,Quartz}} \cdot \frac{(t_1^2/t_2)^{Quartz}}{(t_1^2/t_2)^{2A5NPDP}}} ,$$

where n^{ω} and $n^{2\omega}$ are the refractive indexes of the polarized fundamental and SH beam, I the SH intensity from the sample, t_1 the transmission factor for the amplitude of the fundamental beam at the air-crystal interface, t_2 the one of the SH beam at the crystal-air interface and Δk is defined as:

$$\Delta k = 2\pi \left| \frac{n^{2\omega}}{\lambda^{2\omega}} - 2\frac{n^{\omega}}{\lambda^{\omega}} \right| .$$

The sample is cut at incidence (100) plane. The taper angle between incident and emergent plane is 2.1°. In this experiment, Nd^{3+}:YAG laser (λ=1.064μm) and Forsterite ($Cr:Mg_2SiO_4$) laser (λ=1.34μm) are used. The pulse duration, repetition rate and the energy per pulse of both laser sources are 50ns, 20Hz, and 1.3mJ, respectively. By comparison of SH intensity (Fig.2) between the 2A5NPDP and the α-quartz, the coefficient d_{33} is obtained as 18±2pm/V at 1.064μm and 12±1pm/V at 1.34μm (Table 2).

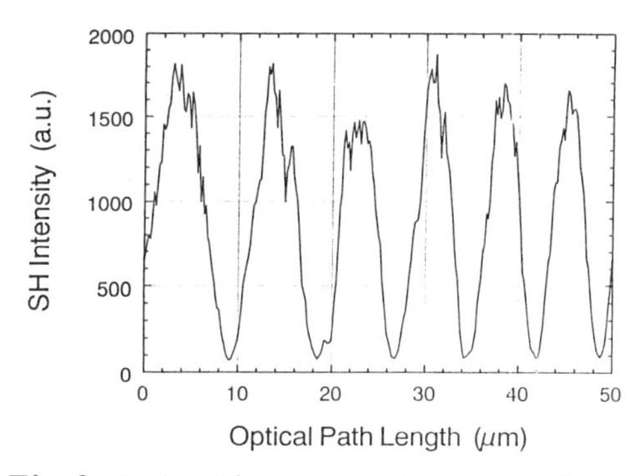

Fig.2. Maker fringe curve of SH intensity of 2A5NPDP at 1.064μm. The sample is moved along the c-axis. The d_{33} coefficient is determined by the maximum intensity.

Table 2. Quadratic Nonlinear Coefficients of 2A5NPDP

a) λ=1.064μm

d_{ij}	*Theoretical Value (pm/V)	Measured Value (pm/V) Sample A	*Sample B
d_{15}	35 ± 1		7 ± 1
d_{24}	16 ± 1		1±0.4
d_{33}	122 ± 4	18 ± 2	

b) λ=1.34μm

d_{ij}	*Theoretical Value (pm/V)	Measured Value (pm/V) Sample A	*Sample B
d_{15}	28 ± 2	**13 ± 3	6 ± 1
d_{24}	12 ± 1	**2.2 ± 0.2	
d_{33}	87 ± 4	12 ± 1	12 ± 1

Sample A: The crystals grown by sol gel technique.
Sample B: The crystals grown by free solution technique.
*See the reference of Kotler 1992.
** See the reference of Horiuchi 1995.

Conclusion

The optimal growth condition of 2A5NPDP in gel is found in terms of the content of TMOS and the molar concentration of H_3PO_4. With these samples, the d_{33} coefficient was obtained by SHG as 18±2pm/V at 1.064μm and 12±1pm/V at 1.34μm. These values are still smaller than the theoretical ones. Since d_{33} component is not involved in the effective quadratic nonlinear optical coefficient d_{eff}, it is impossible to discuss the crystalline quality with the effective interaction length. However, the coherence length was obtained as 10.5μm and this value is close to the calculated one (12.6μm at λ=1.064μm) and the peak intensity of SH was almost constant. Taking into account of these facts, one can say that crystalline quality of the 2A5NPDP is optically good. Considering that the effective interaction length of the gel-grown crystal is at least 88% (Horiuchi 1995), it is difficult to believe that the quadratic nonlinear optical property is reduced due to the poor quality. From this fact, it is supposed that the d_{ij}

coefficients are overestimated. It is inferred that the hyperpolarizability of 2A5NPDP is significantly reduced from the one of 2A5NP due to the strong hydrogen bond coupling between anions and cations. Further studying on the calculation is expected to solve the discrepancy.

References

Horiuchi, N.; Lefaucheux, F.; Robert, M.C.; Josse, D.; Khodja, S.; Zyss, *J. J.Crystal Growth* **1995**, *147*, 361-368.

Kotler, Z.; Hierle, R.; Josse, D.; Zyss, J.; Masse, R. *J. Opt. Soc. Am.* **1992**, *B9*, 534-547.

Lefaucheux, F.; Robert, M.C.; Bernard, Y. *J. Crystal Growth* **1988**, *88*, 97-106.

Masse, R.; Zyss, J. *Mol. Eng.* **1991**, *1*, 141-152.

Singh, S. In *Handbook of Lasers with Selected Data on Optical Technology*; Pressley, R.J., Ed.; Chemical Rubber Co.: Cleveland, OH, 1971.

Sutter, K.; Bosshard, Ch.; Wand, W.S.; Surmely, G.; Günter, P. *Appl. Phys. Lett.* **1988**, *53*, 1779-1781

Zyss, J.; Masse, R.; Bagieu-Beucher, M.; Lévy, J.P. *Adv. Mater.* **1993**, *5*, 120-124.

A Simple Inexpensive Bridgman-Stockbarger Crystal Growth System for Organic Materials

J. Choi, M.D. Aggarwal, W.S. Wang, R. Metzl, and K.Bhat, Department of Physics, Alabama A&M University, Normal, AL 35762
Benjamin G. Penn, and Donald O. Frazier, Space Science Laboratory, NASA Marshall Space Flight Center, AL 35812

Direct observation of solid-liquid interface is important for the directional solidification to determine the desired interface shape by controlling the growth parameters. To grow good quality single crystals of novel organic nonlinear optical materials, a simple inexpensive Bridgman-Stockbarger (BS) crystal growth system has been designed and fabricated. Two immiscible liquids have been utilized to create two zones for this crystal growth system. Bulk single crystals of benzil derivative and n-salicylidene-aniline have been successfully grown in this system. The optimum lowering rate has been found to be 0.1 mm/h for the flat interface. Results on the crystal growth and other parameters of the grown crystals are presented.

Significant efforts have been made in the field of organic nonlinear optical materials because of their potential applications such as second-harmonic generation(SHG), frequency mixing and electro-optic modulation. Organic molecules have been reported to have larger nonlinear optical susceptibilities in many cases than those of most inorganic materials within the 0.5-2.0 micron transparency domain (Sutter, 1988). Laser damage thresholds of some organic single crystals are much higher than conventional inorganic single crystals (Nitti, 1993). Most of the organic single crystals are more difficult to grow than inorganic crystals due to their low thermal conductivity, their large supercooling tendencies and thermal instability. Organic single crystals have been grown from the solution (Kotler, 1992), vapor and melt growth (Aggarwal, 1992). Melt growth is superior to solution growth in that the crystal can be grown faster and the grown crystal is free from the solvent inclusions. Materials that have lower thermal conductivity take more time to solidify from the melt. For this reason, organic crystals are grown at a slower rate. A Bridgman-Stockbarger method(Aggarwal, 1993) has been used to grow single crystal of organic materials because of relatively simple geometry of system and ease of controlling growth parameters during the growth processing. The solid-liquid interface shape and location are the important factors in Bridgman-Stockbarger growth system. The interface shapes strongly influence the dislocation, grain size, and dopant concentration during directional solidification (Corriel, 1979; Favier, 1977). Interface shape and position can be controlled by material properties and furnace control such as temperature gradient and lowering rate. Therefore understanding of growth parameters is important for the Bridgman-Stockbarger system to determine the desired interface shape during the process.

Post-processing techniques have been used to determine interface shape and location for the opaque system (Feigelson, 1980; Capper, 1983). They were able to change the interface shape by changing the growth parameters. However, these techniques do not allow the real-time observation and controlling interface shape and location during crystal growth. The x-ray radiographic techniques have also been developed to enable the direct visualization for opaque system (Barber,1986,1995).

To grow novel organic materials, the control of the growth parameters is more advantageous using real time observation. Therefore a transparent furnace and observing system is required for real-time measurement. A simple, inexpensive transparent BS system is designed and fabricated in our laboratory that is described in the next section. The interface shape, location and growth rate are determined as a function of lowering rate to understand influence on growth factors.

II. Experimental set-up

The Bridgman-Stockbarger melt growth system is shown in Fig.1. It consists of a growth chamber, crystal lowering mechanism and temperature control system. A Glass beaker is used as growth chamber that contains two immiscible liquids. Deionized water is used for lower temperature zone as heat transfer medium and silicon oil is used for higher temperature

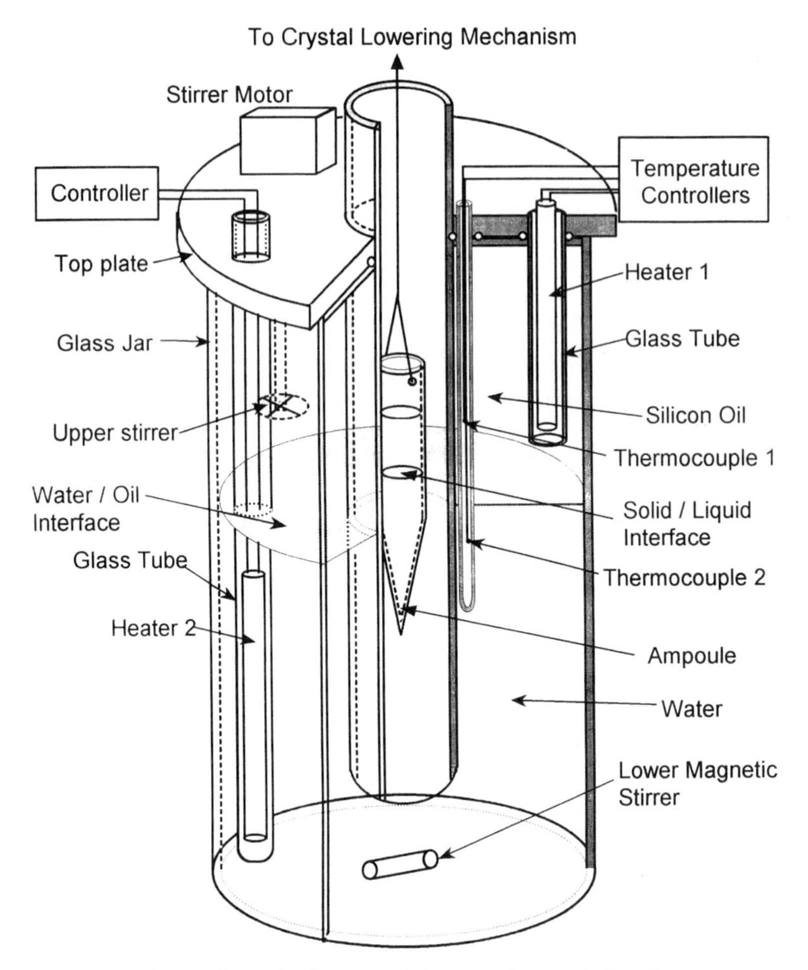

Fig.1 Schematic diagram of the crystal growth chamber

zone. Two tubular shaped heaters that are sealed in glass tubes, fabricated in our laboratory are used for the hot(temperature above the melting point) and cold(temperature below the melting point) zones. Two separate proportional temperature controllers YSI 72 and Eurotherm 818 are used for controlling the temperature of the two zones. The control thermocouples in the two zones are located near the growth ampoule. Both zones are stirred to keep the temperature uniform in the individual zones. A hot plate with stirring function is used as lower zone stirrer and a DC motor is attached at the top of the growth chamber for hot zone stirring. The lowering arrangement is designed and fabricated using Velmex slide that is operated using Hurst stepping motors and gear reducer. Each zone can be set at any desired temperature and the temperature is controlled to within ±0.1°C. For benzil-aniline (melting point 95°C) crystal growth, temperatures of 97.5°C and 82.5°C are set up for hot and cold zones, respectively. The vertical temperature gradient can be obtained up to 25°C to prevent supercooling during growth. Once the seed is formed at the tip of the ampoule then the growth process can be observed through the growth chamber. The lowering rate was controlled by observing interface shape until we could get flat interface. After successful growth of the crystal, the ampoule is slowly cooled at a rate of few degrees per hour. No wetting and cracks induced from difference of thermal expansion between grown single crystal and quartz ampoule were found.

III. RESULTS AND DISCUSSION

Single crystals of benzil, benzil aniline and salicylidene-aniline have been successfully grown in this system. Fig. 2 shows a typical ampoule containing the grown crystal. A typical crystal growth run consists in filling the conical

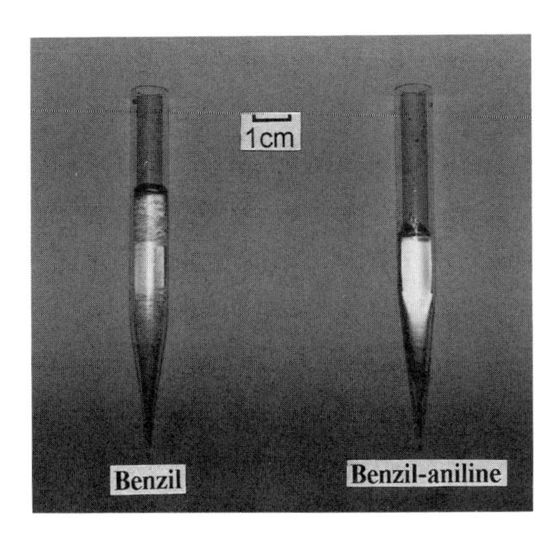

Fig.2 Photograph of benzil and benzil-aniline crystals

ampoule with the organic material, melting it in the hot zone and slowly transporting it to the cold zone thereby crystallizing a small layer of material at a time. Because of the transparency of the system, the temperature of the hot and cold zones are adjusted such that the solid-liquid interface is clearly visible in one of the zones (Fig. 3). The temperature gradient is adjusted to 10 -15°C to initiate the nucleation at the tip of the conical ampoule for benzil and benzil-aniline. However cold zone has to be cooled below the room temperature (~15°C) for nucleation of salicylidene-aniline(melting point 55°C) that has

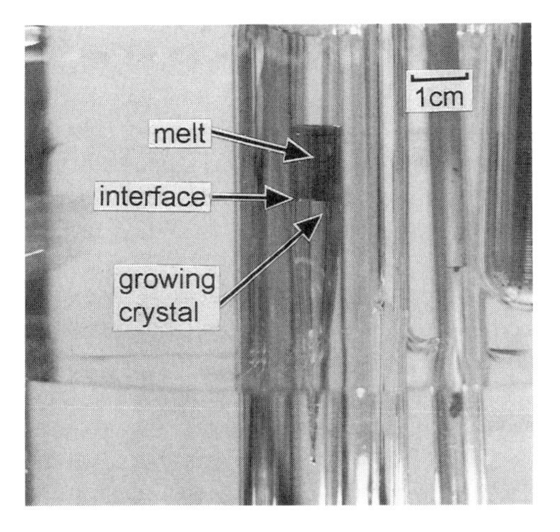

Fig.3 Photograph of a cross-section of the growth chamber showing solid -liquid interface

larger supercooling tendency(> 25°C). Then the cold zone temperature is slowly increased until the flat solid-liquid interface is achieved.

Various lowering rates are used during the crystal growth experiments and the shape of the interface was recorded after viewing through the telemicroscope which is mounted on a x-z translator. Normally the interface is divided in 10 points and the motion of each is monitored as the crystal growth proceeds. It has been found that the faster lowering rate increases the concavity of the solid liquid interface towards the growing crystal. From these experiments the optimum lowering rate has been found to be 0.1 mm/h for the flat interface.

Acknowledgements

The authors gratefully acknowledge the support of the National Aeronautics and Research administration for their grants NAG-125 and MAG-4078 under which this work is performed.

References

Aggarwal, M.D.; Wang, W.S.; Shields, Angela W.; Penn, Benjamin G.; Frazier,Donald O., *Rev Sci. Instrum.* **1992**, 63, 5481-5482.

Aggarwal, M.D.; Wang, W.S.; Choi, J.; Chang, K.J.; Shields, Angela W.; Penn, Benjamin G.; Frazier,Donald O. *Meas. Sci. Technol.* **1993**, 4, 793-795.

Barber, Patrick G.; Crouch, Roger K.; Fripp,Jr., Archibald L.; Debnam Jr., William J., Berry Jr., Robert F. Simchick Richard, J. Cryst. Growth **1986**, 74, 228-230.

Barber, P.G.; Berry, R.F.; Debnam, W.J.; Fripp, A.L.; Woodell, G.; Simchick, R.T. *J. Cryst. Growth* **1995**, 147, 83-90.

Capper, P.; Gosney, J.J.G.; Jones, C.L.; Queleh, M.J.T. *J. Cryst. Growth,* **1983**, 63, 154-160.

Corriel, S.R.; Sekerka, R.F. *J. Cryst. Growth* **1979**, 46, 479-482.

Favier, J.J. and Lesoult, G.; *J. Cryst. Growth* **1977**, 37,76-88.

Feigelson, Robert S.; Route, Roger K. *J. Cryst. Growth* **1980**, 49,261-273.

Kotler, Z.; Hierle, R.; Josse, D.; Jyss, J.;*J. Opt. Soc.Am.* **1992**, B9,534-547.

Nitti, S.; Tan, H.M.; Banfi, G.; Degiorgio, V.; Bailey, R.T.; Cruickshank, F.; Pugh, D.; Shepherd,E.A.; Sherwood, J.N.; and Simpson, G.S. *J.Phys. D: Appl. Phys* **1993**, 26,B225-B229.

Sutter, K.; Bosshard, Ch. and Gunter, P. SPIE **1988,**1017,121-126.

New and Effective Technique to study Supersaturated Solutions by the Pulsed (Spin-Echo) NMR Method

Alexander R. Kessel, Russian Academy of Sciences, Physico-Technical Institute, Kazan 420029, Russia
Aleksey N. Temnikov, Kazan State Technological University, Kazan 420015, Russia
Alexander F. Izmailov and Allan S. Myerson, Polytechnic University, Brooklyn, New York 11201, USA

The specially developed technique of Spin-Echo (SE) measurements is proposed for the determination of solution solubility (binodal line) and its dependence on temperature. Only a few solution samples of known concentrations can be used to obtain phase diagrams in a wide range of temperatures. Determination of any phase diagram state requires only one SE measurement. The SE measurements provide also an opportunity to quantitatively study the formation of solute clusters in supersaturated solutions.

The Nuclear Magnetic Resonance (NMR) is one of the most powerful contemporary methods for the study of molecular structure and mobility in various materials in a condensed state. The NMR consists in the observation of effects of a radio-frequency magnetic field H_1 on nuclear spins placed in a stationary magnetic field H_0. In spite of a specific quantum-mechanical character of interactions in this spin system some of the NMR phenomena can be described classically. Such a description is especially appropriate for the pulsed NMR experiments. In the stationary magnetic field each nuclear spin precesses around H_0 (z-axis) at the resonant frequency $\omega_0 = \gamma H_0$ (γ is the gyromagnetic ratio). Therefore a total macroscopic magnetization M_0 arises along z-axis. A pulse of the resonant radio-frequency magnetic field H_1, applied perpendicularly to H_0, (along x- or y-axis) turns the magnetization M_0 to an angle $\varphi = \gamma H_1 t_p$, where t_p is the pulse duration. In correspondence with the value of the angle φ and direction of the radio-frequency H_1 the applied pulses are called $90°_x$ - pulse, $180°_y$ - pulse and so forth. Usually for the initial $180°_y$ - pulse and so forth. Usually for the initial excitation of spin system the $90°_x$ - pulse is used. This pulse turns the vector M_0 into xy-plane. After switching the pulse off the magnetization starts to rotate in xy-plane with the nuclear resonant frequency ω_0 producing a free induction signal $A(t)$ in a receiving coil of the NMR relaxometer. The signal $A(t)$ is proportional to the magnetization in xy-plane. This signal is observed until the phase coherence in rotation of individual nuclear spins is kept. For a system of equivalent spins placed in the highly homogeneous field H_0 the free induction decay has usually an exponential form:

$$A(t) = A(0)e^{-\frac{t}{T_2}}, \tag{1}$$

where $A(0)$ is the initial amplitude of a signal, which is proportional to the total number of resonant spins in a sample. The time T_2 is the characteristic spin-spin relaxation time known as the transverse relaxation time. Therefore, one can measure $A(0)$ and T_2 observing such a decay.

Unfortunately, precise measurement of T_2 is complicated by inhomogeneity of the external field H_0. This inhomogeneity results in a spatial distribution of the resonant frequency for spins placed in different parts of the sample volume. In addition, there are the fluctuations of the resonance frequency of individual spins due to molecular diffusion. These effects violate the phase coherence and, thus, lead to an additional decay of the free induction signal. To overcome this difficulty the multipulse methods have been developed. Among the numerous refined experimental techniques there is the well-known Carr-Purcell-Meiboom-Gill (CPMG) pulse

1054–7487/96/0266$12.00/0

sequence $90°_x$ - $\tau(-180°_y - 2\tau)_n$ [1,2] which is very convenient for the study of liquids and liquid-like systems (see *Fig.1*). After each $180°_y$ -pulse a Spin-Echo (SE) signal appears. Therefore, by means of this multiple pulse technique it is possible to create such a situation when the decay of amplitudes of the SE signals in an inhomogeneous external magnetic field would coincide with the free induction decay in a homogeneous field (see *Fig. 1*).

Strong connection of the spin relaxation time T_2 with the molecular mobility is a characteristic feature of the NMR. General expression for transverse relaxation time T_2 in liquids is:

$$T_2^{-1} = K\tau_c, \quad \mathbf{(2)}$$

where K is the constant depending on the molecular structure and relaxation mechanism, and τ_c is the correlation time of molecular motion. There is an universal rule: T_2 is longer in systems with faster molecular motion. For instance, T_2 for crystals does not exceed a few tens of a microsecond while for liquids it can be of order of several seconds. In simple liquids the transverse magnetization decay is usually described by the exponential function *A(t)*. In solids the magnetization decay has usually more complicated form given by Gaussian, Pake or Abragam functions [3].

1. Study of Heterogeneous Liquid Systems by Means of the Pulsed NMR

The presented above facts give an idea for a possible application scheme of the pulsed NMR technique to the study of heterogeneous liquid systems consisting of *n* different components. In particular, such systems may include the supersaturated solutions consisting of separate solute molecules and their subcritical aggregates (clusters). The NMR signal of each component has its own relaxation time $T_{2,i}$ ($i = 1,..,n$) since mobility of each component is different. This allows a separation of contributions from different components to the free induction decay. The relative population of each component can be determined by measuring the corresponding initial signal amplitudes $A_i(0)$. For example, if the heterogeneous physical system consists of *n* different components an expression for the free induction decay *A(t)* acquires the form:

$$A(t) = \sum_{i=1}^{n} A_i(0)e^{-\frac{t}{T_{2,i}}}. \quad \mathbf{(3)}$$

The amplitudes of initial signals have to satisfy the condition:

$$A(0) = \sum_{i=1}^{n} A_i(0), \quad \mathbf{(4)}$$

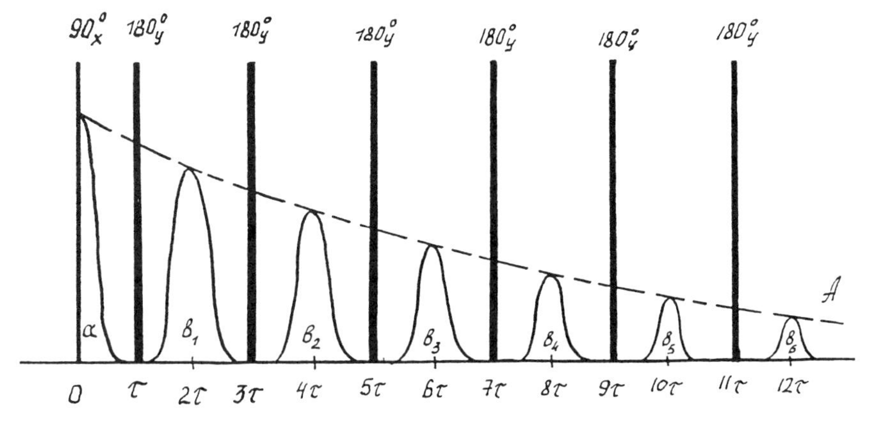

Figure 1. Schematic diagram of the CPMG multiple pulse experiment. The duration of resonant radio-frequency $180°_y$ - pulses is twice greater than that of the $90°_x$ - pulse and significantly shorter than a distance between pulses. ***a*** is the free induction decay in the inhomogeneous external field H_0 ; $\boldsymbol{b_i}$ are the spin-echo signals; ***A*** is the free induction decay in the highly homogeneous field H_0 . The amplitudes *A(t)* correspond to the relaxation of observed spin-echo values. Initial parts of ***a*** and ***A*** coincide.

where $A(0)$ is the initial amplitude of the free induction signal following the first $90°_x$ -pulse. Dividing both part of equation (3) by $A(0)$, one obtains an equation for the relative populations of different components P_i:

$$\frac{A(t)}{A(0)} = \sum_{i=1}^{n} P_i e^{-\frac{t}{T_{2,i}}}, \qquad \sum_{i=1}^{n} P_i = 1, \qquad \textbf{(5)}$$

where $P_i = A_i(0)/A(0)$. This relationship allows the determination of populations P_i $(i = 1,...,n)$ by measuring the experimentally observed quantities $A_i(0)$ and $T_{2,i}$ $(i = 1,...,n)$.

2. *Experimental Procedure for the pulsed NMR*

The measurements of transverse relaxation were performed on the relaxometer operating at the proton resonant frequency *21.5 MHz* by applying the CPMG spin-echo multiple pulse sequence. The repetition time was *15 sec* long to avoid the spin-system saturation and the distance 2τ between the $180°_y$ - pulses was *1 ms*. Samples of urea and citric acid solutions in distilled H_2O were prepared with the solid crystalline powders (by "Reachim" company, purity > 99%). Solutions were first prepared at the lowest temperature of the NMR measurements. Then temperature was increased and samples were kept at a new temperature during 3 hours. After that the next NMR measurements were performed.

3. *The NMR Study of Saturated Solutions*

In *Fig. 2* there are two plottings for the transverse magnetization decay of two aqueous solutions with urea concentrations *40 wt %* and *70 wt %* measured at the temperature *10° C*. The solubility limit for urea in distilled water at this temperature is *45.7 wt%* (*0.202* mole fraction) [3]. Thus, one of these samples (*40 wt %* of urea) was unsaturated whereas an other (*70 wt %* of urea) consisted of two phases: solid crystals of urea and saturated at *10° C* aqueous solution of urea. The transverse magnetization decay for these samples has different forms as it is demonstrated in *Fig. 2.* The experimentally obtained curve for magnetization decay for the

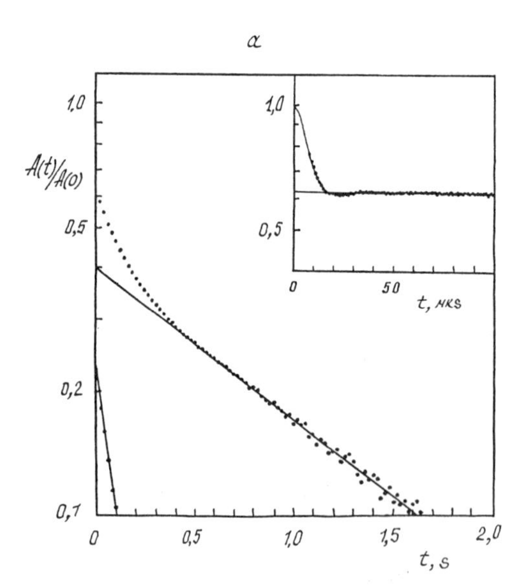

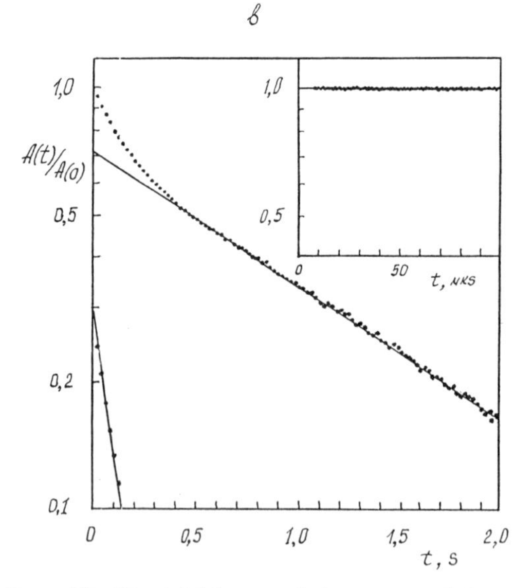

Figure 2. Typical experimental curves of the transverse magnetization decay observed in aqueous solutions of urea (a) above and (b) below the saturation point. Straight lines represent individual components of exponential form (6). The initial part of the presented free induction decay is after the first $90°_x$ - pulse. Relative populations of the magnetization decay are: a) $P_a = 0.40$, $P_b = 0.225$, $P_s = 0.375$; b) $P_a = 0.71$, $P_b = 0.29$, $P_s = 0.00$.

saturated two-phase sample (see *Fig. 2a*) can be described as the following sum of two exponentials and one Abragam function:

$$\frac{A(t)}{A(0)} = P_1 e^{-\frac{t}{T_{2,1}}} + P_2 e^{-\frac{t}{T_{2,2}}} + P_s \frac{\sin(bt)}{bt} e^{-\frac{(at)^2}{2}}. \quad (6)$$

To describe a signal of the less concentrated, unsaturated sample, it is enough to have a sum of only two exponential function (see expression (5) with $n = 2$). In such solutions the fast-relaxing Abragam signal is absent (*Fig. 2b*) since it describes a contribution due to solid phase of urea. Therefore, the first two terms of expression (6) which are also observed in an unsaturated solution (*see Fig.2b*) may be attributed to the protons of water molecules with the relative population P_1 and the protons of solute molecules in liquid phase with the relative population P_2. The signal proportional to P_2 includes contributions from single solute molecules dissolved in the solution and their subcritical aggregates. The relaxation time of solute protons $T_{2,2}$ is shorter than the relaxation time of water protons $T_{2,1}$ due to two reasons. First, the mobility of solute molecules is less (correlation time is longer) than the mobility of water molecules. And second, urea protons interact strongly with the fast-relaxing nitrogen quadruple nuclei (K value in (2) is greater).

The experimentally obtained curves of magnetization decay were fitted by means of expressions (5) and (6) employing the least squares method. This allowed determination of the relative populations P_1, P_2, P_s as well as the relaxation times $T_{2,1}$ and $T_{2,2}$ for a number of urea concentrations and solution temperatures. The same experiments were also carried out with aqueous solutions of citric acid.

4. *Determination of Solubility by Means of the NMR*

One of the interesting problems which can be solved by means of the NMR is the determination of phase (solubility) diagrams. To carry out this study it is necessary to determine the distribution of dissolved solute molecules among different sample phases at various temperatures and concentrations. The fractions of water molecules F_1, solute molecules in liquid phase $F_{2,liq}$ and solute molecules in solid phase $F_{2,sol}$ can be determined from the relative intensities of the transverse magnetization decay components P_1, P_2 and P_s. To perform this one has to take into account that a water molecule contains two protons whereas a urea (solute) molecule - four. The final results for an arbitrary urea concentration c is:

$$F_1 = \frac{2P_1}{1 + P_1},$$

$$F_{2,liq} = F_{2,liq}(c) = \frac{1}{2}P_2(1 + c),$$

$$F_{2,sol} = F_{2,sol}(c) = \frac{1}{2}P_s(1 + c).$$

In *Fig. 3a* the concentration dependence of the function $R(c) = [1 + F_1/F_{2,liq}(c)]^{-1}$ is presented. This quantity gives a fraction of urea molecules in liquid phase. As it follows from our experimental observations for urea aqueous solutions (see *Fig. 3a*) the qantity $R(c)$ increases linearly with urea concentration. When the concentration c reaches the solubility limit $c = c_{sat}$ (*0.202 mole* fraction) the growth of the quantity $R(c)$ ceases. Further increase of solution concentration leads to the formation of growing urea crystalline phase leaving an absolute amount of urea molecules dissolved in saturated solution unchanged. The growth of crystalline phase is demonstrated in *Fig. 3b* where the concentration dependence of $F_{2,sol}(c)$ is presented. As it follows from this figure the solid phase appears just after the crossing of the saturation point ($c=c_{sat}$) and increases linearly with an amount of solute in an equilibrium mixture *"saturated solution + solute crystals"*.

Such a remarkable possibility to measure populations of various phases by the pulsed NMR technique allows the following effective experimental method for the determination of solution phase (solubility) diagrams. The main idea of this method is based on a fact that the experimentally observed dependence of $F_{2,sol}(c)$ on the solute concentration c is given by a straight line which always passes through the points with the coordinates ($c = c_{sat}$, $F_{2,sol}(c_{sat})=0$) and ($c = 1$, $F_{2,sol}(1) = 1$) for $c>c_{sat}$. So for the determination of the solubility

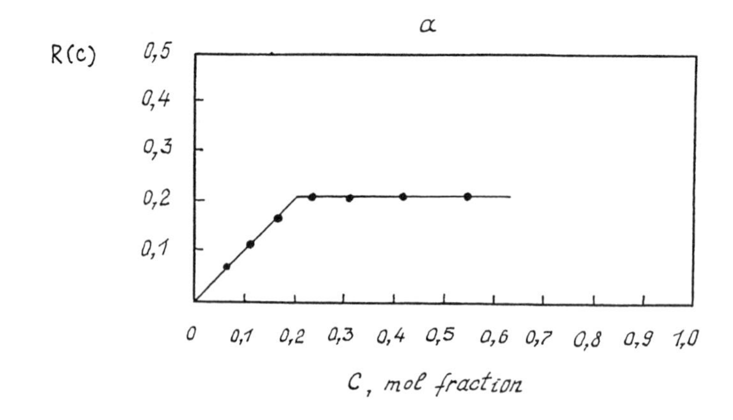

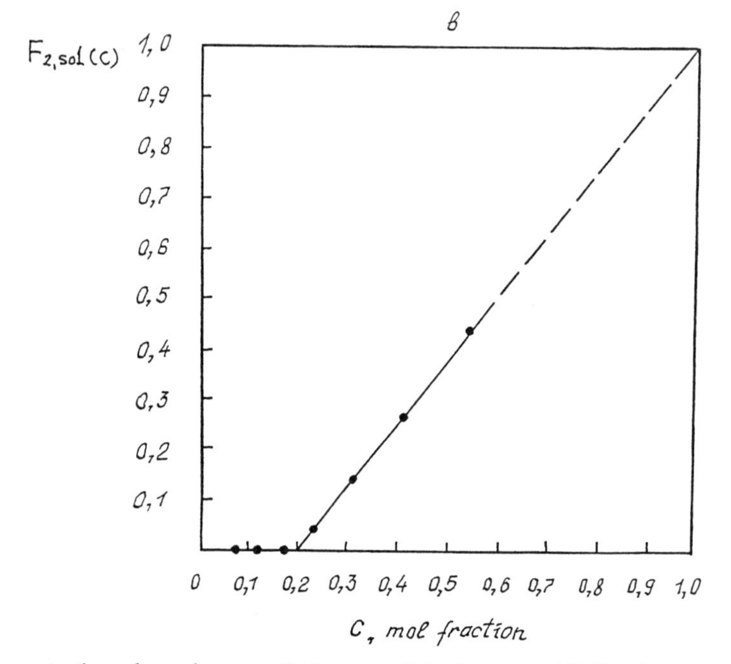

Figure 3. Concentration dependences of the relative fractions of urea molecules in different phases are given: (a) in liquid phase and (b) in solid phase at $10°C$. The solubility for this temperature is $c_{sat} = 0.202$ molal fraction [4].

c_{sat} it is absolutely enough to measure only once the fraction of solute molecules in solid phase $F_{2,sol}(c^*)$ for the mixture *"saturated solution + solute crystals"* with the solute weight fraction $c^* > c_{sat}$. The straight line which passes through the points ($c = c^*$, $F_{2,sol}(c) = F_{2,sol}(c^*)$) and ($c=1$, $F_{2,sol}(1) = 1$) intersects c-axis exactly at $c = c_{sat}$, i.e. in the point with the coordinates ($c=c_{sat}$, $F_{2,sol}(c_{sat}) = 0$). Simple geometrical consideration allows the following expression for the solubility c_{sat}:

$$c_{sat} = \frac{c^* - F_{2,sol}(c^*)}{1 - F_{2,sol}(c^*)}. \tag{7}$$

This expression implies that the determination of the solubility c_{sat} requires only one measurement of the pulsed NMR signal for any solute concentration $c^* > c_{sat}$.

In the case when measurement of the fast-relaxing component is not possible due to technical reasons the alternative version of this method can be suggested. From the normalization condition $F_1(c) + F_{2,liq}(c) + F_{2,sol}(c) = 1$ it follows that:

$$F_{2,sol}(c) = 1 - [F_{2,liq}(c) + F_1(c)] = 1 - L(c), \quad \textbf{(8)}$$

where $L(c)$ is the relative amount of molecules in liquid phase. For the determination of the solubility c_{sat} under these conditions one additional solution sample is needed. There are the following two main requirements imposed on this sample: first, it should contain the same amount of protons as in the initial experiment with the mixture *"saturated solution + solute crystals"* and, second, all protons should be in liquid state. In particularly, pure water can serve as a second sample. In this case the water mass m_w can be calculated by means of the following simple expression:

$$m_w = 3m_{sample}\frac{1 + c^*}{3 + 7c^*}, \quad \textbf{(9)}$$

where m_{sample} is the mass of solution sample with the urea concentration $c^* > c_{sat}$. The value of $L(c^*)$ is determined by measuring the initial amplitudes $A(0)$ and $A_w(0)$ of signals corresponding to liquid phase of urea solution and pure water:

$$L(c^*) = \frac{A(0)}{A_w(0)}. \quad \textbf{(10)}$$

Thus, by means of expressions (7) and (8) the quantities c_{sat} and $F_{2,liq}(c^*)$ can be determined.

The second method has a number of strong advantages as compared with the first one. These advantages, which appear due to the complete analysis of the total magnetization decay, are: (1) There are no restrictive requirements to characteristics of the relaxometer transmitter and receiver. This is very important since for the reliable determination of the fast-relaxing solid component the $90°_x$ - pulse must be extremely short (~ *1 mks*) and the receiver has to be capable of resuming its sensitivity very rapidly after the pulses overload. At the same time the $90°_x$ - pulse duration can be of order of *5-10 mks* and the receiver dead time can be of order

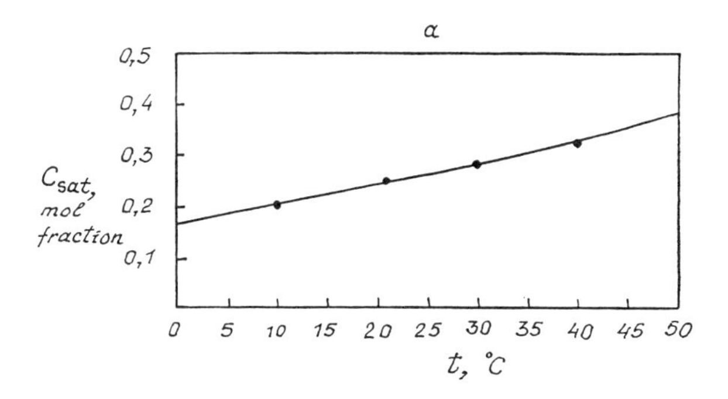

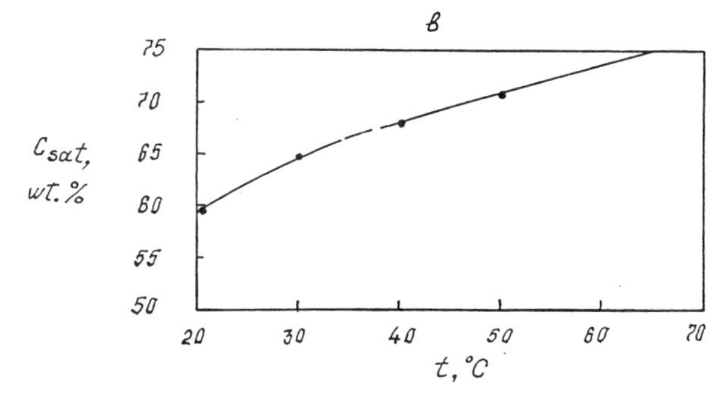

Figure 4. Temperature dependences of c_{sat} for aqueous (a) urea and (b) citric acid solutions. Solid lines corresponds to literature data [4].

of *150 mks* for the measurements of $A_{sample}(0)$ and $A_w(0)$ in liquid phase. (2) The time loss for signal accumulations is avoided. This is also very important since the signal from solid phase in two-phase mixtures is very weak and demands accumulation in the case when solute amount is close to that corresponding to the solubility concentration c_{sat} . (3) There are no time losses for the analysis of the complicated experimental decay curve for solid fraction. (4) The accuracy of determination of the solubility c_{sat} is much higher since experimental measurements of the initial amplitude of slowly relaxing signals and mass of liquid sample are more precise than the calculated approximations of the signal form for the fast relaxing solid component.

In *Fig. 4* our results for the solubility c_{sat} are presented for the urea and citric acid aqueous solutions in a wide range of temperatures. Comparisons of the obtained results with the solubility diagrams drawn from the table data [3] are presented. The deviation of the obtained results for c_{sat} from the cited data does not exceed 2%.

Let us emphasize in conclusion the most significant advantages of the suggested pulse-NMR technique: (1) Only a few samples of the mixture *"saturated solution + solute crystals"* are required to obtain the phase (solubility) diagram in a wide range of temperatures. (2) The total time of measurements is determined mainly by the time T_{eq} needed for the establishment of thermodynamic equilibrium under the change of sample temperature. The time of the NMR measurements is negligible compared with $T_{eq.}$

The authors are deeply grateful to the International Soros Foundation (A.R.K. and A.N.T., ISF Grant NNK000) and NASA (A.F.I. and A.S.M., NASA Grant NAG8-960) for financial support of this investigation.

References:

1. Carr H.Y.; Purcell E.M., Phys. Rev., **1954,** *94,* 630.
2. Meiboom S.; Gill D., Rev. Sci. Instrum., **1958,** *29,* 688.
3. Abragam A., *The Principles of Nuclear Magnetism*, Claredon Press: Oxford,1961.
4. *Solubility Data Series*, Pergamon Press: Oxford, 1965.

Diffusion Coefficient and Viscosity in Crystal Growth in Microgravity

Michael Bohenek and Allan S. Myerson, Department of Chemical Engineering, Polytechnic University, 6 Metrotech Center, Brooklyn, NY 11201

The diffusion coefficient and viscosity of supersaturated water solutions of three crystalline substances were measured. The diffusion coefficient data were used to model the concentration gradients near the crystal for one of the substances.

The growth of crystals in microgravity holds the promise of producing large defect-free crystals. This is because the absence of gravity eliminates gravity driven convection flows at the crystal-liquid interface. Yoo et al. (1986) studied the growth of triglycine sulfate (TGS) in microgravity. They were unable to match their observed concentration profiles around the growing crystal to those predicted by a computer model. One reason for this was the lack of data for the diffusion coefficient and viscosity in the supersaturated region.

It is the purpose of this work to measure the diffusion coefficient and the viscosity of supersaturated solutions of TGS, KDP(potassium dihydrogen phosphate), and ADP(ammonium dihydrogen phosphate) in water.

Diffusion Coefficient

The diffusion coefficient was measured by means of a Guoy interferometer(Sorrell,1982). **Fig. 1** shows data for KDP from a mass fraction of around .05 to .21, which is about 5% above the saturation limit of .20 (Handbook, 1965). **Fig. 2** shows the diffusion coefficient of ADP at from .05 to .295, just above the saturation limit of .29(Linke, 1965). **Fig. 3** shows the diffusion coefficient of TGS from zero concentration to a mass fraction of .255 just above the saturation

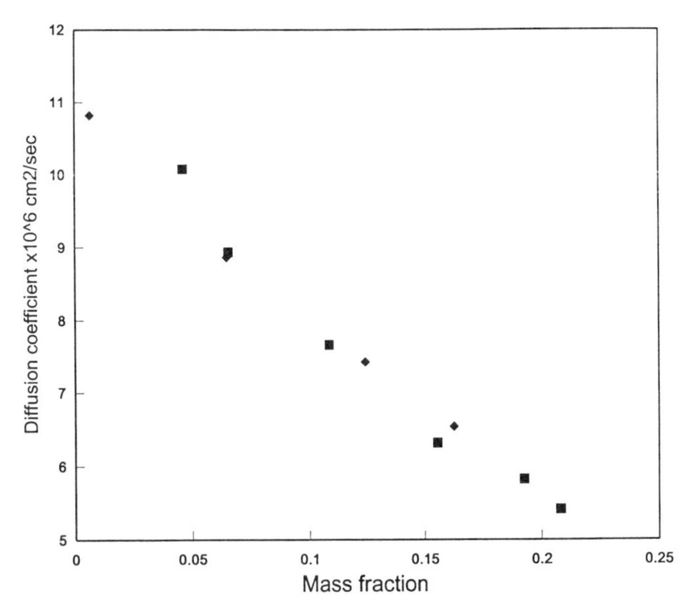

Fig. 1 Diffusion coefficient of KDP in water as a function of mass fraction. saturation=.20 Squares are present data. Diamonds are from Hatfield (1966)

1054–7487/96/0273$12.00/0

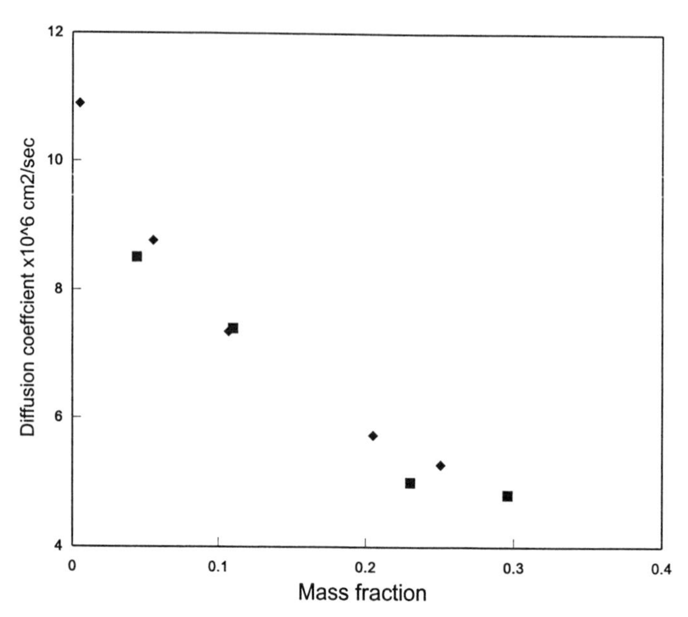

Fig. 2 Diffusion coefficient ADP in water as function of mass fraction. saturation=.29 Squares present data Diamonds are from Hatfield (1966)

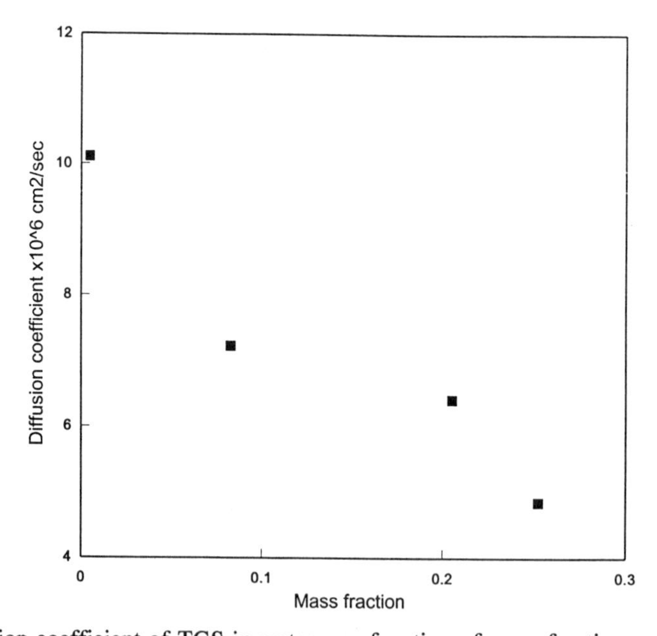

Fig. 3 Diffusion coefficient of TGS in water as a function of mass fraction. saturation=.24

of about .24(Kroes, 1986). Data for most points represent an average of several experiments. For KDP and ADP previous data(Hatfield,1966) are also plotted. For all three substances the diffusion coefficient is about the same order of magnitude, starting at 10^{-6} and declining to about half that value near saturation. The decline with concentration can be approximated by a square root dependence, as suggested by Hatfield. Therefore a square root dependence will be used in the numerical model described later.

It has proven difficult to measure very far into the supersaturated region. The experiments that have successfully penetrated the supersaturated region show the diffusion coefficient to be a continuation of the undersaturated trend. There is no evidence yet of a steep downward trend such as that found for NaCl and KCl(Chang, 1982).Usually high-quality crystals are grown with only a few degrees of supersaturation, so data in this narrow range should still be useful.

Viscosity

The viscosity data were obtained using a Canon capillary viscometer. **Figs. 4-6** show the kinematic viscosity of water solutions vs. concentration for all three substances. The data cover from infinite dilution to supersaturation. The supersaturations obtained here are higher, about 10% for ADP and TGS solutions and 15% for KDP solutions. Higher supersaturations are more easily obtained with the viscometer than the Guoy interferometer. ADP solutions are by far the most viscous, being 150% more viscous at the upper limit of the measured range than at infinite dilution. TGS solutions are about 80% more viscous at the upper limit and KDP solutions around 60% more viscous.

In all three cases, the viscosity monotonically increases with the concentration. The slope of the viscosity vs. concentration line increases with concentration as well, so the slope in the supersaturated regions is very steep. There is no sudden change of slope at saturation suggested by the data.

Numerical Analysis

Due to the experiments unusual location(in space) a very important simplification is allowed in the analysis of crystal growth. Convection is negligible and can be ignored. A rough model for the growth of TGS has been formulated to show in the simplest possible way the effects of the change in diffusion coefficient. Computations were made only for TGS. KDP and ADP have similar diffusion coefficients, so the results should be comparable. The model used was a simple finite difference scheme, mapped in Cartesian coordinates. A one-dimensional profile serves to demonstrate the point adequately. The equation for diffusive flux contains an additional term proportional to the derivative of diffusivity, besides the fact that the diffusion coefficient is now a function of concentration.

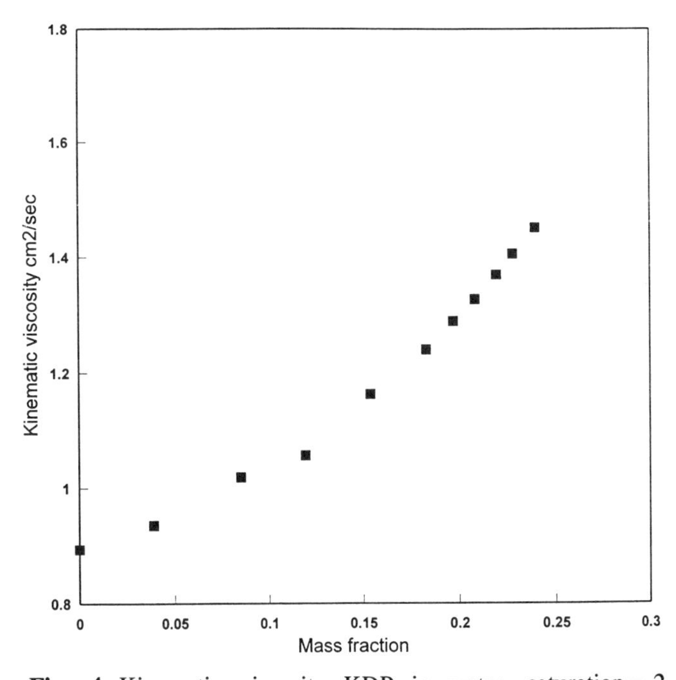

Fig. 4 Kinematic viscosity KDP in water. saturation=.2

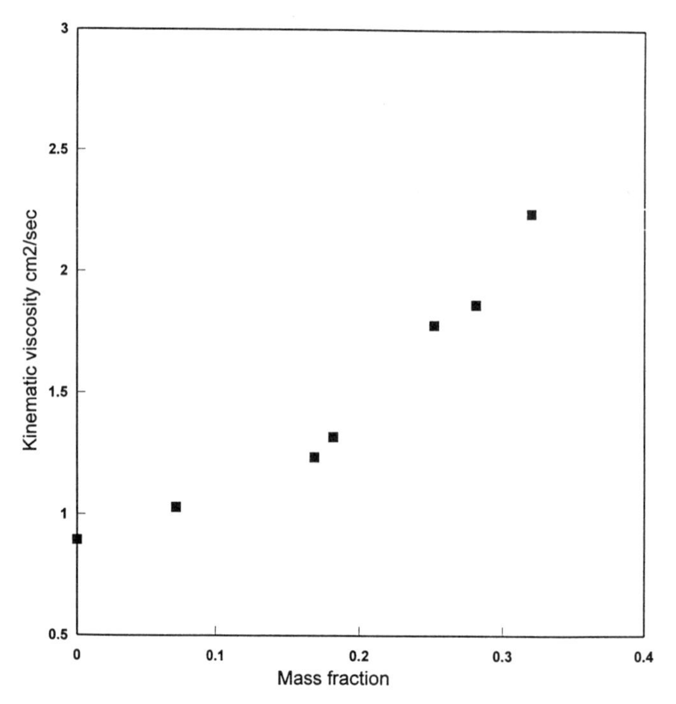

Fig. 5 Kinematic viscosity ADP in water. saturation=.29

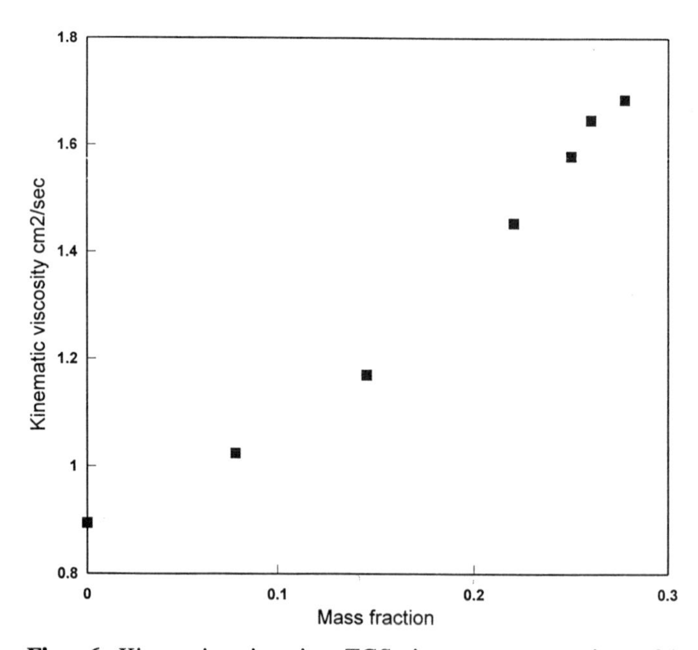

Fig. 6 Kinematic viscosity TGS in water. saturation=.24

$$\frac{dc}{dt} = D(c)\left(\frac{d^2c}{dx^2}\right) + \frac{dD}{dc}\left(\left(\frac{dc}{dx}\right)^2\right) \quad (1)$$

At one end the crystal is growing according to a boundary condition simplified from that found in Yoo(1986). The diffusive flux into the surface, adjusted for surface motion, is proportional to the supersaturation. D is now variable, while the growth rate is expressed in concentration terms for simplicity.

$$A(c - c_{eq}) = \frac{D(c)\frac{dc}{dx}}{Z_c(1 - c/Z)} \quad (2)$$

Zc is the molar volume of crystal. Z is the molar volume of the solution. A is a surface velocity(cm/sec) proportional to supersaturation. These values are derived from those given in Nadarajah(1990). The diffusion coefficient is taken to be a square root correlation in molarity equal to $10^{-6}(10.7-6.2\sqrt{c})cm^2/sec$. This fits the data with a standard deviation of 13%. The other boundaries consist of solid walls with zero mass flux. The initial concentration is set to some high value and is allowed to decay. **Fig.7** is for a supersaturation in molar of 22%, which corresponds to about 10C. **Figs. 8** is for 111% or about 50C. The figures show the concentration profile extending outward from the crystal. For short times the crystal grows only a small amount, therefore the boundary motion itself can be ignored. Each graph also shows profiles for two constant diffusivities, the average over that region and the diffusivity at infinite dilution. For the lower initial supersaturation, there is negligible difference in the concentration profile between the variable diffusivity case and the case where the diffusivity is constant at the average value. There is some effect of variable diffusivity at very high supersaturations. At all concentrations the values are significantly different from the values computed by taking the diffusion coefficient at infinite dilution. Usually for growing single high quality crystals the supersaturation is equivalent even less than 10C. Therefore, the diffusion coefficient changes very little over the concentration range.

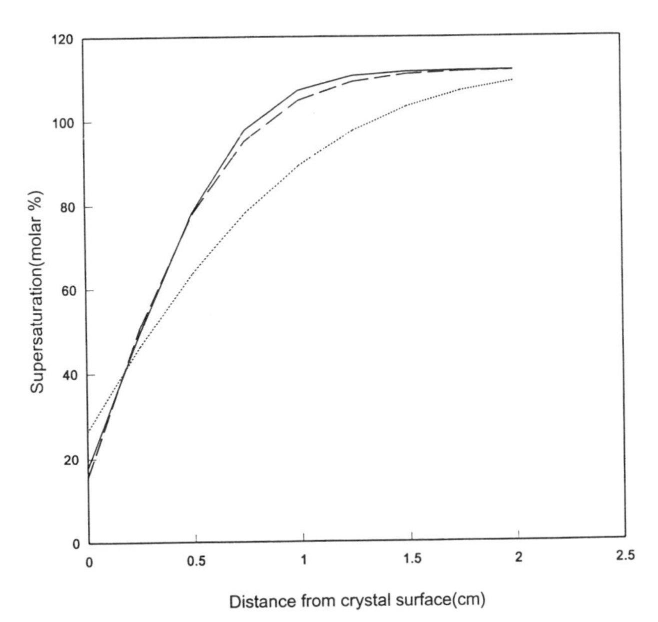

Fig. 7 Calculated supersaturation profile near crystal for TGS solution with initial supersaturation=112% at 12 hours Straight line-variable diffusivity Long dashes-constant diffusivity at average concentration (3.7E-6 cm2/sec) Short dashes-constant diffusivity infinite dilution(10.8E-7)

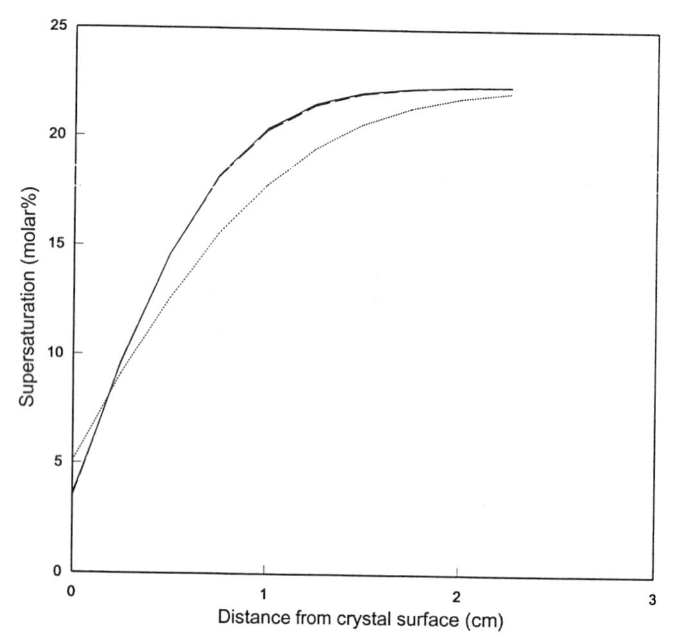

Fig. 8 Calculated supersaturation profile near crystal for TGS solution with initial supersaturation =23% at 12 hours Straight line-variable diffusivity Long dashes-constant diffusivity at average concentration(4.8E-6) Short dashes-constant diffusivity infinite dilution (10.8E-6)

Conclusion

The diffusion coefficient and viscosity of supersaturated solutions of KDP, ADP and TGS are significantly different than those of low concentration solutions. The variation with concentration must be included in any analysis of crystal growth.

Acknowledgement

Support of this work through NASA grants NA68-960 and NAG8-1060 is gratefully acknowledged.

References

Chang,Y.C.;Myerson,A.S. *AIChE Journal* **1982**, 28(5), 772.

Linke,William F.;Siedell,Atherton *Handbook of Solubilities of Inorganic and Metal-Organic Compounds*; ACS:Washington,D.C., 1965.

Hatfield,J.D.;Edwards,O.W.;Dunn,R.L.,*J. Phys. Chem.* **1966**,70(3),2554.

Kroes,R.L.;Reiss,D.;*J. Crystal Growth*, **1986**, 65,414

Nadarajah,A.;Rosenberger,F.;Alexander J. *J. Crystal Growth*, **1990**,104,218.

Sorrell,L.;Myerson,A.S. *AIChE Journal*, **1982**, 28, 772.

Yoo,Hak-Do;Wilcox,W.R.;Lal,Ravindra; Trolinger,James D., *J. Crystal Growth*, **1986**, 75, 591.

INDEX

Index

A

Ab initio approach, crystal structure determination, 22–27
p-Acetamidobenzoic acid, incorporation in acetaminophen crystal lattice, 101–102
Acetaminophen
gel crystallization, 97, 99
incorporation of structurally related substances, 95–104
molecular modeling, 96
molecular structures of related additives, 101*f*
morphology, 80*f*, 81*f*, 97
solvent effects on crystal habit, 78–84
structural studies on individual crystal surfaces, 99
unit cell, 100*f*
Acetanilide (AA), incorporation in acetaminophen crystal lattice, 102
Acetic acid, additive in caprolactam study, 60–63
Acetone
acetaminophen crystals, 80*f*, 81*f*
solvent polarity, 83–84
Acetoxyacetanilide, incorporation in acetaminophen crystal lattice, 102, 103
Achiral molecules
generation of chirality, 239–245
stereoview of crystals, 244*f*
Acridine, generation of chirality, 239–245
Additive(s)
crystal growth rate and morphology, 44, 53
incorporation in acetaminophen crystal lattice, 99, 100*f*, 101–102
segregation, 103
Additive bonding analysis, naphthalene with biphenyl additive, 49
Adipic acid crystallization
agglomeration, 146–148
batch cooling, 145–150
crystal quality, 146, 147*f*, 148*f*
experimental conditions and results, 146*t*
Agar, effects on nucleation and habit modification, 103
Agglomeration
adipic acid crystallization, 146–148
batchwise crystallization of adipic acid aqueous solution, 145
Agitation
crystallization of LCT, 185
effect on nucleation and average crystal size, 191
effect on stearin yield and dropping point, 190–191
Alkanoic acids, binding energies, 55
Alkyl chain length, crystal quality, 132
Alprazolam
crystal structure and habit, 85–94
molecular formula, 87*f*
X-ray powder diffraction patterns, 89*f*, 90*f*
2-Amino-5-nitropyridinium dihydrogen phosphate, crystals, 260*f*
Ammonium dihydrogen phosphate
diffusion coefficient and viscosity of supersaturated solutions, 273–278
diffusion coefficient as function of mass fraction, 274*f*
kinematic viscosity, 276*f*
Ammonium sulfate
crystallization concentration, 252*t*
effect of chromium ion in solution, 249–255
water activity, 254*f*,*t*
Amorphous polymers, crystallization, 196
Amphiphillic molecules
emulsifiers for pure *m*-CNB in water, 128
template for *m*-CNB nucleation, 134–135
Amylose and amylopectin, starch granules, 196
Analgesic, acetaminophen, 78–84
Aspect ratios, crystals prepared by evaporation or cooling, 97*t*, 98*f*
Atom–atom bonds
identification, 49
naphthalene–biphenyl, 51*t*
naphthalene–naphthalene, 52*t*
Attachment energy
biphenyl additive in naphthalene lattice, 47*t*
calculation, 32–33
caprolactam and additives, 60–61*t*
connected nets, 29
crystal morphology calculation, 7, 43–44
definition, 29
growth morphology of crystals, 36
habit predictions, 79
predicted acetaminophen crystal morphologies, 82*f*

B

Batch cooling crystallization, adipic acid aqueous solution, 145–150
Benzamide, benzoic acid as additive, 45–46
Benzil and benzil aniline, growth of single crystals, 264–265
Benzoic acid
additive in caprolactam study, 60–63
additive to benzamide, 45–46
computation of F faces, 30–32
growth habit of caprolactam with benzoic acid, 61*f*, 62*f*
PBC analysis, 28–35
visualization of a connected net, 29*f*
Benzophenone, toluene solution vs. morphology, 8, 9*f*
Binding energy
alkanoic acids on surfaces of β-succinic acid, 55*t*
biphenyl additive in naphthalene lattice, 47*t*
crystal surfaces and additives, calculation, 53–55
molecular weight of alkanoic acids, 55
Bioavailability, polymorphic form, 70
Biphenyl
additive to naphthalene, 47, 48*f*
crystal shape, 40–41
growth morphology of crystals from toluene solution, 40–41
interatomic interactions, 51*f*
intermolecular bonding analysis, 49, 50*t*
Blocker additive
biphenyl in naphthalene crystal, 48*f*
calculation, 47
crystal faces, 8
modification of crystal habit, 101
Bond energy
calculation models, 18–19
interface between crystal and mother phase, 17
Bonding analysis, calculations, 48–52
Bonds in crystal graph, monomers and dimers, 31*t*
Bridgman–Stockbarger crystal growth system, organic materials, 263–265
Bulk pharmaceutical chemicals, regulatory considerations in crystallization processes, 66–71
Bulk water, glucose solution, 180
Butter fat, fractionation by crystallization, 185–195

C

Cambridge Crystallographic Database, intermolecular bonding patterns, 5–7
Caprolactam
comparison of theoretical and experimental results, 63*f*
crystal habit modification in presence of carboxylic acids, 59–63
habit in presence of benzoic acid, 61*f*, 62*f*
modeling of crystal habit modification, 60
Carboxylic acids
crystal habit of modification of caprolactam, 59–63
influence of hydrogen bond strength on caprolactam, 59–60
S-Carboxymethylcysteine (SCMC)
composition of grown parts of seeds, 108
concentration vs. operation time, 106*f*, 107*f*
crystal growth rate, 72–76, 107–108
crystal shape, 73, 77*f*
effect of pH on solubility, 72–77
operational conditions for observation of solubility, 72*t*
optical purity vs. operation time, 108*f*, 109*f*
optical resolution in cooled type batch operation, 105–109
Carminic acid
chemical structure, 179*f*
electronic spectrum influenced by pH, 178
emission spectra, aqueous solutions, 181*f*, 183*f*, 184*f*
fluorescent probe of sugar solution composition, 178–184
sucrose concentration, 180
Carr–Purcell–Meiboom–Gill
liquid systems, 266–272
multiple pulse experiment, 267*f*
Chamber solvent vapor, dependence of relative humidity on solute concentration, 229*f*
Chemical potential, driving force of crystallization, 157
Chemistry review, crystallization, 68–70
Chiral drugs
crystallization, 67–68
regulatory concern, 66
Chiral molecules
crystal data of X-ray structure analysis, 240*t*
diphenylacetic acid and 2,2-diphenylpropionic acid, 244*f*
host–guest molecules in inclusion crystal, 246–248
spontaneous crystallization, 239–245
stereoview of crystal, 244*f*
Chiral seed, crystallization of racemic drug, 68
Chirality
generation in molecular crystals composed

of two different achiral molecules, 239–245
source, 239–241
2-Chlorobenzoic acid, additive in caprolactam study, 60–63
Chloronitrobenzene (CNB), crystallization of meta and para isomers, 127–136
m-Chloronitrobenzene (*m*-CNB)
crystallization from component mixtures, 128
crystallization temperature of emulsion, 133*t*
growing crystals, 130*f*
transfer from melt droplets to crystals, 133
Chromium(III) ion
effect on nucleation of supersaturated ammonium sulfate solutions, 249–255
final concentration in ammonium sulfate solution, 252*t*
Cinnamic acids, *m*-CNB templates, 135
Circuit boards, EDTA precipitation from solutions used in electroless copper deposition, 122–126
Citric acid, temperature dependence of concentrated solution, 271*f*
Cloud point, normal aklane precipitation, 151
Cocoa butter
classification of crystalline forms and melting points, 211*t*
crystallization and polymorphism, 209–215
major triglycerides, 209
six crystalline forms, 210
X-ray diffraction data, 213*f*–214*f*
Cohesive energy, *See* Lattice energy
Complex organic compounds, crystal growth kinetics, 116–121
Computational chemistry, application to study of molecular materials, 2–14
Conglomerate crystallization
definition, 67
racemic chiral drugs, 68
Connected net
crystal morphology, 16
F slice, 29
order–disorder phase transition, 18
reasons for multiplicity, 20
roughening transition, 17–18
user-friendly software, 19–20
Constant crystal dissolution, sucrose crystals, 217
Containerless levitation of microdroplets, no dust impurities and dirt, 222
Continuous pulling method, crystal growth from vapor, 232*f*
Cooled type batch crystallizer, concentration of L-SCMC, 105–109
Cooling curves, LCT without agitation, 187, 188*f*
Cracks, crystalline layers, 114
Crystal
torsion angles, 242*t*
types of faces, 28
Crystal binding energy, *See* Lattice energy
Crystal chemistry, molecular materials, 2–3
Crystal faces
effect of impurities on reduction of growth rate, 201–203
hydrogen bonding potential, 82–84
Crystal graph
dimer and monomer, 31*f,t*
set of centers of crystal mass and bonds, 28–29
Crystal growth
effects of impurities on kinetics, 200
influence of fluid phase, 36–42
kinetics, 159–161, 200
L-SCMC seeds in a DL-SCMC solution of pH 0.5, 72–77
microgravity, 273–278
numerical analysis, 275–277
operational conditions for tests, 73*t*
solute diffusional flux, 226
thermodynamics of metastability, 222–230
Crystal growth chamber, schematic diagram, 264*f*
Crystal growth kinetics
complex organic compounds, 116–121
too fast, 113
use of supersaturated solutions in determination, 157–162
Crystal growth modification, structurally related additives, 95
Crystal growth system, Bridgman–Stockbarger, 263–265
Crystal habit
alprazolam, 85–94
based on dimer PBC analysis, 35
modeling, 79–80
modification by caprolactam in presence of carboxylic acids, 59–63
theoretical and experimental, 34–35
Crystal lattice
molecular arrangements, 2
morphological simulations, 7
Crystal materials, inclusions, 163–170
Crystal morphology, attachment energy methods, 78–84
Crystal packing patterns, construction stages, 11
Crystal perfection
assessment, 232
dislocations, 233–234
Crystal purity
crystallization temperature, 129*t*
droplet size, organic and surfactant weight fraction, 132*t*
Crystal shape
calculation, 7–8
change with growth, 76*f*
changing growth rates, 207
presence of impurities, 207
Crystal size

agitation, 193*f*
crystallization procedure, 149
flow rate, 192*f*
temperature, 191*f*
time, 118*f*
Crystal structure
ab initio approach to determination, 22–27
molecular modeling, 25
prediction, 10–13
single crystal and powder analysis, 13*f*
solution from powders, 12–13
temperature effect, 124
Crystal surfaces and additives, binding energy, 53–55
Crystalline layers, structure, 112–113
Crystallization
cocoa butter fat, 209–215
D- and L-SCMC, 107–108
driving force, 157, 158
emulsion solidification, 127–136
gelatinized starch, 196
L-SCMC in DL-SCMC supersaturated solution, 105–109
mechanism, 134
molecular crystals between nitroanilines and nitrophenols, 256–258
New Drug Application, 70
optimization, 28
process controls, 68
secondary recovery procedures, 66–67
starch, 196–199
two unavoidable processes, 115
Crystallization energy, growth morphology of crystals, 36
Crystallization in foods, control, 172–177
Crystallization inhibitor, CNB para isomer, 129
Crystallization kinetics
expression, 158–159
LCT, 191
milk fat, 137
nucleation and growth, 173–174
Crystallization procedures, FDA regulation, 66–71
Crystallization process, growth of the layer, 140–141
Crystallization rates, organic weight fraction and droplet diameter, 131
Crystallized starch, peak temperatures of melting endotherm, 198–199
Crystallographic orientation factor, definition, 36
Crystallographic stability, diltiazem hydrochloride, 85–94

D

Dendritical growth
efficiency of separation by crystallization, 163
melt crystallization, 164*f*
Density
free energy, 68
role of concentration dependence in crystal growth, 223–227
Designer impurities, *See* Tailor-made additives
Dewaxing, lubricating oil basestocks, 151
6,13-Dichlorotriphendioxazine
bond lengths, 26*t*
crystal structure determination, 23–25
nonhydrogen atomic coordinates, 26*t*
results of Rietveld refinement, 24
Differential scanning calorimetry (DSC), stearin and olein fractions, 187
Diffusion boundary layer, supersaturation and thermodynamics of metastability, 222
Diffusion cell, schematic description, 165*f*
Diffusion coefficient
growth of crystals in microgravity, 273–278
saturation limit, 273–275
Diffusion washing, postcrystallization treatment of crystalline layers, 114
Diltiazem hydrochloride
crystal structure, 86*t*
crystallographic stability, 85–94
molecular formula, 87*f*
molecular packing, 93*f*
X-ray powder diffraction patterns, 92*f*
Diphenylacetic acid, generation of chirality, 239–245
Dislocation, introduced during crystal growth, 234
Dislocation theory, elastic energy, 237
Disruptor additive
calculation, 45–46
growth of crystals, 8*f*
intermolecular bonding sequence, 7–8
metacetamol, 102
modification of crystal habit, 101
Dissolution kinetics, single sucrose crystals, 216–219
Dreiding force field, dimer and monomer morphology, 33*f*
Driving force
crystallization, 157, 158
dimensionless supersaturation, 159*f*
Dropping point, selection of fats in food-fat applications, 187, 190–191
Drops, crystallization, 133–134

E

Electrodynamic levitation

electrically charged microdroplets of supersaturated solutions, 222
properties of aqueous solution at high concentration, 249
Electroless copper plating, EDTA recovery process, 122–126
Emission peak intensity ratio, carminic acid emission in glucose and buffer solutions, 181*f*, 182*f*
Emulsifiers, effect on crystallization of *m*-CNB, 132
Emulsion composition
droplet size, organic and surfactant weight fraction, 132*t*
time dependence, 131*f*
Emulsion solidification, purification and crystallization, 127–136
Enantiomeric crystals, ratio in a conglomerate, 68
Etch pits
distribution around indentations, 236*f*
habit faces of C_{60} crystals, 234*f*
typical distributions, 234
Ethanol
acetaminophen crystals, 80*f*
crystallization of β-succinic acid, 55–58
solubility, growth behavior, and sucrose crystal shape, 200–208
Ethylenediaminetetraacetic acid (EDTA)
crystals showing agglomerated particles, 124*f*
particles consisting of β-EDTA, 125*f*
precipitation from solution, 122–126
precipitation tests, 123*t*
two crystal forms, 124–126
volume average diameter vs. precipitation time, 125*f*
Exchangeable protons, aqueous sucrose solutions, 182
Extinction condition, connected nets of benzoic acid dimers, 30

F

F faces, computation, benzoic acid, 30–32
F slice, determination of morphological importance, 29
F units, adsorbed growth units, 37–38
Face growth
rates, caprolactam and additives, 62*f*
retardation, 81
Falling film crystallizer, experimental setup, 138*f*
Fast crystallization, problems with solid–liquid separation, 112
Fat crystallization, chocolates, compound coatings, butter, margarine, and peanut butter, 172
Fatty acid composition, LCT and its stearin and olein fractions, 191, 193*f*, 194
Flat box method
derivation of F slices from crystal structure, 28–29
determination of connected nets, 16
PBC analysis, 28–35
Flat face, cut crystal, 28
Flow improvers
commercial, 152
waxy components of oil or fuel, 151
Fluid phase
influence on crystal growth, 36–42
prediction of growth morphology of crystals, 41
Fluorescence emission spectroscopy, carminic acid in sugar solutions, 178–184
Fondant, crystallization, 174–175
Food, controlling crystallization, 172–177
Food and Drug Administration (FDA)
Good Manufacturing Practice regulations, 66–67
regulation of bulk pharmaceutical chemicals, 66–71
Force fields
crystal habit parameters, 34–35
crystal structure, 16
habit predictions, 79
predicted acetaminophen crystal morphologies, 82*f*
Fractionation process
butter fat, 185
milk fat, 137–144
Free energy, liquid-to-solid phase change, 68
Fructose, effect on solubility, growth behavior, and sucrose crystal shape, 200–208
Fuel additives, fullerenes, 151–156
Fullerenes
development and use, 152–153
growth, perfection and defects in C_{60} and C_{70} single crystals, 231–238
growth of wax crystals, 153–154
single crystals grown from vapor, 233*f*, 235*f*
structural types for pour-point depressants, 155*f*

G

Gel crystallization, related substances or additives, 99*t*
Gel growth, organomineral crystal 2-amino–5-nitropyridinium dihydrogen phosphate, 259–262
Gelatinization, heating of native starch, 196
Gelatinized starch, extent of crystallization and melting behavior of crystallized starch, 199

Glass transition
effect on crystallization in starch, 196–199
metastable limit, 173
Glucose, effect on solubility, growth behavior, and sucrose crystal shape, 200–208
Glucose solutions
carminic acid, 178, 180
emission peak intensity ratio for carminic acid, 181*f*
Glyceride composition, LCT and its stearin and olein fractions, 194*f*
Glycine
crystal growth rates, 157–162
fundamental driving force vs. dimensionless supersaturation, 161*f*
growth kinetics, 118–120
growth rates as functions of dimensionless driving force, 160*f*
Good Manufacturing Practice regulations, FDA, 66–67
Growth environment
external shape of crystals, 52
molecular crystal, 43–52
Growth faces, identification, 44
Growth instability, crystallization procedure, 149
Growth kinetics
glycine precipitated by 2-propanol, 121
organic solutes, 120–121
single sucrose crystals, 216–219
Growth morphology
biphenyl and naphthalene crystals grown from toluene solutions, 40–41
influence of fluid phase, 36–42
urea crystals grown from aqueous solution, 40
Growth rate, liquid inclusions in solid layers, 163–170
Growth rate dispersion
definition, 216
single sucrose crystals, 217–218
Growth units
crystalline, 28
effective, 39*f*
monomers and dimers, 30
orientations and molecular conformations, 37
quasi and genuine, 38

H

Habit modification
blocker and disruptor tailor-made additives, 44–47
growth inhibitor, 152
Hardness, test by microindentation, 232–233
Hartman–Perdok theory
crystallographic and statistical mechanical foundations, 15–21
growth morphology of crystals, 36
Hb-additive, caprolactam crystal habit, 59
High resolution powder diffraction, crystal structure determination, 22–27
Homogeneous nucleation, emulsified systems, 127
Host–guest molecules, role of MeOH in chiral combination, 246–248
Hydrodynamics of natural convection, thermodynamic metastability, 224
Hydrogen bonding
crystal faces, 82–84
crystal growth rate, 81
habit predictions, 79
methylparaben, 102
molecular packing, 3, 59
within crystal lattice and to solvent water molecules, 83*f*
p-Hydroxyacetanilide, *See* Acetaminophen

I

Ice cream, size of ice crystals, 176
Ice crystals
ice cream and frozen foods, 172
manufacture and storage of frozen desserts, 176–177
Impurities
axial ratio, 206*f*
crystal growth, morphology and nucleation, 53, 249
solubility, growth behavior, and sucrose crystal shape, 200–208
sucrose solutions, 207
Impurity inclusions
mechanism and measures for process improvement, 113*f*
places in layer melt crystallization, 114
Impurity sources, crystallization, 112–113
In-process blending, synthesis lots, 67
Inclusion crystal
chiral combination of host–guest molecules, 246–248
materials, 163–170
Induction time
effect of additives on nucleation processes, 55–58
β-succinic acid in ethanol solution in presence of alkanoic acids, 57*t*
Inhomogeneous cell model, shape of crystals, 19
Inspection, crystallization processes, 67
Interaction energy
atom–atom approach, 44
packing effects in organic crystals, 3
Interfacial structure analysis
generalization, 38
surface scaling factor, 37–38
Intermolecular bond, effect of impurities on crystal growth, 54–55

Intermolecular bonding vectors, β-succinic acid, 497
Intermolecular interactions
calculated lattice energy, 3–4
Cambridge Crystallographic Database, 5–7
Isothermal solute diffusion, natural convection, 226

K

Kinetically controlled discrimination, emulsion system, 134
Kinked face, cut crystal, 28

L

Lattice energy
calculated and experimental, 3, *4t*, *5f*
definition, 44
function of summation limit for anthracene, urea, succinic acid, and glycine, *4f*
growth morphology of crystals, 36
Lattice geometry
habit predictions, 79
predicted acetaminophen crystal morphologies, *82f*
Lifson force field, crystal habit, 34–35
Lipid crystallization, chocolates, compound coatings, butter, margarine, and peanut butter, 172
Liquid inclusion
explanation for migration, 164–165
migration in a crystalline layer, 114, *165f*, *168f*
migration in solid layers, 163–170
migration through a diffusion cell, 165–166
sources of experimental uncertainties, 169
Liquid melt droplets, purity of solid layer melt crystallization process, 112
Lock and key mechanism, AA, 102
Long chain triglycerides (LCT)
cooling curves, *188f*
crystallization behavior, 185–195
solid fat content, *189f*, *190f*
Lubricating oils
treated with substituted fullerenes, 154–155
waxes, 151–152

M

Maltose, effect on solubility, growth behavior, and sucrose crystal shape, 200–208
Mechanics minimization, binding energy calculation, 54
Melt crystallization
advantages, 112
butter fat, 185
dendritical growth, *164f*
experimental setup, *186f*
milk fat triglycerides, 185–195
separation process, 112–115
solid layer, 137–144, 163
Melting enthalpy, relationship with storage temperature, *198f*
MeOH, role in chiral combination of host–guest molecules, 246–248
Metacetamol, incorporation in acetaminophen crystal lattice, 101–103
Metal plating, EDTA as chelating agent, 122
Metastability, solution, 223–224
Metastable zone width
effect of additives on nucleation processes, 55–58
β-succinic acid in ethanol solution in presence of alkanoic acids, *56t*
Methylparaben, incorporation in acetaminophen crystal lattice, 102
Microdroplets, supersaturated levitated, 227–230
Microgravity, crystal growth, 273–278
Microindentation
measurement of hardness, 232–233
slip system, 235
Migration
impurity concentration vs. rate and distances, 166–169
liquid inclusions in solid layers, 163–170
rates vs. temperature gradients, *167f*, *168f*
Milk fat
crystal structure, *139f*
fractionation process, 137–144
phase change, 137
surplus, 185
triglycerides, melt crystallization behavior, 185–195
Modeling, crystal habit, 79–80
Modeling techniques, morphological, 43–44
Molecular coordination spheres, crystal structure, 11
Molecular crystals
crystal data, *258t*
crystallization, between nitroanilines and nitrophenols, 256–258
generation of chirality, 239–245
growth environment, 43–52
Molecular dynamics
binding energy calculation, 54
spin relaxation time, 267
Molecular modeling
application to study of molecular materials, 2–14

levels of operation, 2
Molecular nuclei, crystal structure prediction, 11
Molecular packing, chiral crystal of acridine and diphenylacetic acid, 241*f*
Molecular pairs, chiral and achiral, 243*f*
Molecular pharmaceutics, prediction of material characteristics, 86
Molecular structure and mobility, NMR study, 266
Molecular trickery, simulation of crystal morphology, 44–47
Momany force field, crystal habit, 34–35
Monte Carlo cooling approach, packing of one-dimensional stacked aggregates, 11–12
Monte Carlo simulated annealing process, crystal packing problem, 12
Morphological modeling, tailor-made additives, 45*f*
Morphological modification, causes, 99, 101
Morphology
  agreement between predicted and grown crystals, 80
  influence of fluid phase, 36–42
  prediction, 15–20
  theoretical, construction, 33–34
Mother liquor, inclusions in crystalline layers, 163–170
Mother phase morphology, bond energy calculation, 17
Multilinear regression analysis, sublimation enthalpies, 9–10
Multistep fractionation
  milk fat crystallization, 142–143
  solid layer technology, 143

N

Naphthalene
  biphenyl as additive, 47, 48*f*
  crystal shape, 41
  growth morphology of crystals from toluene solution, 40–41
  interatomic interactions, 51*f*
  intermolecular bonding analysis, 49, 50*t*
Natural convection
  isothermal solute diffusion, 226
  solution metastability, 224
Neural network, sublimation enthalpies, 9
New Drug Application (NDA), crystallization, 70
Nitroanilines and nitrophenols, crystallization of molecular crystals, 256–258
Nucleation
  D-SCMC crystals, 105
  factors affecting rate, 53, 57, 192*f*
  generalized kinetic curve, 173*f*
  ice cream and frozen desserts, 176
  impacts between seeds and drops, 133
  incorporation of impurities, 112
  milk fat, 137
  supersaturated ammonium sulfate solution, 249–255
Numerical analysis, crystal growth, 275–277
Nylon, adipic acid, 145

O

Olein fraction
  crystallization conditions, 187
  stearin yield and dropping point, 190*t*
On-line processing, in situ X-ray cell, 210
Optical resolution, guest compounds, 247*f*
Order–disorder phase transition, connected net, 18
Organic materials
  Bridgman–Stockbarger crystal growth system, 263–265
  crystal growth kinetics, 116–121
Organic supersaturated solutions, theory and experiment, 222–230
Orthocetamol, incorporation in acetaminophen crystal lattice, 102–103

P

Packing coefficient, efficiency of solid state arrangement, 3
Paclobutrazol, intermediate molecular structure and morphologies, 7*f*
Palmitic acid, interactions with crystal surfaces, 55
Paracetemol, *See* Acetaminophen
Particle population density, residence times, 117*f*
Peak intensity, diltiazem hydrochloride commercial batches, 91
Periodic bond chain (PBC)
  automated analysis method applied to benzoic acid, 28–35
  crystal lattice faces, 15
  crystal morphology calculation, 7
  underlying assumptions for analysis, 28
Pharmaceutical materials
  determination of physical properties, 85–86
  primary production methods, 85
  regulatory considerations in crystallization processes, 66–71
Phase change, milk fat, 137
Phase diagrams, determination by NMR, 269–272
Phase equilibria, driving force for crystallization, 172
Physical state, effect on crystallization in starch, 196–199

Plastic deformation, one of origins of hardness, 237
Plating of surfaces by metals, EDTA as chelating agent, 122
Polar morphology, urea, 47–48
Polymorphism
cocoa butter fat, 209–215
crystallization studies, 70
pharmaceutical materials, 85
reproducible production, 69
Pores, cracks, 114
Postcrystallization treatments, final purity of crystalline materials, 112
Postprocessing techniques, interface shape and location, 263
Potassium dihydrogen phosphate
diffusion coefficient and viscosity of supersaturated solutions, 273–278
diffusion coefficient as function of mass fraction, 273*f*
kinematic viscosity, 275*f*
Pour-point depression
alkylamino- and alkylfullerenes, 151
all-hydrocarbon, 153–154
fullerene additives, 155*t*
syntheses of model, 154*f*
Powder diffraction patterns
crystal structure determination, 22–27
experimental and theoretical, 25*f*
molecular packing calculations, 25
Precipitation
EDTA recovery process, 122–126
pH vs. rate of crystal growth, 124
Process conditions, effect on product properties, 122–126
Processing cell, tempering process of cocoa butter fat, 214
Pulsed NMR method
experimental procedure, 268
heterogeneous liquid systems, 267–268
supersaturated solutions, 266–272
Pure crystals, effect of impurities, 201*f*, 202*f*
Purification, emulsion solidification, 127–136

Q

Quadratic nonlinear coefficients, 2-amino–5-nitropyridinium dihydrogen phosphate, 261
Quadratic nonlinear optical effect, 2-amino–5-nitropyridinium, 259–262

R

Racemic compounds, definition, 67
Ranitidine hydrochloride, patents and recrystallization, 69
Recrystallization
ice cream and frozen foods, 172, 176
ice cream stored at different temperatures and storage conditions, 177*f*
ranitidine hydrochloride, 69
secondary recovery procedures, 66–67
type of sweetener, 176–177
Relative binding energy, definition, 45
Relative growth rate, habit controlling parameters, 36
Relaxation time, transformation from F to S unit, 38
Residue, impurities in crystallization, 113
Rietveld method, crystal structure determination, 12–13, 23
Roughening phenomena
determination of morphological importance, 29
energy within connected net, 17
flat crystal face, 39–40
methods for determining bond energy, 18

S

Salicylidene aniline, growth of single crystals, 264–265
Saturated solutions, NMR study, 268–269
Scaling experiments, surface scaling factor, 38–40
Scheraga force field, crystal habit, 34–35
Search/structure generation algorithm, crystal structure, 11
Seeded exponential cooling, adipic acid, 147–148
Seeding
crystal size, 149
unexpected forms, 69
Separation process, melt crystallization, 112–115
Shear solution viscosity, dependence on bulk solute concentration, 225*f*
Shear stress cell, in situ experiments, 212*f*
Single crystal
fullerenes, 231–238
microhardness, 231
Single crystal X-ray structural data, solid state properties of alprazolam and diltiazem hydrochloride, 85–94
Slice energy
calculation, 32
crystal morphology calculation, 7
growth morphology of crystals, 36
Slip system
dislocation theory, 237
microindentation, 235
Software
AMPAC, calculation of point charges, 33
Cambridge Crystallographic Database, 5–7
CERIUS, calculated habits influenced by glucose and fructose, 206
CERIUS^2
binding energy calculation, 54

force field parameters, 33
individual crystal surfaces, 96
molecular modeling, 29, 32, 79
morphology prediction using attachment energy, 87
CERIUS3.2, crystal habit modification, 60
Chem3D Plus, crystal lattice, 93*f*
development, 19–20
DREIDINGII, interaction energies and optimum geometries, 60
F GRAPH
connected nets using appropriate crystal graphs, 31
implementation of flat box method, 29
HABIT, crystal form construction, 15
HABIT95, morphological investigations on molecular materials, 43–52
Lazy Pulverix, prediction of X-ray powder patterns, 87
mechanical force fields, 16
MM2, molecular mechanics package, 24
MOPAC, atomic electrostatic potential charges, 87
PKFIT, peak shape function, 23
PODSUM, beam decay, 23
REFCEL, unit cell dimensions, 24
SHAPE, crystal drawing program, 43
Solid layer melt crystallization
fractionation of milk fat, 137–144
multistep fractionation, 143
separation process, 112–115
Solid–liquid interface
cross-section of growth chamber, 265*f*
shape and location, 263
Solid state arrangement, intermolecular interactions, 2
Solid state properties, diltiazem hydrochloride and alprazolam, 85–94
Solubility, determination by NMR, 269–272
Solute activity
aqueous solution, 158
coefficient ratio vs. dimensionless supersaturation, 159*f*
Solute mass diffusivity, role of concentration dependence in crystal growth, 223–227
Solution density, dependence on solute concentration, 225*t*
Solution shear viscosity, role of concentration dependence in crystal growth, 223–227
Solvation water, glucose solution, 180
Solvent–crystal interactions, effect on crystal habit, 78
Solvent effects, acetaminophen crystal habit, 78–84
Solvent polarity
crystallization, 81
habit shift, 83–84
Speciality chemicals, crystal structure determination, 22–27
Spherical void electrodynamic levitator trap
effect of impurity on metastable zone width, 249
experimental setup, 228*f*
microdroplets of supersaturated solutions, 227–230
schematic, 251*f*
Spin-echo measurements, supersaturated solutions, 266–272
Spontaneous crystallization, chiral two-molecule crystal, 239–245
Starch
crystallization, 196–199
melting enthalpies as function of storage time, 197*f*
onset, peak, and endset temperatures and enthalpy of melting, 197*f*
Stearin
differential scanning calorimetry melting profiles, 188*f*, 189*f*
effect of temperature on yield, 190
selection of fats in food-fat applications, 187, 190–191
solid fat content, 189*f*, 190*f*
Stearin fraction
crystallization conditions, 187
stearin yield and dropping point, 190*t*
Stepped face, cut crystal, 28
Storage, changes in crystalline structure of foods, 172
Storage temperature
effect on onset, peak, and endset temperatures, 198*f*
starch crystallization, 196–198
Strom direct method, crystal graph PBCs, 17
Structurally related substances, incorporation into acetaminophen crystals, 95–104
Sublimation enthalpy
calculation vs. experiment, 10*f*
estimation, 8–10
predicted results, 10*t*
β-Succinic acid
atom–atom bonds, 49*t*
binding energy, metastable zone width and nucleation induction time, 53–58
binding energy vs. induction time, 57*f*
bonding analysis, 48–49
crystal growth rates, 157–162
dominant atom–atom interactions, 50*f*
fundamental driving force vs. dimensionless supersaturation, 161*f*
growth rates as functions of dimensionless driving force, 160*f*
intermolecular bonding vectors, 49*t*
metastable zone width and induction time in ethanol solution, 55–58

particular atom contributions to lattice energy, 50*t*
solubility in ethanol, 56*f*
theoretical morphology using Donnay Harker method and attachment energy method, 54*f*
Sucrose
concentration during drying and crystallization in thin sugar films, 175*f*
crystal in presence of ethanol, fructose, glucose, and maltose, 204*f*, 205*f*
crystal shape modification, 201*f*, 202–203
crystallization in a fondant, 174–175
growth and dissolution rate dispersion of crystals, 216–219
impurities effect on solubility, growth behavior, and crystal shape, 200–208
initial concentration, effect on crystal growth during drying, 174*f*
normal crystal, 203*f*
size independent growth and dissolution, 218
Sucrose and lactose mixtures, phase/state diagram, 173*f*
Sucrose concentration, carminic acid, 180
Sucrose solutions
carminic acid, 178
emission peak intensity ratio for carminic acid, 183*f*
Sugar, crystallization during drying of thin films, 175–176
Sugar coating, process, 175
Sugar crystals, candy, cereals, and icings, 172
Sugar solutions, fluorescence emission spectra of carminic acid, 178–184
Superlattice, diltiazem hydrochloride, 93
Supersaturated solution
concentration of L- and D-SCMC, 105–107
crystal growth, 223–227
spin-echo measurements, 266–272
thermodynamic properties, 157–162
Supersaturation
crystal growth rates, 75*f*, 159
experiments, 116–117
growth instability, 149
heterogeneous nucleation, 222
nucleation of ammonium sulfate solution, 249–255
rigorous test of available theories for crystal growth, 116
tendency to crystallize, 172
thermodynamic investigations, 227–230
Surface scaling factor
calculation, 37–40
definition, 36
Surfactant, emulsion stabilization and crystal dispersion, 134
Suspension crystallization, fractionation of milk fat, 137
Suspension technique, milk fat crystallization, 142–143
Sweating
diffusion washing, 114
postcrystallization treatment of crystalline layers, 113
Synchrotron radiation X-ray diffraction, in situ, 209–215

T

Tailor-made additives
caprolactam, 62
carminic acid, 179
industrial point of view, 59
simulation of crystal morphology, 44–47
Temperature, crystal growth rate, 203, 206
Template, *m*-CNB nucleation, 134–135
Tensagex, purification above critical micelle concentration, 134
Texture, crystallization, 172
Theoretical crystal morphology, construction, 33–34
Thermodynamics of metastability, crystal growth, 222–230
Three-winged propeller, torsion angles, 240
Toluene, biphenyl and naphthalene crystal growth morphologies, 40–41
Trace fluorescent probe, carminic acid, 183–184
Transverse magnetization decay, urea, 268–269
Triacylglycerols, milk fat phase change, 137
Triglycine sulfate
calculated supersaturation profile near crystal, 277*f*, 278*f*
diffusion coefficient and viscosity of supersaturated solutions, 273–278
diffusion coefficient as function of mass fraction, 274*f*
kinematic viscosity, 276*f*
numerical analysis, 257–277
Twinning, diltiazem hydrochloride, 93

U

Unit cell
dimer and monomer, 30*f*
positions of centers of mass in fractional coordinates, 30*t*
Urea
concentration dependence in liquid and solid phase, 270*f*
crystal shape, 40
hydrogen bond clusters, 4
intermolecular interactions, 6*t*
modelling polar morphologies, 47–48
NMR solubility study, 269–272

temperature dependence of concentrated solution, 271*f*
three-dimensional structure of cluster, 6*f*
transverse magnetization decay, 268–269

V

Vacuum morphology, bond energy calculation, 17
Validation, crystallization step, 67
Vickers microhardness, temperature dependence, 236–237
Viscosity
growth of crystals in microgravity, 273–278
infinite dilution to supersaturation, 275

W

Washing, postcrystallization treatment of crystalline layers, 113–114
Water activity, effect on ammonium sulfate solution, 254*f*,*t*
Wax(es), fuels and lubricating oils, 151–152
Wax crystal(s)
fullerene-based modifiers, 153–154
habit modification by growth modifier, 152*f*
interaction with model fullerene-based growth inhibitor, 153*f*
with and without growth inhibitor, 152*f*
Wax crystal modifiers, petroleum industry, 151
Wiggler beamline, experimental layout, 213*f*
Wulff construction, growth morphology of naphthalene and biphenyl crystals, 41*f*

X

X-ray structural data, prediction of solid state properties of alprazolam and diltiazem hydrochloride, 85–94
X-ray studies, polymorphism in cocoa butter, 209–210

Y

Yield, droplet size, organic and surfactant weight fraction, 132*t*

Z

Zantac, *See* Ranitidine hydrochloride

Production: Margaret J. Brown
Acquisition: Anne Wilson
Index: Colleen P. Stamm
Cover design: Amy J. O'Donnell

Printed and bound by Maple Press, York, PA